TRANSNATIONAL BLACKNESS

THE CRITICAL BLACK STUDIES SERIES

CENTER FOR CONTEMPORARY BLACK HISTORY
COLUMBIA UNIVERSITY

Edited by Manning Marable

The Critical Black Studies Series features readers and anthologies examining challenging topics within the contemporary black experience—in the United States, the Caribbean, Africa, and across the African Diaspora. All readers include scholarly articles originally published in the acclaimed quarterly interdisciplinary journal Souls, published by the Center for Contemporary Black History, the research and publications center of the Institute for Research in African-American Studies at Columbia University. Under the general editorial supervision of Manning Marable, the readers in the series are designed both for college and university course adoption, as well as for general readers and researchers. The Critical Black Studies Series seeks to provoke intellectual debate and exchange over the most critical issues confronting the political, socio-economic, and cultural reality of black life in the United States and beyond.

Titles in this series published by Palgrave Macmillan:

Racializing Justice, Disenfranchising Lives: The Racism, Criminal Justice, and Law Reader (2007)

> Edited by Manning Marable, Keesha Middlemass, and Ian Steinberg

Seeking Higher Ground: The Hurricane Katrina Crisis, Race, and Public Policy Reader (2008)

> Edited by Manning Marable and Kristen Clarke

Transnational Blackness: Navigating the Global Color Line (2008)

> Edited by Manning Marable and Vanessa Agard-Jones

The Islam and Black America Reader (2009)

> Edited by Manning Marable and Hishaam Aidi

Beyond Race: New Social Movements in the African Diaspora (2010)

> Edited by Manning Marable

The New Black History: The African-American Experience since 1945 Reader (2010)

> Edited by Manning Marable and Peniel E. Joseph

TRANSNATIONAL BLACKNESS

NAVIGATING THE GLOBAL COLOR LINE

Edited by Manning Marable and Vanessa Agard-Jones

palgrave
macmillan

TRANSNATIONAL BLACKNESS

Copyright © Manning Marable and Vanessa Agard-Jones, 2008.

First published in 2008 by PALGRAVE MACMILLAN® in the US—a division of St. Martin's Press LLC, 175 Fifth Avenue, New York, NY 10010.

Where this book is distributed in the UK, Europe and the rest of the world, this is by Palgrave Macmillan, a division of Macmillan Publishers Limited, registered in England, company number 785998, of Houndmills, Basingstoke, Hampshire RG21 6XS.

Palgrave Macmillan is the global academic imprint of the above companies and has companies and representatives throughout the world.

Palgrave® and Macmillan® are registered trademarks in the United States, the United Kingdom, Europe and other countries.

ISBN-10: 0-230-60267-3 (hardcover)
ISBN-13: 978-0-230-60267-0 (hardcover)

ISBN-10: 0-230-60268-1 (paperback)
ISBN-13: 978-0-230-60268-7 (paperback)

Library of Congress Cataloging-in-Publication Data

Transnational blackness : navigating the global color line / edited by Manning Marable and Vanessa Agard-Jones.
        p. cm.—(Critical black studies series)
    Includes bibliographical references and index.
    1. Blacks—Race identity. 2. Race relations—History. 3. Race awareness. I. Marable, Manning, 1950- II. Agard-Jones, Vanessa.

HT1581.T73 2008
305.896–dc22   2008012343

A catalogue record of the book is available from the British Library.

Design by Scribe Inc.

First edition: October 2008

10 9 8 7 6 5 4 3 2 1

Printed in the United States of America.

Contents

BLACKNESS BEYOND BOUNDARIES

NAVIGATING THE POLITICAL ECONOMIES OF GLOBAL INEQUALITY

MANNING MARABLE

The advance guard of the Negro people . . . must soon come to realize that if they are to take their just place in the van of Pan-Negroism, then their destiny is *not* absorption by the white Americans. That if in America it is to be proven for the first time in the modern world that not only Negroes are capable of evolving individual men like Toussaint, the Saviour, but are a nation stored with wonderful possibilities of culture, then their destiny is not a servile imitation of Anglo-Saxon culture, but a stalwart originality which shall unswervingly follow Negro ideals.

—W. E. B. Du Bois, 1897[1]

ON MARCH 5, 1897, THE NEWLY FORMED AMERICAN NEGRO ACADEMY MET FOR its inaugural sessions in Washington, D.C., W. E. B. Du Bois, then a twenty-nine-year-old social scientist and recent PhD graduate of Harvard University, delivered the second paper to this gathering of black American intellectuals, "The Conservation of Races," that would foreshadow much of his future life's work. The paper centered in part on the question of what constituted "blackness," or the construction of black identity within the challenging contexts of white-dominated societies. Inside the United States, Du Bois argued, each African American must struggle to determine "what, after all, am I? Am I an American or am I a Negro? Can I be both?" Du Bois then sought to delineate the boundaries between Africanity, race, and citizenship that constantly confronted black Americans:

> We are Americans, not only by birth and by citizenship, but by our political ideals, our language, our religion. Farther than that, our Americanism does not go. At that point, we are Negroes, members of a vast historic race that from the very dawn of creation has slept, but half awakening the dark forests of its African fatherland. . . . We are that people whose subtle sense of song has given America its only American music, its only American fairy tales, its only touch of pathos and humor amid its mad money-getting plutocracy. As such, it is our duty to conserve our physical powers, our intellectual endowments, our spiritual ideals; as a race we must strive by race organization, by race solidarity.[2]

For Du Bois at this time, the boundaries of blackness were defined largely by aesthetics, culture, and the highly charged construction of "race." But as the twentieth century unfolded, Du Bois expanded his understanding about the common grounds that people of African descent shared throughout the colonial and segregated world. This led him to embrace the politics of Pan-Africanism, and efforts by black activists in the Caribbean, the United States, and Africa itself to overthrow white minority regimes. Intellectually, it gave Du Bois a truly global concept of what today would be termed "Black Studies." Part of the mission of Black Studies as an intellectual project has been the remapping of collective identity and memory, in part by using Du Bois's criteria. But it should also combine theory with collective action, in the effort not simply to interpret but to transform the world, empowering black people in the process.

During the 1960s, when Black Studies departments were first being launched within predominantly white academic institutions, an ideological debate subsequently developed over the appropriate geopolitical and cultural boundaries for what the study of "blackness" should comprise. Many prominent African American cultural nationalists, such as Kwanzaa-founder Maulana Karenga, vigorously argued that Black Studies must trace its intellectual lineage back to classical Egyptian civilization. The black experience in the United States, in this Afrocentric interpretation, was a small subsidiary of a much grander African civilizational saga. Other black studies scholars noted the destructive effects of the transatlantic slave trade, and focused on the cultural and political resistance of African Diasporic populations scattered across North and South America, the Caribbean, Europe, and Asia as the decisive elements in the making of the modern world. Scholars largely trained in the United States often had a more parochial vision of Black Studies, emphasizing the local struggles waged by African Americans to achieve political rights and equality against the American nation-state. As a measurement of the lack of theoretical and conceptual consensus among these scholars, departments and programs dedicated to Black Studies still call themselves by various names: "Afro-American Studies," "African American Studies," "Africana Studies," "African and African American Studies," "African Diasporal Studies," and "Comparative Race and Ethnicity Studies."

At Columbia University, when I founded the "Institute for Research in African-American Studies" (IRAAS) in July 1993, the precise name of the program was the result not of a theoretically grounded academic discussion but rather a pragmatic political compromise. The "Institute of African Studies" had been established in Columbia University's School of International and Public Affairs approximately one-quarter century earlier, and its director and small faculty were deeply concerned that IRAAS would colonize and incorporate their curricula into our own program. The decision was made to keep African American and Caribbean Studies distinctly separate from African Studies. Over subsequent years, as the fortunes of Columbia's African Studies Program rose and fell, I came to regret that decision administratively, as well as intellectually.

It is impossible to relate the full narrative of the experiences of people of African descent in the United States, and throughout the Caribbean and the Americas, without close integration and reference to the remarkable history of the African continent, its many peoples, languages, and diverse cultures. The South Atlantic and especially the Caribbean were "highways" for constant cultural, intellectual, and political exchange between people of African descent, especially during the past three centuries. Pan-Africanist-inspired social protest movements like Marcus Garvey's Universal Negro Improvement Association and African Communities League (UNIA) started in Jamaica but accelerated into hundreds of chapters across the United States as a mass movement, and then grew hundreds of new chapters throughout Central America and Africa. Documenting the UNIA's complex story by focusing solely on the events of one nation, such as the United States, distorts the narrative and cripples our understanding of fundamental events. Similarly, South Africa's "Black Consciousness Movement" of the 1970s and the brilliant protest writings of Steven Biko cannot be interpreted properly without detailed references to the "Black Power Movement" in the United States during the 1960s, and to the influential speeches and political writings of Malcolm X of the United States and Frantz Fanon of Martinique.

"Blackness" acquires its full revolutionary potential as a social site for resistance only within transnational and Pan-African contexts. This insight motivated W. E. B. Du Bois to initiate the Pan-African Congress Movement at the end of World War I. George Padmore, Kwame Nkrumah, Du Bois, and others sponsored the Fifth Pan-African Congress, in Manchester, England, in October 1945, out of the recognition that the destruction of European colonial rule in Africa and the Caribbean, and the demise of the Jim Crow regime of racial segregation in the United States, were politically linked. Any advance toward democracy and civil rights in any part of the black world objectively assisted the goals and political aspirations of people of African descent elsewhere. An internationalist perspective, from a historian's point of view, also helped to explain the dynamics of the brutal transnational processes of capitalist political economy—the forced movement of involuntary labor across vast boundaries; the physical and human exploitation of slaves; the subsequent imposition of debt peonage, convict leasing, and sharecropping in postemancipation societies; and the construction of hypersegregated, racialized urban ghettoes, from Soweto to Rio de Janeiro's slums to Harlem. As this edited volume illustrates, the twentieth century was full of examples of "blackness beyond boundaries as praxis"—intellectual-activists of African descent who sparked movements of innovative scholarship, as well as social protest movements, throughout Africa and the African Diaspora.

In 1900, Du Bois had predicted that the central "problem of the twentieth century" would be the "problem of the color line," the unequal relationship between the lighter versus darker races of humankind.[3] Du Bois's color line included not just the racially segregated Jim Crow South and the racial oppression of South Africa but also the British, French, Belgian, and Portuguese colonial domination in Asia, the Middle East, Latin America, and the Caribbean among indigenous populations.[4] Building on Du Bois's insights, we can therefore say that the problem of the twenty-first century is the problem of global apartheid: the racialized division and stratification of resources, wealth, and power that separates Europe, North America, and Japan from the billions of mostly black, brown, indigenous, undocumented immigrant, and poor people across the planet. The term "apartheid" comes from the former white minority regime of South Africa; an Afrikaans word, it means "apartness" or "separation." Apartheid was based on the concept of *herrenvolk*, a "master race" that was predestined to rule all non-Europeans. Under global apartheid today, the racist logic of *herrenvolk* is embedded ideologically in the patterns of unequal economic and global accumulation that penalizes African, South Asian, Caribbean, Latin American, and other impoverished nations by predatory policies.

Since 1979–80, with the elections of Ronald Reagan as U.S. president and Margaret Thatcher as prime minister of the United Kingdom, America and Great Britain embarked on domestic economic development strategies that are now widely known by the term "neoliberalism." Neoliberal politics called for the dismantling of the welfare state; the end of redistributive social programs designed to address the effects of poverty; the elimination of governmental regulations and regulatory agencies over capitalist markets; and "privatization," the transfer of public institutions and governmental agencies to corporations. Journalist Thomas B. Edsall has astutely characterized this reactionary process of neoliberal politics within the United States in these terms: "For a quarter-century, the Republican temper—its reckless drive to jettison the social safety net; its support of violence in law enforcement and national defense; its advocacy of regressive taxation, environmental hazard and probusiness deregulation; its 'remoralizing' of the pursuit of wealth—has been judged by many voters as essential to America's position in the world, producing more benefit than cost."[5]

One of the consequences of this reactionary political and economic agenda, according to Edsall, was "the Reagan administration's arms race" during the 1980s, which "arguably drove the Soviet Union into bankruptcy." A second consequence, Edsall argues, was America's disastrous military invasion of Iraq. "While inflicting destruction on the Iraqis," Edsall observes, "[George H. W.] Bush multiplied America's enemies and endangered this nation's military, economic health and international stature. Courting risk without managing it, Bush repeatedly and remorselessly failed to accurately evaluate the consequences of his actions."[6]

Edsall's insightful analysis significantly did not attempt to explain away the 2003 U.S. invasion of Iraq under President George W. Bush and subsequent military occupation as a political "mistake" or an "error of judgment." Rather, he located the rationale for the so-called "war on terrorism" within the context of U. S. domestic, neoliberal politics. "The embroilment in Iraq is not an aberration," Edsall observes. "It stems from core [Republican] party principles, equally evident on the domestic front." The larger question of political economy, left unexplored by Edsall and most U.S. mainstream analysts, is the connection between U.S. militarism abroad, neoliberalism, and macro-trends in the global economy. As economists, such as Paul Sweezy and Harry Magdoff, noted decades ago, the general economic tendency of mature, global capitalism is toward stagnation. For decades, in the United States and western Europe, there has been a steady decline in investment in the productive economy, leading to a decline in industrial capacity and lower future growth rates. Profit margins inside the U. S. have fallen over time, and corporations have been forced to invest capital abroad to generate higher rates of profitability. There is a direct economic link between the deindustrialized urban landscapes of Detroit, Youngstown, and Chicago with the expansion of industries in China, Vietnam, Brazil, and other developing nations.

Since capitalist economies are "based on the profit motive and accumulation of capital without end," observed Marxist author Fred Magdoff, "problems arise whenever they do not expand at reasonably high growth rates."[7] Since the 1970s, U.S. corporations and financial institutions have relied primarily on debt to expand domestic economic growth. By 1985, total U.S. debt—which is comprised of the debt owed by all households, governments (federal, state, and local), and all financial and nonfinancial businesses—reached twice the size of the annual U.S. gross domestic product (GDP). By 2005, the total U.S. debt amounted to nearly "three and a half times the nation's GDP, and not far from the $44 trillion GDP for the entire world," according to Fred Magdoff.[8]

As a result, mature U.S. corporations are forced to export products and investment abroad, to take advantage of lower wages, weak or nonexistent environmental and safety standards, and so forth to obtain higher profit margins. Today about 18 percent of total U.S. corporate profits come from direct overseas investments. Partially to protect these growing investments, the United States has pursued an aggressive, interventionist foreign policy across the globe. As of 2006, the United States maintained military bases in fifty-nine nations. The potential for deploying military forces in any part of the world is essential for both political and economic hegemony.[9] Thus the current Iraq War was not essentially a military blunder caused by a search for "weapons of mass destruction" but rather an imperialist effort to secure control of the world's second largest proven oil reserves; it was also the first military step of the Bush administration's neoconservatives to "remake the Middle East" by destroying the governments of Iraq, Iran, and Syria.

Although the majority of nations in the international community either openly opposed, or at least seriously questioned, the U.S. military occupation of Iraq, the neoliberal economic model of the United States has been now widely adopted by both developed and developing countries. Governments across the ideological spectrum—with the important exception of some Latin American countries in recent years—have eliminated social welfare, health, and education programs; reduced governmental regulations on business activity; and encouraged the growth income inequality and entrepreneurship. Even noncapitalist countries like Cuba have revived the sex-trade-oriented tourism business, which has contributed to new forms of gender and racial prejudice in that country. As a result, economic inequality in wealth has rapidly accelerated, reinforcing traditional patterns of racial and ethnic domination.

A 2006 study by the World Institute for Development Economic Research of the United Nations University established that, as of 2000, the upper 1 percent of the globe's adult population, approximately 37 million people, averaged about $515,000 in net worth per person, and collectively controlled roughly 40 percent of the world's entire wealth. By contrast, the bottom one-half of the planet's adult population, 1.85 billion people, most of whom are black and brown, owned only 1.1 percent of the world's total wealth. There is tremendous inequality of wealth between nations, the UN report noted. The United States, for example, comprised only 4.7 percent of the world's people, but it had nearly one-third, or 32.6 percent, of global wealth. By stark contrast, China, which had one-fifth of the world's population, owned only 2.6 percent of the globe's wealth. India, which has 16.8 percent of the global population, controlled only 0.9 percent of the world's total wealth. Within most of the world's countries, wealth was disproportionately concentrated in the top 10 percent of each nation's population. It comes as no surprise that in the United States, for example, that as of 2000 the upper 10 percent of the adult population owned 69.8 percent of the nation's total wealth. However, Canada, a nation with much more liberal social welfare traditions than the United States, nevertheless still exhibited significant inequality. More than one-half (53 percent) of Canadian assets, were owned by only 10 percent of the population. European countries such as Norway, at 50.5 percent, and Spain, at 41.9 percent, had similar or slightly lower levels of wealth inequality.[10]

The most revealing finding of the World Institute for Development Economics Research was that similar patterns of wealth inequality have come to be prevalent throughout the developing world. In Indonesia, for example, 65.4 percent of the nation's total wealth belonged to the wealthiest 10 percent in 2000. In India, the upper 10 percent owned 52 percent of all Indian wealth. Even in China, where the ruling Communist Party still maintains vestiges of what might be described as "authoritarian state socialism," the wealthiest 10 percent owned 41.4 percent of the national wealth.[11]

But even these macroeconomic statistics, as useful as they are, obscure a crucial dimension of wealth concentration under global apartheid's neoliberal economics. In the past quarter century in the United States, where deregulation and privatization have been carried to obscene extremes, we are presently witnessing a phenomenon that the media has described as "the very rich" who are leaving "the merely rich behind." One study by New York University economist Edward N. Wolff found that 1 out of every 825 households in the United States in 2004 earned at least $2 million annually, representing nearly a 100 percent increase in the wealth percentage recorded in 1989, adjusted for inflation. As of 2004, 1 out of every 325 U.S. households possessed a net wealth of $10 million or more. When adjusted by inflation, this is more than four times as many wealthy households as in 1989. The exponential growth of America's "super-rich" is a direct product of the near elimination of capital gains taxes and the sharp decline in federal government income tax rates.[12]

Inside the United States, the processes of global apartheid are best represented by the "New Racial Domain" (NRD). The NRD is different from other earlier systemic forms of racial domination inside the United States—such as slavery, Jim Crow segregation, and ghettoization or strict residential segregation—in several critical aspects. These earlier racial formations, or exploitative racial domains, were grounded or based primarily, if not exclusively, in the political economy of U.S. capitalism. Antiracist or oppositional movements that blacks, other ethnic minorities, and white antiracists built were largely predicated upon the confines or realities of domestic markets and the policies of the U.S. nation-state. Meaningful social reforms such as the Civil Rights Act of 1964 and the Voting Rights Act of 1965 were debated almost entirely within the context of America's expanding, domestic economy, and influenced by Keynesian, welfare-state public policies. The political economy of America's NRD, by contrast, is driven and largely determined by the forces of transnational capitalism, and the public policies of state neoliberalism. From the vantage point of the most oppressed U.S. populations, the NRD rests on an unholy trinity, or deadly triad, of structural barriers to a decent life. These oppressive structures are mass unemployment, mass incarceration, and mass disfranchisement. Each factor directly feeds and accelerates the others, creating an ever-widening circle of social disadvantage, poverty, and civil death, touching the lives of tens of millions of people in the United States.[13]

Transnational Blackness presents examples of individuals and organizations of African descent, primarily originating in the United States, that challenged the legitimacy and power of the global color line and its oppressive political economies of inequality. Such examples varied widely in the tactics and strategies for social change they employed. What they held in common was a long memory of resistance to human exploitation, and the knowledge of African-descendant cultural heritages and rituals that connected the diverse peoples of the African Diaspora. For Du Bois over a century ago, there were certain "Negro ideals" worth fighting to preserve, which challenged the hegemonic materialism of Europe and America. Similarly, as the twenty-first century unfolds, and as the global color line's struggles for social justice intensify, the role of black activist-intellectuals and social protest movements will assume even greater significance transnationally.

NOTES

1. W. E. B. Du Bois, "The Conservation of Races," Occasional Paper No. 2 (Washington, DC: American Negro Academy, 1898).
2. Ibid.
3. W. E. B. Du Bois, *The Souls of Black Folk* (New York: Oxford University Press, 2007), 8.
4. See W. E. B. Du Bois, "The Color Line Belts the World," *Collier's Magazine* 20 (October 1906): 20.
5. Thomas B. Edsall, "Risk and Reward," *New York Times*, December 5, 2006.
6. Ibid.
7. Fred Magdoff, "The Explosion of Debt and Speculation," *Monthly Review* 58, no. 6 (November 2006): 24–26.
8. Ibid.
9. The Editors, "U.S. Military Bases and Empire," *Monthly Review* 53, no. 10 (March 2002): 1–14.
10. Eduardo Porter, "Study Finds Wealth Inequality Is Widening Worldwide," *New York Times*, December 6, 2006.
11. Ibid.
12. Louis Uchitelle, "Very Rich Are Leaving the Merely Rich Behind," *New York Times*, November 27, 2006.
13. I have outlined in greater detail the political economy of the New Racial Domain in Manning Marable, *Living Black History: How Reimagining the African-American Past Can Remake America's Racial Future* (New York: Basic Civitas, 2006), 214–300.

PART I

THEORIZING RACE IN A GLOBAL CONTEXT

CHAPTER 1

RACE AND GLOBALIZATION

RACIALIZATION FROM BELOW

LEITH MULLINGS

IN 1903 THE GREAT AFRICAN AMERICAN SCHOLAR AND PAN-AFRICANIST W. E. B. Du Bois noted, "The problem of the twentieth century is the problem of the color line . . . the relation of the darker to the lighter races of men in Asia and Africa, in America and the islands of the sea."[1]

In the twenty-first century we are still confronted with an international color line, the racialized consequences of massive impoverishment, and displacement integral to globalized capitalism. The challenges of structural racism create simultaneous spaces for oppression and resistance. The modern color line is not only imposed from above but also becomes a site for contestation from below. This book testifies to the complexities of the paradox of race in the contemporary world. As these papers demonstrate, globalization has resulted in new forms of racialization but has also created new, transnational forms of resistance to racism. Race making—the construction of race as a way to rationalize global inequalities—also creates a basis for global collective action. These innovative new movements take race as a space for organizing global social movements against the inequities of globalization and have the potential to transcend both the scope and the reach of earlier Pan-Africanist movements. Four centuries of the transatlantic slave trade and racialized subordination of people of African descent produced a construction of race throughout much of the world. As a result, many regions of the world were dominated by what one could call a racial mode of production—involving not only exploitation of labor but also the skills of Africans and their descendants—to build the modern world system. In many areas of the world, race became a worldview that rationalized domination and privilege, on one hand, and dispossession of land, labor, wealth, and rights, on the other. "Scientific racism," which emerged in the eighteenth century, provided a pseudoscientific patina for a set of beliefs that categorized people into different races, each endowed with unequal capacities, and alleged not merely that biological and social differences were fixed, inheritable, and unchangeable, but also that races could be ranked hierarchically, with the white race as the pinnacle of civilization.

Yet the imposition of race also created the structural context for producing sites of resistance and creative spaces for the articulation of subaltern opposition. The twentieth century saw magnificent mass struggles whose objective was to overturn powerful structures of racial hierarchies: the diverse and sometimes contradictory anticolonial struggles throughout Africa and the Third World, the civil rights movement in the United States, and the anti-apartheid movement in South Africa. These struggles brought about powerful worldwide transformations in politics and culture. The people of Africa, Latin America, and the Caribbean, who lived under European colonization, have now acquired nominal sovereignty in nation-states, defeated the apartheid state in South Africa, and dismantled legal segregation in the United States. Today, most anthropologists reject the notion of biological race.[2]

Powerful new forces now configure how race is lived. With the fall of socialism, globalized capitalism became the dominant world system. Old and new forms of "accumulation by dispossession"[3] have increasingly impoverished much of the world's population, creating new forms of racialized and gendered stratification and new sites of racialization, such as the prison-industrial complex. U.S. phenomena including rising unemployment, the dismantling of the welfare state, the privatization of previously public services (such as education), and the growth of incarceration as a way of controlling dissent[4] all have their counterparts in Africa, Latin America, and the Caribbean. In the areas of the greatest oppression, the legacies of colonialism, international debt, structural adjustment policies, and the enormous circulation of arms have resulted in major setbacks for the progressive thrusts of the post–World War II liberation movements. Black women frequently become the first victims of violence during structural adjustment, warfare, ethnic strife, and domestic conflict.

In these new conditions, the meaning of race is constantly reconfigured as new forms of exclusion built upon the continuing consequences of enslavement, colonialism, and imperialism. As capitalism incorporates elements of racialized populations into its neoliberal global project, the familiar rigid stereotypes and polarizing discourses built during earlier epochs become less useful to capital's needs, and new ways of managing race and inequality are manufactured and fostered. In the United States and South Africa, for example, the implementation of policies of "color blindness" may serve to repackage white supremacy by denying the continuing significance of racism.[5] Racial stratification now works not only through frameworks alleging biological differences between populations but also through an emphasis on individual meritocracy or group culture. In Europe, right-wing anti-immigration ideologues no longer openly claim incontrovertible biological differences between races in opposing immigration from Asia, Africa, the Caribbean, and the Middle East but rather assert that there are unbridgeable cultural differences between immigrant groups and Europeans—a neoracism, a "racism without race."[6] In the United States, "culture of poverty" or "underclass" frameworks function pseudobiologically as mutually enforcing paradigms that provide explanations for—and ultimately deflect attention from—the structural forces that have produced savage inequalities of race, class, and gender.[7]

It is also true that the forces of globalization previously described also create spaces for new mobilizations and counterhegemonic movements that can be transnational in theory and practice; Brecher, Costello, and Smith describe this as "globalization from below."[8] Similarly, we can see the emergence and acceleration of counterhegemonic social movements framed in the language of race and racism in order to signal dispossession, make claims on resources, form transnational alliances, and challenge racialization from above—a process we might call "racialization from below." This development is particularly striking in areas such as Latin America where ideologies glorifying race mixture and the lack of legal segregation have inhibited the growth of such organizations. The expansion of these movements is assisted by the use of new information technology,[9] hemispheric and global conferences, such as the UN World Conference against Racism, Racial Discrimination, Xenophobia and Other Forms of Intolerance (WCAR) and their associated documents.

For example, in Latin American countries with sizeable populations of African descent, racially based social movements have emerged and gained strength in the 1980s and 1990s. This is particularly significant because in many of these countries, though the dominant ideology projects a national ideal of racial mixture, pervasive discrimination coexists with an official denial of the existence of racism.[10] In the last two decades, many Latin Americans of African descent have increasingly rejected the options of mixed race categories and the "mulatto escape hatch" in order to assert, on their own terms, an oppositional racial—or more precisely, racialized—identity. In some cases, these movements have won land titles, a commitment to implement antidiscriminatory policies (such as affirmative action), and official government recognition of their historical distinctiveness.

Brazil, the country with the largest black population in the hemisphere, is an interesting case in point. Observers frequently note the paradoxical coexistence in Brazil of persistent racial inequality with the failure to generate mass antiracist movements. The generally accepted national ideology that Brazil is a "racial democracy"—a concept that vigorously denies racial inequality yet aggressively subordinates the status and identity of blackness—has historically limited organizing on the basis of racial discrimination despite the existence of a longstanding, if fragmentary, black consciousness movement.[11] Recently, the articulation and mobilization of a militant black identity has emerged as a potentially powerful political concept unifying those who self-identify as being of African descent and has begun to comprise a major component in the national debate about the future of the Brazilian state.[12] As Sheila Walker observes, although "Africanity" has always been a significant feature of Brazilian culture, Afro-Brazilians are only now unambiguously claiming their blackness.[13] As a result of the growing capacities of that movement in the late 1990s, the Brazilian centrist government under former President Fernando Henrique Cardosa officially acknowledged the pervasive existence of racial discrimination in Brazil and appointed a national commission to explore it. Militant Afro-Brazilians were prominent in the official delegation to the WCAR and were able to negotiate a major shift in Brazil's dialogue on race.[14] In May 2001, then President Cardosa signed a decree initiating the National Program for Affirmative Action. Currently, there is a growing (and strongly contested) movement led by Afro-Brazilian members of the ruling Workers' Party to push for a much more comprehensive racial equality statute that would aggressively enforce affirmative action and similar policy reforms in order to redress Brazil's racial disparities. President Luiz Inacio Lula da Silva of the Workers' Party has responded positively to these demands by proposing measures addressing discrimination in government. Recently, he appointed the first black Supreme Court justice and four Afro-Brazilian ministers to his cabinet, including a newly created cabinet-level Special Secretariat for Promotion of Racial Equality.[15] In the face of implacable resistance to the enforcement of affirmative action measures in the form of legal actions, the current government has not only supported affirmative action but also instituted innovative school attendance grants that allow poor families to keep their children in school rather than pulling them out to work.[16] In theory, the policy is designed to address class inequality, but in practice, within the racialized social hierarchy of the country, such reforms also address the profoundly racialized effects of marginalization.

Colombia is another important state in which "racialization from below" is developing as a strategy of resistance. Colombia contains the third highest African-descended population in the Americas, after Brazil and the United States. Afro-Colombians have historically inhabited lands on the Pacific Coast that are rich in timber, gold, farming potential, and biodiversity. Beginning in the 1980s, national and multinational corporations have been attracted to this potential market and have invested heavily in this region.[17] As anticorporate protests grew all over the country, a nascent Afro-Colombian social movement crystallized in the course of organizing to promote legislative reforms.[18] Afro-Colombians were able to win some recognition of territorial rights and cultural distinctiveness in the new 1991 National Constitution and, through subsequent legislation in 1993, gained the right to apply for land titles. However, after winning these rights, Afro-Colombians have been subject to massive displacement from their ancestral lands.[19] Afro-Colombians have increasingly been victims of assassinations of activists and massacres of villagers, primarily by right-wing paramilitary forces,[20] and several observers suggest that levels of violence directed against them have been significantly escalated by the billions of U.S. dollars pouring into the area through the "War on Drugs."[21] In this context, Asale Angel-Ajani describes the Afro-Colombian strategy of organizing into "peace communities," which adopt a general policy of neutrality regarding the armed conflict, as an attempt to separate themselves from the escalating violence and the inevitable ravages of constant warfare. To assert their approach of neutral disengagement from violence, Afro-Colombians have increased their organizing efforts in the national and international arenas. The first Afro-Colombian national conference was held in 2002, and Afro-Colombians have also mounted efforts to mobilize international solidarity by traveling to Europe and the United States to raise awareness about their situation and create alliances.

In this process, culture plays a strategic role in creating and sustaining an Afro-Colombian identity, on the one hand, and constructing a larger transnational movement, on the other. Joseph Jordan describes this reimagining blackness, African consciousness, and Afro-Colombian ethnicity as a process "that provides a means for extra-national citizenship connecting Afro-Colombians to other communities of African descent throughout the Americas." The adoption and indigenization of popular cultural forms, such as hip-hop, provide transnational cultural matrices for the articulation of new forms of identity. These function as a strategy of resistance, with the potential to reimagine an African descendant identity, thereby connecting Afro-Colombians to other communities of African descent, as well as making group demands on the nation-state.[22] In Europe, there are more than ten million people of African, Asian, and Middle Eastern descent. Clarence Lusane describes how, drawing heavily on UN declarations and resolutions, "black" communities have been centrally involved in Europe-wide campaigns that call for the implementation of antidiscrimination policies. According to Lusane, Europe is now poised to make "the most sweeping changes in anti-discrimination legislation since the modern effort to come together as a region."[23]

From August to September of 2001, the WCAR convened in Durban, South Africa. The WCAR and its preparatory conferences were an important point at which these nascent and often contradictory movements began to converge. The WCAR and its accompanying Non-Governmental Organizations (NGO) Forum was attended by thousands of representatives of governments, international agencies, NGOs, academia, policy makers, media, and communities from over 160 countries. In addition to indigenous and African-descendant populations from all over the world, other groups such as the Burakumin of Japan, the Roma from Europe, and the Dalits from the Indian subcontinent were all in attendance and pressed their claims for addressing comparable forms of discrimination. The more than eight thousand attendees participated in formal and informal meetings, exchanged information and forged alliances. Both meetings produced policy documents, declarations, and programs, and the WCAR adopted a Declaration and Program of Action that commits member states to undertake a wide range of measures to combat racism and racial discrimination at the international, regional, and national levels. These have been used by movements all over the world as organizing tools, for popular education and, as models for legislation.

The various movements represented at Durban emerged from extremely diverse conditions and find themselves at various levels of development, but all are intensified by the transnational consequences of globalized capitalism and the imposition of market-based structural adjustment "reforms" on exploited populations. The recognition of common issues transcending boundaries and the interrogation of the role of profits from slave trade, colonialism, and more recent forms of domination play a part in current conditions of poverty; are important steps in building a transnational movement to challenge these conditions; and at the same time accelerate the ability of these movements to see themselves as positioned both within and beyond the nation-state.

Samir Amin observed that one of the most significant features of the conference was the development of an analysis that situated racism and discrimination as "generated, produced and reproduced by the logic and expansion of capitalism," which in its contemporary globalized form "can only result in 'apartheid on a global scale.'"[24] The concept of global apartheid was an important theme of the conference. In a 2001 article in the *Nation*, Salih Booker and William Minter described global apartheid as "an international system of minority rule whose attributes include: differential access to basic human rights; wealth and power structured by race and place; structural racism, embedded in global economic processes political institutions and cultural assumptions; and the international practice of double standards that assume inferior rights to be appropriate for certain 'others,' defined by location, origin, race or gender."[25]

Amin hailed WCAR as "a people's victory" in which "the spirit of Bandung has breathed again." He compared it to the historic Bandung Conference held in Indonesia in 1955, attended by leaders from Africa and Asia.[26] However, WCAR is potentially a more advanced development than the meeting in Bandung, which was comprised essentially of representatives of states or states-in-waiting—that is, government officials and representatives from national liberations organizations who possessed official credentials. The NGO forum of WCAR, on the other hand, encompassed a range of sometimes disorganized and often contradictory popular currents. At the same time, WCAR transcended the more traditional category of Pan-Africanism by incorporating subaltern groups who are not of African descent but find themselves similarly affected by the weight of global apartheid.

These counterhegemonic movements nevertheless face considerable challenges, and their success in confronting global apartheid will depend on the extent to which they are able to build on the dominant approach that emerged from WCAR—one that attempts to transcend an essentialized notion of race and move toward a perspective linking subaltern populations, not by race but by racialization. There are strong pressures, often from neoliberal organizations such as the World Bank, to emphasize issues of cultural difference at the expense of more radical efforts toward fundamental social change.[27] If history is any guide to the extent that this nascent movement succumbs to such pressures and retreats toward a more parochial vision of their political tasks, they risk developing fissures along lines of class, gender, and other differences and even eventually imploding. As these divergent movements progress to the next step, it will be essential for them to continue to build solidarities across identities and boundaries, developing coalitions among the working class, antiracists, and feminist constituencies and alliances with progressive national liberation movements. In so doing, place-based efforts will have to transform themselves from movements of integration, autonomy, or reform into more ambitious movements confronting all forms of inequality. To achieve these goals, it will be necessary to construct a new language of human emancipation that has the capacity to project a new vision of an alternative global social order in which "difference" does not inevitably convey the reality of structural inequality.

NOTES

1. W. E. B. Du Bois, *The Souls of Black Folk* (1903, repr., Boulder, CO: Paradigm, 2004).
2. Leonard Lieberman, "How 'Caucasoids' Got Such Big Crania and Why They Shrank," *Current Anthropology* 42, no. 1 (2001): 69–95.
3. David Harvey, *The New Imperialism* (New York: Oxford University Press, 2003).
4. Leith Mullings, "Losing Ground: Harlem, the War on Drugs, and the Prison Industrial Complex," *Souls: A Critical Journal of Black Politics, Culture, and Society* 5, no. 2 (2003): 1–21.
5. Howard Winant, *The World Is A Ghetto: Race And Democracy Since World War II* (New York: Basic Books, 2001); Lee Baker, *From Savage to Negro: Anthropology and the Construction of Race 1896–1954* (Berkeley: University of California Press, 1998); George Fredrickson, *Racism: A Short History* (Princeton, NJ: Princeton University Press, 2002).
6. Etienne Balibar, "Is There a "Neo-Racism?" in *Race, Nation, Class: Ambiguous Identities*, ed. Etienne Balibar and Immanuel Wallerstein (New York: Verso, 1991), 15–28.
7. Leith Mullings, *On Our Own Terms* (New York: Routledge, 1997).
8. Jeremy Brecher, Tim Costello, and Brendan Smith, *Globalization From Below: The Power of Solidarity* (Cambridge, MA: South End, 2000).
9. Francis Njubi, "New Media, Old Struggles: Pan Africanism, Anti-Racism and Information Technology," *Critical Arts* 15, no. 1–2 (2001): 117–34.
10. Marisol De la Cadena, "Reconstructing, Race, Racism, Culture and Mestizaje in Latin America," *NACLA Report on the Americas* 34, no. 6 (2001): 16–28.
11. Michael Hanchard, *Orpheus and Power: The Movimiento Negro of Rio de Janeiro and Sao Paulo, Brazil, 1945–1988* (Princeton, NJ: Princeton University Press, 1994); Kimberly Jones-de Oliveira, "The Politics of Culture or the Culture of Politics: Afro-Brazilian Mobilization, 1920–1968," *Journal of Third World Studies* 20, no.1 (2003): 103–20; Frances Windance Twine, *Racism in a Racial Democracy: The Maintenance of White Supremacy in Brazil* (New Brunswick, NJ: Rutgers University Press, 1998).
12. Michael Hanchard, ed., *Racial Politics in Contemporary Brazil* (Durham, NC: Duke University Press, 1999); A. S. Guimaraes, "Behind Brazil's 'Racial Democracy,'" *NACLA Report on the Americas* 34, no. 6 (2001): 38–39.
13. Sheila Walker, "Africanity vs Blackness: Race, Class and Culture in Brazil," *NACLA Report on the Americas* 35, no. 6 (2002): 16–20.
14. J. Michael Turner, "The Road to Durban—and Back," *NACLA Report on the Americas* 35, no. 6 (2002): 31–35.
15. Marion Lloyd, "In Brazil, a New Debate Over Color," *Chronicle of Higher Education* 50, no. 23 (2004): 38–40.
16. Thomas Boston and Usha Nair-Reichert, "Affirmative Action: Perspectives from the United States, India and Brazil," *Western Journal of Black Studies* 27, no. 1 (2003): 3–14.
17. Arturo Escobar, "Displacement, Development, and Modernity in the Colombian Pacific," *International Social Science Journal* 55, no. 175 (March 2003): 157–67.
18. Peter Wade, *Blackness and Race Mixture: The Dynamics of Racial Identity in Colombia* (Baltimore, MD: Johns Hopkins University Press, 1993); Arturo Escobar, "Displacement."
19. Arturo Escobar, "Displacement."
20. Ulrich Oslender, "Communities in the Crossfire," in *Hemisphere: A Magazine of the Americas* 11, no. 3 (2002): 24–27.

21. Forrest Hylton, "An Evil Hour: Uribe's Colombia in Historical Perspective," *New Left Review* 23 (2003): 51–93.

22. Peter Wade, "Working Culture: Making Cultural Identities in Cali, Colombia," *Current Anthropology* 40, no. 4 (1999): 449–71.

23. Clarence Lusane, Chapter 15 of this volume.

24. Samir Amin, "World Conference Against Racism: A People's Victory," *Monthly Review* (December 2001): 21.

25. Salih Booker and William Minter, "Global Apartheid," *Nation* 273, no. 2 (2001): 11–16.

26. Samir Amin, "World Conference against Racism: A People's Victory," *Monthly Review* (December 2001): 21.

27. Charles Hale, "Does Multiculturalism Menace? Governance, Cultural Rights and the Politics of Identity in Guatemala," *Journal of Latin American Studies* 34 (2002): 485–524.

CHAPTER 2

GLOBAL APARTHEID, FOREIGN POLICY, AND HUMAN RIGHTS

FAYE V. HARRISON

HUMAN RIGHTS—"THE REASONABLE DEMANDS FOR PERSONAL SECURITY AND BASIC well-being that all individuals can make on the rest of humanity by virtue of their being members of the species Homo sapiens"[1]—are in increased jeopardy in this era of globalization. Small, poor countries increasingly are dominated by imposed economic controls that make a mockery of their rights to self-determination. For about two decades, this neoliberal regime—in which developed nations aid poorer nations on the condition that they restructure their economies and political systems to accommodate maximum wealth accumulation by multinational corporations—has arrived packaged as so-called free trade. This phenomenon is more than an idea or ideology. It is a cultural system, "a paradigm for understanding and organizing the world and for informing our practices within it."[2] It is "an approach to the world which includes in its purview not only economics but also politics, not only the public but also the private, not only what kinds of institutions we should have but also what kinds of subjects we should be."[3]

The reasons for this assault on human rights—political and socioeconomic— are complex. In many parts of the world, however, it can be attributed, at least in part, to the relative immunity with which transnational corporations and agencies dictate social, political and economic issues within nation-states, especially smaller nations. These nations' ability to protect rights to education, health care, and humane work standards is drastically compromised by internationally mandated policies and programs that give higher priority to corporate rights and the rights of transnational capital than to the basic needs and dignity of ordinary human beings. Although the social contract that more democratic states once had with their citizens is disappearing, the repressive role of state power clearly is not. In many cases, Western, particularly the United States, foreign aid packages include generous provisions for police and military upgrading. Thanks to this free market in arms, intergroup tensions within smaller nations now are more apt to escalate into militarized conflicts.

For example, the militarized condition of life in Jamaica provides a prime example of how U.S. foreign aid for fighting drugs and crime impacts developing nations. During the 1980s, after the politically orchestrated demise of the democratic socialist administration, the policing and military capacity of the conservative Jamaica Labor Party government was substantially upgraded with a sizable security aid package, the largest ever given by the United States to any country in the Commonwealth Caribbean. The aid enabled the government of Edward Seaga, former prime minister, to act more punitively against the "dangerous elements"—crime, labor discontent, and political unrest—that threatened law and order on the island and threatened the United States' strategic interests in the region. The well-funded war against crime was led by the Special Operations Squad, popularly dubbed "Seaga's eradication squad." In the mid-1980s, Americas Watch issued a human rights report that decried the growing pattern of extrajudicial executions responsible for half of the nation's total homicides. The militarization of the state and the often indiscriminate deployment of repressive police tactics remain a problem today. Last year, these problems prompted Amnesty International to censure the government in a special report.[4]

These problems are not confined to the southern hemisphere; comparable trends are also in evidence in the north. In the United States alone, Reaganomics, Contract with America, welfare reform, the dismantling of affirmative action, Proposition 187, policing by racial profiling, and the prison-industrial complex are variations on the same theme. Note that they closely resemble the structural adjustment programs that the International Monetary Fund (IMF) stipulates and the war on drugs and crime underwritten by the United States in debt-ravaged "developing" countries. Common themes emerge upon examination of these tactics to regulate the global economy and police the crises that regulation engenders. This neoliberal method results in processes that might be called capitalism's second primitive accumulation and a "recolonization" of markets in a world fraught with dilemmas of postcolonialism and the postmodern condition.[5]

David L. Wilson, an activist with the Nicaragua Solidarity Network, provides another example of this dynamic in his analysis of *maquiladoras* (assembly plants operating as subsidiaries or subcontracted firms of transnational corporations) as a site for the workings of neoliberalism. Neoliberalism, he writes, is a regime of development and a phenomenon of primitive accumulation, which in many ways is comparable to the classic case that Marx described for the transition into capitalism. Primitive accumulation then and now creates "a vast labor pool of people desperate for jobs, even at wages below subsistence levels." Wilson argues: "[W]hat is new about neoliberalism is a sort of primitive accumulation against capitalist and post-capitalist economic forms—against industrial production for the domestic market, against small-scale capitalist or cooperative agriculture (often the result of agrarian reform), and against the tenuous but crucial safety net that has developed in many third world countries."[6] These conditions of change, mediated by the IMF, World Bank, and World Trade Organization (WTO)—all multinational but strongly U.S.-influenced—have been intensified by the geopolitical and politico-economic realignments engendered in today's post–cold war milieu in which alternatives to capitalism are widely discredited.

Subsequently, capitalism's conflicts with communism and socialism have—except for the brutal U.S. embargo against Cuba—given way to wars between competing ethnic groups and to a U.S.-funded "war against drugs" that is driven by the contradictory foreign policy of the remaining superpower.[7]

In this context, one of the gravest human rights problems is the intensification of discrimination and violence that target people on the basis of race.[8] Race is a socially constructed distinction, material relation, and dimension of social stratification that intersects with and is mutually constituted by class, gender, ethnicity, nation, and increasingly transnational location and identity. Although culturally variable, it encodes social differences often presumed to be hereditary—differences that, if not carefully managed and policed, are considered threats to a nation's social structure.

Although historically, racial differences were considered to be rooted in biological variations, today these differences are increasingly expressed not in racial terms but in cultural terms. These trends in reconfiguring race are evident across a wide array of international settings, from European zones of ethnic cleansing (where, through mass rapes, women became permanently partitioned racial subjects) to African contexts (in which ethnonational conflicts are racial and, in extreme cases, intensify to the point of genocide).[9] The conflict between the Hutus and Tutsis in Rwanda and Burundi is a tragic instance of this. Anthropologist Lisa Malkki has pointed out that members of Burundi's Hutu community has crafted mythohistorical narratives that define their differences from Tutsis in terms of "moral essences," which operate as powerfully as the biological distinctions that operated during an earlier era.[10]

In many places around the world, race is being reconfigured in more acceptable ideological codes and rhetoric, which some scholars view as a new form of racism without races. Social critics in France, Germany, and Austria point out that even right-wing xenophobes in their countries formally acknowledge that blatant racism is widely discredited and that "races" do not "really exist." Although this may sound progressive, this cursory, one-dimensional awareness does not mean that racism has withered away or is not being reproduced in modern and postmodern guises. Despite the nominal no-race stance taken by some western European neofascists, their punitive assaults against Third World immigrants and eastern European refugees (e.g., the Roma in Bosnia) effectively demonize ethnonational outsiders and subject them to conditions so oppressive that a new form of apartheid may be emerging. Encoded in the notions of *immigrant* and *refugee* are meanings of ethnic absolutism that invent or renew racial identities on reconfigured landscapes of national inclusion and exclusion. Paradoxically, although certain categories of immigrants are viewed as troublesome parasites whose cultures threaten the purity of European nations, their economic participation in ethnically and sexually segmented labor keeps their host economies thriving and enriches their employers.

This ambivalence is also present in the United States, where nativist campaigns target immigrants. Californians supported Proposition 187, which barred children of illegal immigrants (mainly Mexican) from educational and health services, even as California's agribusiness and service sectors became increasingly dependent on the exploitable labor of the children's parents. Propositions such as 187 are not intended to create an inhospitable atmosphere for immigrants, thereby urging them to return to their native countries; it is really about keeping them in their (exploited and vulnerable) place within the United States by restricting their legal rights. In other words, these measures perpetuate a deskilled and stigmatized labor force that cannot make credible human rights demands like those increasingly made by Americans and legal residents of color—demands that are eroding white privilege and engendering a crisis of white identity.

Alongside, and in some instances interacting with, these culturalist essentialisms, though, is the relentless resurgence of biology-based accounts about the nature and roots of social difference. This is clearly the case in North America where Richard Herrnstein's and Charles Murray's 1994 book, *The Bell Curve: Intelligence and Class Structure in American Life*, and research funded by neoconservative foundations such as the Pioneer Fund have revitalized discussions on measurements of intelligence, athleticism, fertility patterns, and criminal violence.

These disturbing patterns are reemerging despite the decades-old perspective of such scholars as Ralph Bunche, who noted in his *A World View of Race* that racial distinctions lack any real biological basis. In his bold analysis of imperialism, global intergroup conflicts, and the threat racism posed to world peace, Bunche—who was awarded the Nobel Prize in 1950 for his role as UN mediator in Palestine and Israel's conflicts with neighboring Arab states—underscored the economic basis of the global racial hierarchy and its fundamental intersection with class exploitation.[11]

Bunche was influenced by Du Bois, who in 1915 published a seminal essay, "African Roots of the War," in which he theorized about imperialism and global conflict before Lenin published his *Imperialism, The Highest Stage of Capitalism* in 1917. Historian and biographer David Levering Lewis writes that the essay, which articulated Du Bois's "mature ideas about capitalism, class, and race" in the workings of colonialism and the causes of World War I, was "one of the analytical triumphs of the early twentieth century."[12] Nearly ninety years after Du Bois's analysis and seventy-six years after Bunche's, political scientists and others who study international relations still need to be urged to include race and racism in their analyses of global politics and political economy.

Anyone who reads the newspaper—and knows how to read between the lines—is aware that racism and the interlocking injustices of xenophobia, class exploitation, and gender oppression are escalating global phenomena. If they read the alternative media, they know that some people think the globalization of free-market ideas and policies, especially those imposed on vulnerable nations (i.e., neoliberalism) have something do with this trend. If they read or heard broadcast news reports in late August and early September 2001, they are well aware that these problems were foci around which the fraternal twin meetings, the World Conference against Racism (WCAR) and its parallel Non-Governmental Organizations (NGO) Forum, convened in Durban, South Africa. The meetings marked the year 2001 as the International Year of Mobilization against Racism, Racial Discrimination, Xenophobia, and Related Intolerance, one of the highlights of the Third Decade to Combat Racism and Racial Discrimination (1993–2003).[13] South Africa's symbolic power as a postapartheid society and, previously, as the setting for a protracted struggle for African liberation and multiracial democracy, resonates deeply with the political sensibilities and yearnings of antiracists the world over. As a sort of secular "Mecca," the Durban meetings attracted "pilgrims" from all over the world. Not surprisingly, quite visible among them were NGO representatives and country delegates from the African continent and Diaspora. In the spirit of optimism, we might say that the pilgrims who gathered in Durban participated in symbolically charged and substantively meaningful rituals of rebellion and solidarity. On this hopeful note, let us also assume that some of them—by virtue of their experience and by virtue of their critique of those experiences—underwent a significant rite of passage that led them to a new phase of critical knowledge, consciousness, and struggle. Their expanded social action and political mobilization toolkits may have enabled them to better respond to today's volatile atmosphere of restructuring, which is an atmosphere that seems particularly resilient in the face of many of the resistance tactics employed in the past.

As they police the crises that neoliberalism unleashes, the managers of today's global economy insist that there are no alternatives to the market liberalization, privatization, and cuts in government spending—domestic and foreign—being mandated by the IMF, World Bank, WTO, and U.S. policy. These neocolonial ideologies are informed by transnational interests that force vulnerable nations to redefine their national priorities. Neoliberal ideology and policy directives, which cross national boundaries with impunity, have promoted free market rights at the expense of human rights. It is crucial to note that globalized politics and policies, particularly those of such post–World War II institutions as the IMF and World Bank, are now largely controlled by a single superpower: the United States. Although, as Sherle R. Schwenninger writes, "the perception of U.S. power and influence has in many cases exceeded its reality," the United States dominates, especially in the area of finance. Owing to the "unusual circumstances of the post-cold war period—Europe's preoccupation with the European monetary union, Japanese deflation, Russian weakness, low oil prices, geopolitical inertia in East Asia," the United States controls "world monetary policy in a way not seen since the 1950s." As a consequence, it has been able to "[push] financial liberalization (without adequate safeguards) on such unprepared countries as South Korea, Thailand and Russia."[14] In other spheres as well, the United States sets and limits the policy in accordance with its interests—interests that often disregard the threatened life chances, health, and subsistence security of most human beings.

Will post-Durban thinking and organizing lead a critical mass of antiracists beyond the formal trappings of democratization and its mystifying public relations rituals and selective enforcement of human rights? Will new mobilization strategies move activists beyond the state-centeredness of most UN programs toward more effective ways of combating the rights violations for which transnational interests are responsible? In light of the forces that circumscribe and often dictate what states (particularly peripheral states) can do, will the pilgrims push for principled implementation of substantive, concrete change, and true human rights?

Although South Africa was certainly an ideal site for the UN-sponsored human rights conference and antiracist pilgrimage, in all honesty, the country can only be characterized as *postapartheid* in the narrowest terms. De facto justice has yet to be achieved. Thus the concerted struggle against apartheid must continue in South Africa *as well as in the world at large*. The biggest threat to human rights and to human *life* and life chances, particularly those of racially subjugated people, is the structural violence that emanates from *global apartheid*.[15] Structural violence is the symbolic, psychological, and physical assaults against human psyches, physical bodies, and sociocultural integrity that emanate from situations and dominant institutions. This broader range of symbolic, psychological, economic, and environmental assaults is neglected, because the conventional human rights system has mainly focused on liberal notions of individuals' political and civil rights within nation-states. Yet ultimately and ironically, these structurally induced forms of violence set the stage for the very political abuses that have traditionally been the focus of human rights monitoring. A central issue that has not received its due is the question of social and economic rights, which are controversial in their potentially profound implications for income and wealth redistribution. These rights are highly contested and effectively sabotaged by structures of power and privilege dictated primarily by transnational forces, such as the IMF, the World Bank, the WTO, and U.S. foreign policy—entities whose politico-economic purview, as mentioned earlier, transcends that of individual nation-states.

Apartheid is a policy of enforced separation and disparities between races. Other than through the deployment of state repression in explicitly racially coded laws, oppressively segmented labor markets, and brutal policing, apartheid's enforcement can also be accomplished through subtler covert means that evade and disguise race while reproducing it nonetheless. Consequently, apartheid persists although de jure forms of racism officially ended with the historic 1994 election of Nelson Mandela as South Africa's president. Beyond the specifics of South Africa, the term *apartheid* can be applied to the global order, that is, the so-called New World Order, and not simply as an effective metaphor. As Salih Booker and William Minter point out, the "concept captures fundamental characteristics of the current world order missed by such labels as 'neoliberalism,' 'globalization' or even 'corporate globalization.'" Global apartheid is a reality marked by the operation of "undemocratic institutions [that] systematically generate economic inequality." Booker and Minter define it as "an institutional system of minority rule whose attributes include: differential access to basic human rights; wealth and power structured by race and place; structural racism, embedded in global economic processes, political institutions and cultural assumptions; and the international practice of double standards that assume inferior rights to be appropriate for certain 'others', defined by location, origin, race or gender."[16]

Peace studies researcher Gernot Köhler reinforces this view by pointing out that, in the current world situation, a minority race of whites and "honorary whites" dominates the majority of humanity, which is composed disproportionately of peoples who were once defined by former colonial authorities as racial outsiders and are now—at least implicitly—treated as racial subjects by neocolonial powers and their political economy.[17] Related to this view is anthropologist Ann Kingsolver's concept of "strategic alterity," which refers to "the practice of shifting between strategic assertions of inclusion and exclusion (or the marking and unmarking of 'selves' and 'others') to both devalue a set of people and to mask that very process of strategic devalorization"—the present-day transnational world is organized around such strategic differences.[18] As Köhler writes, apartheid in its global form is even more severe than what existed in predemocratic South Africa in that the disparities in wealth, power, military control, health, and life expectancy are even more extreme and are still growing.

Sociologist Howard Winant correctly underscores that "the contemporary international hierarchy (i.e., capitalism, a system that necessarily creates and perpetuates racial hierarchies) works through varied experiences, identities, and conflicts rather than through any overarching uniformity. . . . [E]ach nation-state, each political system, each cultural complex necessarily constructs a unique racialized social structure, a particular complex of racial meanings and identities." Thus "the . . . increasing internationalization of race can only be understood in terms of prevalent patterns, general tendencies, [but] in no sense can such generalizations substitute for detailed analyses of local racial formations."[19]

Scholarly research on issues of race and racism is producing a rich canon on the diversity of racism and the culturally specific ways that race—whether marked or unmarked—is socially and politically constructed (and reconstructed) as an institutional or structural basis for identity. Many of these studies elucidate how the social processes that give rise to new race-centered identities are reconfiguring the sociocultural terrains of ethnicities, nationalisms, religious allegiances and conflicts, and gender politics. Throughout the global order, new identities and movements have emerged organized around intensified and often primordialized distinctions—that is, differences assumed to stem from a people's beginning. Increasingly, territorially anchored struggles over the meaning and control of place are emerging, alongside struggles over newly reconfigured criteria for group membership within deterritorialized, diasporic, and transnational space.[20] These apparently contradictory yet complementary tendencies are occurring within, and as divergent responses to, a world that has "become much more tightly integrated into a nexus of . . . global fields . . . [across which] sophisticated telecommunications, an accelerated mobility of capital and labor, and rapid flows of commodities and culture compress both time and space [over] fractured technoeconomic, geopolitical, and sociocultural landscapes."[21]

Although it is clear that various forms of racism exacerbate global apartheid, critical race analysts must also examine how disparate, culturally specific social constructions of race interact with the ideological and structural forces of race-making that emanate from international and transnational forces. The ensuing portion of this chapter will investigate the racial politics of one of those culturally specific forces: U.S. foreign policy. Upon that foundation, I will then compare the various interplays between specific domestic configurations and transnational spheres of race and power.

To a considerable extent, neoliberalism reflects the cultural logics of the dominant nation-states of the north, particularly the United Sates. In other words, certain cultural logics have more sway than others do in the market of cultural exchange and in the international hierarchy of cultures, peoples, and nations. The ideological underpinnings of U.S. policy—domestic and foreign, geopolitical and economic, overt and covert—rest upon racist presuppositions. These presuppositions, which have shifted over time, include social Darwinist, eugenicist notions of difference and more liberal ideas that distance themselves from discredited, biodeterminist discourses. A multinational network of policies rooted in white supremacy fuels global apartheid. These policies, however, are being refashioned as formerly colonized people scatter across the globe and force multicultural awareness in white folks' backyards. These changes are especially evident in the tensions that arise when dominant western nations interact and build alliances with—as well as compete with—East Asian, primarily Japanese, entities.

Anthropologist Aihwa Ong elucidates the role that the transnational mobility of Japanese capital plays in racializing the international division of labor. Others have examined the liminal location the Japanese occupy in the global racial hierarchy, sandwiched between the "civilized white" and the "barbarous black." John Russell argues that, by denigrating blackness in their mass culture, the Japanese align themselves with whiteness and all that it symbolizes in terms of wealth, power, cultural capital, and racial supremacy.[22] As evidenced by the staggering increase in Caucasian-Asian interracial marriages in the United States, this alignment is increasingly acknowledged by Caucasian and Asians who imagine a somatic norm based on a mixture of Asian and European features. One could argue, however, that the Japanese do not simply consent to white hegemony in their appropriations of symbols of whiteness but also contest and undermine its normativity by co-positioning themselves at the apex of the global hierarchy.[23]

International relations scholar Robert Vitalis notes that foreign policy and the study of international relations are driven by a longstanding unspoken "norm against noticing" race.[24] A dramatic example of this occurred when the United

States refused to participate in the WCAR, because it objected to two key issues on the conference's agenda: reparations for slavery and colonialism (an issue of considerable international scope with implications for African and Caribbean debt relief and for post–affirmative action policy struggles in the United States); and the Israeli-Palestinian conflict, particularly discussions of Zionism as racism. The U.S. State Department was especially adamant in its insistence that an international conference focused on racism was not an appropriate venue for discussing Israel and for implicating U.S. foreign policy in the Middle East. From the U.S. perspective, the WCAR violated an important international norm by simply acknowledging the potential racial implications of its policies.

At the Durban forum and conference, however, Palestinian NGOs and solidarity groups from other parts of the world, including South Africa and Brazil, resisted the United States' attempt at political censorship and asserted that discrimination and oppression comparable in many ways to that of predemocratic South Africa exist in Palestine. This characterization appears in documents distributed at the WCAR NGO Forum in Durban.[25] Additionally, although neither the words "apartheid" nor "race" appear in this document, a report by the Palestinian Coalition for Women's Health titled "Israeli Violations of Women's Health Rights in Palestine During the Al Aqsa Intifada" details violations to the rights to life, medical care; safe education, residence, and work, and well as mental and social welfare that are consistent with examples of global apartheid in other parts of the world.

It is important to note that criticism of Israel's human rights violations against Palestinians is not limited to Muslims or Gentiles. As evidenced by the small but growing cadre of Israeli army officers who are going to jail rather than enforcing repressive Israeli policies in the West Bank, there is Jewish dissent as well, both inside and outside of Israel. In a call for financial support for the progressive Jewish journal, *Tikkun*, Cornel West wrote in the October 1, 2001, issue of *The Nation* that the magazine is on the verge of bankruptcy because pro-Israel Jews are retaliating against their liberal and progressive counterparts for criticizing Israeli policies toward Palestinians.[26] Lest the views of the Palestinian NGOs be considered unfairly biased, evidence from social scientific research on Israel—apparently unmotivated by anti-Semitism or a deliberately propagandized pro-Palestinian position—is emerging to counter such assertions. For example, political scientist Stanley Greenberg conducted a comparative analysis situating Israel alongside the United States, specifically comparing Alabama, South Africa, and Ireland.[27] In this work, Greenberg points to Israel as a settler colonial regime organized around race. This organization is manifested clearly in patterns of land alienation, labor control, and state policy.

Although Greenberg does not equate Israel's situation to South Africa's, he does, however, underscore similarities between the two societies that place them within the same critical framework. Such a comparison illuminates varying forms and instances of racialization, including the more implicit varieties in which salient social distinctions are publicly marked in culturalist or political categories of religion, ethnicity, and ethnonationalism.

The apartheid metaphor, therefore, may be useful and appropriate for inter-rogating the current Palestinian predicament. In a recent article on the peace process's effect on Palestinian geography, John Simon, who directs the Monthly Review Foundation, draws on data, including maps documenting the "Swiss cheesization" of the West Bank, to illustrate recent attempts by the Israeli state to maintain control over 57 percent of the land mass of the West Bank for security reasons. This demand, he writes, "makes a mockery not only of the 'peace process' but of the very notion of an independent Palestine," and it condemns Palestine to "a kind of *Bantustan*-like arrangement yielding a pseudo-state both politically and economically dependent on Israel."[28] Although there was a great deal of polarized debate over the Palestinian question and whether the WCAR was an appropriate venue for it, according to the broad definition of "racial discrimination" found in the International Convention on the Elimination of All Forms of Racial Dis-crimination (ICERD)—namely that it covers discrimination on the basis of race, color, descent, national, or ethnic origin—the Palestinian delegation had every right to bring their case to Durban and to argue their claim that there is a racial-ized dimension to their oppression. The legitimacy of their claim was recognized and supported by many African and African-descended delegates who contested the U.S.-defined and U.S.-enforced norms for labeling racial discrimination.

A clear instance of a foreign policy with racial outcomes and subtexts, similar in many ways to domestic drug policies that criminalize many African Americans and Latinos, is the multibillion-dollar U.S.-led "war on drugs" in Colombia and elsewhere in Latin America. This "war" is propelled by U.S. economic and mili-tary aid that has escalated the violence of Colombia's army, paramilitary forces, guerrillas, and narco-traffickers. Colombians of African descent are most often dislocated and devastated by the civil unrest: Although they make up only 25 percent of the population, Afro-Colombians are 70 percent of those forcibly dis-placed by violence.

Louis Gilberto Murillo, a political leader from the predominantly Black Chocó Province was forced into exile because of his efforts to mobilize a peace plan. He links the struggles of African Americans to Afro-Colombians. He said on a radio broadcast in 2001, "I would like African Americans to note that their tax money is used to support a U.S. policy, including Plan Colombia, which is detrimental to African Colombians. And not just detrimental to their standard of living, but to their lives. It is a policy that kills them."[29] (Plan Colombia is the Colombian government's term for its massive, widespread drug-control program funded largely by foreign governments, including more than $1 billion from the United States.)

Another activist, Marino Cordoba, president of the Association of Displaced Afro-Colombians (AFRODES), echoes Murillo with his report of assassinations and forced removals from rural farming communities. These graphic reports have been corroborated by *NACLA Report on the Americas*, which issued a recent commentary saying that U.S. aid is designed "to give the [Colombian] military 'rapid mobility capability' against guerrillas as well as to accelerate drug plant fumigation. Powerful herbicides . . . rain down on the . . . countryside" causing serious health problems and environmental damage as well as other human rights violations by indiscriminate security forces.[30]

Paradoxically, the war on drugs has destabilized the Colombian state and civil society, undermining the business investments, free markets, and democracy that U.S. officials claim is their goal—a goal used, in part, to justify their military presence in the region. Consequently, what is euphemistically called a low-intensity conflict (because of the restricted deployment of U.S. ground troops), with high-intensity impact, cannot be justified in economic terms or, for that matter, any rational terms. Though the realpolitik of drug control appears to conflict with the economic goals of U.S. foreign policy, it "reestablishes [the United States'] primacy of place by defining the hemispheric security agenda as a struggle against the corrosive influence of drug production, trafficking, and to a lesser extent consumption."[31] Moreover, it manufactures a domestic climate of hysteria over the dangers of illegal drugs. This climate, as Daniel Lazare writes, is built on the manipulation of fear and enables the state, in its domestic and transnational guises, "to operate in such a way that [is] 'free of any normal restraints from the 'bureaucracy,' from congressional subcommittees, and from the press." The "war on drugs" is ultimately a war on reason and a war against political democracy, both at home and abroad that "[enlists] Congress, the media, and ultimately the public itself in an all-consuming jihad. . . . [A] people unable to distinguish truth from falsehood when it [comes] to drugs [is likely to be] unable to distinguish truth from falsehood when it [comes] to global warming, energy policy, separation of church and state, or tax cuts for the rich and famous. Because it [is] unable to assess where its true interests [lie], it [is] all the more subject to domination and control."[32]

According to sociologist James Petras, Plan Colombia's first agendum, above all else, is to protect the U.S.'s geopolitical interests and imperial power in Latin America and beyond. In the late 1990s, the principal locus of both leftist and nationalist "resistance to the U.S. empire . . . shifted to northern South America—namely Colombia, the Eastern highlands of Ecuador, and Venezuela," which are called "the radical triangle." The Colombian insurgency, its systemic threat, and its appeal in other Latin American countries, although they are key factors, are only part of the larger geopolitical matrix that is contesting U.S. hegemony. Adding another element to the matrix resisting U.S. domination, the president of Venezuela, Hugo Chávez Frías (who publicly acknowledges his African ancestry) has instituted nonaligned policies regarding oil production, supply and prices, and is trading freely with Cuba—which has undermined the U.S. strategy of relegating Cuba to the status of a pariah state. Petras further points out that President Chávez's subsidized oil deals "have strengthened the resolve of the Caribbean and Central American regimes to resist Washington [DC]'s efforts" to turn the Caribbean into an exclusive U.S. lake: "While the guerrillas and popular movements represent a serious social and political challenge to U.S. supremacy in the region, Venezuela represents a diplomatic and [political-economic] challenge in the Caribbean basin and beyond. In more general terms, the radical triangle can contribute to undermining the mystique surrounding the invincibility of U.S. hegemony and the notion of the inevitability of free market ideology."

In other words, Plan Colombia's strategic geopolitical aim is to "reconsolidate [imperial] power in northern South America, secure unrestricted access to oil, [the primary source of energy in the United States,] and enforce the 'no alternatives to globalization' ideology for the rest of Latin America." To these strategic ends, the U.S. covertly supports right-wing terrorists through its military aid to the Colombian armed forces, which, in turn, are in alliance with paramilitary forces. Paramilitary terror is an "any means necessary" tactic "to empty the countryside and deny the guerrillas logistical support, food, and new recruits."[33]

As indicated earlier, the impoverished peasants and workers of Colombia, who are most vulnerable to this brutal practice of sociopolitical cleansing, are disproportionately of African descent. This problematic pattern of racial violence is the most recent expression of an already established practice of rampant social cleansing—that is, a "genocide of the poor"—which has long targeted categories of persons presumed to be intrinsic sources of danger and disorder: "prostitutes, homosexuals, drug users and street children. . . . [M]ost of the adolescent victims are black."[34] In light of this bias, it is not surprising that Afro-Colombians, especially young men, are perceived as the usual suspects in the war against drugs and guerrillas. Particularly in the Pacific and Caribbean coastal regions, which are home to the country's highest concentrations of Afro-Colombians, paramilitary and guerrilla activity has intensified and capitalist "development" interests are encroaching. It is not happenstance that this is occurring with increased intensity as African descendants, more than ever, claim their collective rights to the lands they cultivated since emancipation and assert their distinct cultural and political voice as *las comunidades negras* (the black communities).[35]

The repressive consequences of Plan Colombia have seriously constricted civil society, limiting, for example, citizens' rights to make trade union and civic demands or exercise political rights. It has increased the size, centralization, and military capacity of the state, which sees any measure of public debate and civil opposition as "subversive to the war effort" and as a "fifth [column] acting on behalf of the guerrillas."[36] Consequently, alarming—and increasing—numbers of journalists, intellectuals, labor activists, and community leaders have been assassinated. In the short term, U.S. policy in Colombia undermines the climate for foreign investment and free trade. Nonetheless, the hemispheric and global superpower is well aware that its long-range business interests depend on the outcome of political struggles in Latin America and beyond. The American empire is trying to stack the geopolitical cards in its favor, assuming that a pro-imperial resolution of the war will recolonize the region and create the most conducive conditions for the transnational accumulation of U.S. capital.

In hegemonic discourse, the United States is touted as the leading paragon of democracy and freedom. In George W. Bush's current rhetoric, with its appropriation of religious themes, the nation-state represents goodness in the struggle against evil. Yet the United States' human rights record at home and abroad is marginal, and it has long played an obstructionist role in the UN human rights system. U.S. refusal to participate in the WCAR was consistent with the U.S. government's three decades of noncompliance with ICERD even though it signed it. In the wake of the horrendous events of September 11, 2001, Bush publicly stated that evil terrorists hate Americans because we love freedom. The record shows, however, that U.S. policy has disregarded many people's human rights by countering struggles for freedom, self-determination, and economic and environmental justice all over the world. Indigenous and ethnonational people; marginalized nations and nation-states struggling to make their way down nonaligned and (in some cases) noncapitalist paths; racial minorities, immigrants, and refugees abused as strangers, scapegoats, and criminalized dangerous classes in countless nation-states; and women, whose rights the United States refuses to recognize as human rights (as defined in CEDAW, the UN Convention for the Elimination of All Forms of Discrimination against Women, which it has refused to sign), have to confront obstacles created or reinforced by U.S. policy's greater interest in free markets than in free human beings.

In many respects, U.S. policy promotes the recolonization of markets and labor and, hence, helps to create a vast pool of men and women desperate for jobs and forced to work at below-subsistence wages. Women, disproportionately women of color concentrated in the southern hemisphere, have been designated as a "new colonial frontier" for flexible capital accumulation.[37] The neoliberal regime of development depends on gender-dependent dichotomies, such as "women's work" and "men's work," that are supported by patriarchal assumptions that sewing, assembling electronic components, and pursuing home- and community-based informal activities are extensions of women's natural activities requiring no special skills, training, or compensation.

Aihwa Ong, drawing on postcolonial theorist Gayatri Spivak, refers to these women, who are compelled to work for slave wages under unhealthy conditions, as "paradigmatic subjects" of the international division of labor.[38] In Haiti in 1996, women workers in the Disney factory earned only twenty-eight cents an hour, which is less than a living wage even by Haitian standards.[39] It is no coincidence that those super-exploited Haitian workers are black women, second-class citizens in the poorest nation in the western hemisphere. The gross violation of their right to just terms of employment and fair working conditions must be viewed as a consequence of their bottom-level location within intersecting hierarchies of gender, race, class, and nation.

Labor exploitation that intersects with gender and racial oppression are inevitable consequences of "the logic and operation of capital in the contemporary global arena."[40] In a world of growing disparities of wealth, power, and privilege, women comprise 70 percent of the poor, and they are particularly vulnerable to the ideological and physical assaults of nationalist militarization, economically induced environmental degradation, and the economic restructuring and political realignments mediated by policies, such as structural adjustment. Structural adjustment depends on the cultural production of discourses and images about masculine dignity and feminine sacrifice, especially that of women of color. Furthermore, the policy is able to operate in the first place because of widespread expectations and role hierarchies ensuring that women will have to take up the slack when jobs and social safety nets are slashed or eliminated. Underscoring precisely this point, a UNICEF report on Latin America and the Caribbean stated, "If it were not for poor women working harder and longer hours, the poorest third of the population in that region would not be alive today."[41]

Increasingly, the subsistence security and human rights of these women and their families are being eroded, while they subsidize the production and accumulation of the world's wealth, which, more than ever before, is being concentrated at the top of the transnational ladder. Structural adjustment and other neoliberal policies, often fused with the cultural politics of local settings, contribute to the superexploitation of women's labor. Racially subordinate women, of course, bear the heaviest burdens and are the most vulnerable targets.[42]

These superexploited women, however, are not merely victims but also agents who actively negotiate the conditions of their everyday life and work. Through mostly covert acts of resistance and rebellion, they expose and contest the postcolonial industrial logic that institutionalizes inequalities of age, gender, class, and race. Although some of their negotiations may appear to be universal or transcultural (e.g., strikes), these actions are informed, at least in part, by culturally specific meanings, values, and experiences. Whether through social criticism encoded in songs and oral poetry in North Africa, through the covert language of protest expressed by spirit possessions on factory floors in Malaysia, or through conventional strikes that women in South Korea and Jamaica have organized against the repressive regimes of free trade or export-processing zones, women struggle to reclaim their human dignity. They do this in the face of domination that makes gender, race, class, and national—or transnational—identity socially and economically salient in a globalizing world.[43]

Although structural adjustment is a specific policy of the IMF working in conjunction with the World Bank, USAID, and other institutions, the term also refers to a general development orientation and policy climate driven by neoliberal assumptions about economic growth and change. In other words, *structural adjustment* can also serve as metonym for the restructuring and realignments that define present-day globalization.[44] Hence, in the case of Cuba, although the IMF, World Bank, WTO, and USAID do not directly intervene in the Cuban economy, neoliberal policies—the most coercive and punitive being the U.S. embargo—have indeed reshaped the nation during its so-called Special Period since the end of USSR and Eastern Bloc economic support, and they have undermined its revolutionary achievements in ensuring rights to employment, education, and health services. This has occurred even though, when the embargo began in the early 1960s, the politico-economic climate in the world was Keynesian rather than neoliberal. Nonetheless, for the past two decades, this punitive policy has been enforced within a politico-economic matrix of neoliberal globalization.

In the wake of the disintegration of the USSR and the demise of the Eastern Bloc, the Cuban government undertook an "internal readjustment" and "rationalization" to sustain its economy and meet basic subsistence needs. These changes allowed for greater economic diversification, a partial process of privatization, foreign investment, and "dollarization."[45] The effects of these drastic changes— Cuba's own structural adjustment—on everyday life have been considerable.

Cuba's status as a socialist sanctuary is being destabilized under dollarization and the conditions of economic austerity that led to it. Social inequalities are reemerging and becoming conspicuous and crime is becoming a problem. A red flag signaling the changing times can perhaps be found in a troubling December 2001 incident in which five members of a family, including an eight-year-old child and a couple visiting from Florida, were murdered in a robbery in Matanzas Province. This heinous incident was unusual in that murders are extremely rare in Cuba and mass murders "are unheard of."[46] The economic crisis that brought about this unprecedented crime wave has caused escalating unemployment and has reduced safety net provisions—trends that have impacted African-descended Cubans, and Afro-Cuban women in particular, more than any other segment of the population. With less access to kin-mediated remittances from the disproportionately white émigré communities overseas, there is more pressure on Afro-Cuban women, who are more likely than white Cubans to live in female-headed households, stand in long lines for rations, stretch the devalued peso, and make ends meet by any means necessary.[47] *Any means necessary* has come to include doing work on the side—*trabajo por cuenta propia*—in the underground economy aligned with the growing tourist sector. For younger women, particularly those who fit the culturally constructed stereotype of *la mulata* (mulatto woman), this is increasingly being translated into working as *jineteras* (sexual jockeys). This line of work reflects Cuba's historical race, gender, and class boundaries.[48] Desperate to lure foreigners to the country's beaches and hotel resorts, the Cuban government itself has resorted to manipulating prerevolutionary racial clichés by "showcasing 'traditional' Afro-Cuban religious rituals and art, 'traditional' Afro-Cuban music, and Afro-Cuban women," who are foregrounded as performers in these commodified contexts."[49]

The sexual exoticization of African-descended women has a long history in Cuba as well as throughout the African Diaspora and the West, where variations on the theme of black hypersexuality are rampant as either a positively valued essentialism or a fertility or health-related social problem. Nadine Fernandez questions the assumption that black and mulatto women predominate in Cuba's sex tourism by highlighting the role of a racially biased gaze in attributing Afro-Cuban women's interactions with male tourists to prostitution while perceiving white women's interactions in terms of alternative interpretations, including that of "romance." Because of their greater access to dollars and to jobs in the tourist sector, white women are more likely to have privileged access to tourists in restricted venues (shops, restaurants, and nightclubs) where Afro-Cubans are not generally permitted to enter. Consequently, Afro-Cubans interact with tourists "outside tourist installations, making their meetings much more visible and scrutinized by the public eye."[50] In the context of Cuba's current crisis, traditional racial narratives of gender, race, and sexuality are being reasserted and rewritten to fit with recent restructuring.[51]

The U.S. embargo is a flagrant form of foreign intervention. Like official structural adjustment policies, it has been premised on an ideology of power, recolonization, and ranked capitals that assumes that Cubans are expendable troublemakers—perhaps even harborers of terrorism—who deserve to be starved out of their defiant opposition to U.S. dominance. The same ideology that rationalizes the unregulated spread of commodification into all spheres of social life implies that Cuban women's bodies, especially *Afro-Cubanas'* hypersexualized bodies, can be bought and sold on the auction block of imposed economic austerity without any accountability on the part of the *papiriquis*, or sugardaddies, of global capital. The implication of these policies is that Afro-Cuban families and communities can be sacrificed so that northerners can enjoy privileges— including that of living in a "good" and "free" society—that southern workers and peasants subsidize. Cuba's current crisis is being negotiated over the bodies of its women, with African-descended women, *las negras y mulatas, las chicas calientes* (black and mulatto women, hot sexy chicks), expected to bear the worst assaults against what remains in many ways a defiant socialist sanctuary.[52]

Analyses of the heightening of racial conflict and the attendant increase in human rights abuses around the world point to the confluence of several factors. The world has become more integrated in ways that lead to more decentralized and flexible capital accumulation, dramatic disparities in social and life expectancy, a reconcentration of wealth in the hands of a minority, and a decline in subsistence security and environmental sustainability for the broad masses of humanity. Diminished socioeconomic security and increased social stress, often precipitated by free but clearly not fair market mandates, best explain the escalating tensions that provoke volatile politics of racial marginalization. Destabilizing economic forces are helping to deepen fragmentations of identity. As a consequence, once-established national identities are weakened and often give way to new identities in which culturally concrete forms of community, language, and ethnic loyalties become more salient and respond defensively to an increasingly globalized world.

Domestic racism and global apartheid also are intensified by:

- the economic and political destabilization engendered by post–cold war realignments and restructuring;

- the circumscription of state (especially peripheral state) sovereignty under the influence of transnational forces;
- the crisis of social welfare resulting from the diminished ability of the state to provide a safety net and to protect rights to education, health care, and humane work conditions;
- the destabilizing effects of structural adjustment and stabilization programs on civil society;
- the instability and decline of international markets in export commodities (e.g., coffee in Rwanda), especially those around which less diversified, extroverted economies are organized;
- major demographic shifts leading to the increasing internationalization of work forces and societies;
- the elaboration of competing mythicohistorical and mythicosocial accounts that construct differences and identities in essentializing terms drawn from fundamentalist racial ideologies that are centered on biology, morality, and cultural alterity;
- and the globalization of the arms trade, which adds fuel to the fire ignited by intergroup tensions.

The structural violence of the IMF, the WTO, and the superpower politics of the United States are threads visibly woven through several of these factors. Whether expressed through foreign policy or international communications media, the United States' culture of race and its relations of racism have a profoundly global impact in light of the superpower's position in the global hierarchy. For example, the internationalization of American media carries with it powerful representations of black Americans (including images of popular culture and identity politics) that have influenced perceptions of blackness all over the world. The United States' construction of race has had a profound impact on a minority of Afro-Latin Americans, particularly those involved in racially cognizant black movements. Latinos of African descent traditionally are politically fragmented by national ideologies of *mestizaje* (mixedness with the goal and implication of whitening) and by complex racial classification systems that delineate elaborate socioracial continuums. Resisting the hegemonic notion of a "mulatto escape hatch,"[53] advocates of black consciousness increasingly are appropriating the U.S. principle of hypodescent—the "one drop rule"—to build united fronts against racism among African descendants.

The globalization of U.S.-led multilateral interests, however, does not nullify the culpability of domestic forces within nations that contribute to racism. Although they may be situated in contexts shaped by the interests of transnational capital, it is crucial to remember that real people on uneven playing fields make the choices that eventually lead to human rights violations.

Antiracist activists must be attentive to the workings of complicit forms of individual and collective agency—with their political, economic, and psychological dimensions. A holistic understanding of the dynamics of culture, power, and political economy at multiple levels—local, national, and transnational—within a global matrix of domination is needed to craft better means of coalition-building.

The events of September 11, 2001, underscored the urgency of exposing and strategizing about U.S. domestic and foreign policy and of repositioning the United States on the terrain of international relations. The well-being of ordinary Americans is at stake: our very lives depend on it. Will the current war against terrorism degenerate into yet another war against reason and democracy? This would sacrifice civil liberties and human rights on the altar of "superpowerdom" built by the architects of global apartheid.

There were many pilgrims at the WCAR who were well aware that their expanded antiracist repertoires and arsenals must include weapons suitable for combating the negative and dehumanizing aspects of globalization. They also know, now more than ever, that they must help forge a more humane globalization informed by more just ways of imagining and mobilizing global communities united against racism, xenophobia, and all related intolerance. The rite of passage that the WCAR represented for human rights educators, researchers, and activists may lead them in the direction of new tools, strategies, and opportunities for organizing in ways that are at once local, national, regional, and transnational. It will take an enormous amount of conviction and critical imagination, but it is up to diverse agents of many civil societies and receptive sectors of governments to make the WCAR more than an empty public relations symbol for the UN. All antiracists who regard the Durban Plan of Action as an inspiring tool in the struggle for transformation should translate the plan into real victories—and shape a twenty-first century truly reconfigured by change.[54]

NOTES

1. Ellen Messer, "Anthropology and Human Rights," *Annual Review of Anthropology* 22 (1993): 222.

2. David L. Wilson, "Do Maquiladoras Matter?" *Monthly Review* 49 (October 1997): 30.

3. Catherine Kingfisher and Michael Goldsmith, "Reforming Women in the United States and Aotearoa/New Zealand: A Comparative Ethnography of Welfare Reform in Global Context," *American Anthropologist* 103 (September 2001): 716, 727.

4. See Faye V. Harrison, "Crime, Class, and Politics in Jamaica," *TransAfrica Forum* 5 (Fall 1987): 30–32. Also see "Human Rights in Jamaica" (Americas Watch/Human Rights Watch, 1986); "Jamaica," in *Amnesty International Annual Report* (Amnesty International, 2001); "Jamaica: Police Killings—A Human Rights Emergency" (Amnesty International Press Release, October 4, 2001); "Killings and Violence by Police: How Many More Victims?" (Amnesty International, October 4, 2001): 1–72.

5. David Harvey, *The Condition of Postmodernity: An Inquiry into the Origins of Cultural Change* (Cambridge, MA: Basil Blackwell, 1989); Fredric Jameson, *Postmodernism or the Cultural Logic of Late Capitalism* (Durban, South Africa: Duke University Press, 1990); Jean-François Lyotard, *The Postmodern Condition* (Minneapolis.: University of Minnesota Press, 1984).

6. Wilson, "Maquiladoras," 30.

7. In a historic turn of events, the U.S. Congress agreed to a limited relaxation of the embargo following Hurricane Michelle, which inflicted mass destruction on Cuba in November 2001, resulting in widespread food shortages and medical emergencies. As a result, the U.S. State Department authorized the sale of grain and medicines to Cuba. It was the first trade deal between the two countries since 1962, when President Kennedy established the embargo. In spring 2002, however, the Bush administration violated its agreement with Cuba when the State Department denied travel visas to Cuban food import officials who were scheduled to visit the U.S. to buy grain. "Cuba Reaches Historic Trade Deal with U.S.," *Latin American Monitor: Caribbean* 19 (January 2002): 8; "U.S.–Cuban Food Sales in Jeopardy," *Caribbean Update* 18 (May 2002): 8–9.

8. Faye V. Harrison, "The Persistent Power of 'Race' in the Cultural and Political Economy of Racism," *Annual Review of Anthropology* 24 (1995): 47–74; Faye V. Harrison, "Facing Racism and the Moral Responsibility of Human Rights Knowledge," *Annals of the New York Academy of Sciences* 925 (December 2000): 1, 45–69.

9. For analyses of ethnic cleansings and other violently marked racial conflicts in Eastern Europe, Rwanda, and Oromo (a region in Ethiopia that is a site for a nationalist movement for self-determination), see Vesnas Kexic, "Never Again a War: Women's Bodies Are Battlefields," in *Look at the World through Women's Eyes: Plenary Speeches from the NGO Forum on Women, Beijing '95*, ed. Eva Friedlander (New York: Women, Ink, 1995), 51–53; Mahmood Mamdani, *When Victims Become Killers: Colonialism, Nativism, and the Genocide in Rwanda* (Princeton, NJ: Princeton University Press, 2001); and Asafa Jalata, "The Impact of a Racist U.S. Foreign Policy on the Oromo National Struggle," *Journal of Oromo Studies* 6 (1999): 49–90.

10. Lisa Malkki, *Purity and Exile: Violence Memory, and National Cosmology among Hutu Refugees in Tanzania* (Chicago: University of Chicago Press, 1995), 227–28.

11. Ralph Bunche, *A World View of Race* (1936; repr. Port Washington, NY: Kennikat, 1968).

12. David Levering Lewis, *W. E. B. Du Bois: Biography of a Race, 1868–1919* (New York: Henry Holt and Company, 1993), 504; Robert Vitalis, "The Graceful and Generous Liberal Gesture: Making Racism Invisible in American International Relations," *Millennium: Journal of International Studies* 29 (2000): 354. For a discussion of Du Bois's UN petition, participation in early UN conferences, and lobbying to have the UN Charter acknowledge "dependent peoples'" rights to self governing, see Lewis, *W. E. B. Du Bois: The Fight for Equality and the American Century, 1919–1963* (New York: Henry Holt and Company, 2000): 507; also see Amiri Baraka, "Paul Robeson and the Theater," *Black Renaissance* 2 (1): 39, for a discussion on African American and UN activism to protect Black people's human rights.

13. For reflections on the conference, see Samir Amin, "World Conference against Racism; A People's Victory," *Monthly Review* 53 (December 2001): 20–23; and Faye V. Harrison, "Imagining a Global Community United against Racism," *Anthropology News* (December 2001): 22–23.

14. Sherle R. Schwenninger, "America and the World: The End of Easy Dominance," *The Nation* 20 (November 2000): 20.

15. For a detailed analysis of the structural violence in urban Jamaica, see Faye V. Harrison, "The Gendered Politics and Violence of Structural Adjustment: A View from Jamaica," in *Situated Lives: Gender and Culture in Everyday Life*, ed. Louise Lamphere, Helena Ragoné, and Patricia Zavella (New York: Routledge, 1997), 451–68.

16. Salih Booker and William Minter, "Global Apartheid," *The Nation* 9 (July 2001): 11.

17. Gernot Köhler, "Global Apartheid," in *Talking about People: Readings in Contemporary Cultural Anthropology*, ed. William A. Haviland and Robert J. Gordon (Mountain View, CA: Mayfield, 1978), 262–68.

18. Ann Kingsolver, *NAFTA Stories: Fears and Hopes in Mexico and the United States* (Boulder, CO: Lynne Rienner, 2001), 110.

19. Howard Winant, *Racial Conditions: Politics, Theory, Comparisons* (Minneapolis: University of Minnesota Press, 1994), 123.

20. Linda Basch, Nina Glick Schiller, and Cristina Blanc Szanton, *Nations Unbound: Transnational Projects, Postcolonial Predicaments, and Deterritorialized Nation-States* (Langhorne, PA.: Gordon and Breach, 1994).

21. Faye V. Harrison, "Introduction: Expanding the Discourse on Race," Contemporary Issues Forum: Race and Racism, *American Anthropologist* 103 (September 1998): 609.

22. Aihwa Ong, *The Spirits of Resistance and Capitalist Discipline: Factory Women in Malaysia* (Albany: State University of New York Press, 1987); and Ong, "The Gender and Labor Politics of Postmodernity," *Annual Review of Anthropology* 20 (1991): 279–309. Russell, "Race and Reflexivity: The Black Other in Contemporary Japanese Mass Culture," *Cultural Anthropology* 6 (1991): 3–25.

23. A further discussion on this topic is found in Faye V. Harrison's "Unraveling 'Race' for the 21st Century," in *Exotic No More: Anthropology on the Front Lines*, ed. Jeremy McClancy (Chicago: University of Chicago Press, 2002), 145–66. Also see "Everyday Neoliberalism in Cuba: A Glimpse from Jamaica," in *Outsider Within: Reworking Anthropology in the Global Age* (Urbana: University of Illinois Press, 2007).

24. Vitalis, "The Graceful and Generous Liberal Gesture," 333. This norm against noticing race is consistent with the tendency to ignore gender and, in particular, the problematic international politics of masculinity, of which feminist political scientist Cynthia Enloe's work has made us more aware: *Bananas, Beaches, and Bases: Making Feminist Sense of International Politics* (Berkeley: University of California Press, 1989); and *The Morning After: Sexual Politics at the End of the Cold War* (Berkeley: University of California Press, 1993).

25. *Israel's Brand of Apartheid: The Nakba [Catastrophe] Continues*, booklet summarizing the Palestinian NGO Position Paper for the WCAR; *Occupied Jerusalem, a New Soweto?*, study by the Jerusalem Center for Social and Economic Rights; *Amandla Intifada*, newsletter of the Palestine Solidarity Committee, South Africa; and *Down with the Nazi-Israeli Apartheid*, statement by the Afro-Brazilian National Congress.

26. Other dissenters among U.S. Jews include Jewish Mobilization for a Just Peace (Junity), which is a Philadelphia-based international network working to support Gush Shalom (a leading Israeli- Palestinian peace organization); Jews for Peace in Palestine and Israel, based in Washington, DC; Not In My Name, a Chicago-based organization; the Coalition of Jews for Justice, of the San Francisco Bay Area; Jews Against the Occupation, New York City; Boston's Women in Black; and Jewish Women for Justice in Israel and Palestine. This information is found in a October 29, 2001, Jonathan Fremont, e-mail message from the office of Congresswoman Cynthia McKinney (D-GA) regarding the October 25, 2001, congressional news conference on the Middle East Peace Process.

27. Stanley B. Greenberg, *Race and State in Capitalist Development: Comparative Perspectives* (New Haven, CT: Yale University Press, 1980), 41.

28. John Simon, "Palestinian Geography and the Peace Process," *Monthly Review* (October 2001): 30–34 (emphasis added).

29. Louis Gilberto Murillo, "I Survived a Paramilitary Death Squad," *Pacific News Service*, June 4 2001, http://www.pacificnews.org/content/pns/2001/june/ 0604survived.html (accessed July 2001).

30. TIRN [Tennessee Industrial Renewal Network] Fair Trade Committee, "Afro-Colombian Speaker in Tennessee"; "Whither the War in Colombia?" *NACLA Report on the Americas* 35 (September–October 2001): 7.

31. William Walker III, "A Reprise for 'Nation Building': Low Intensity Conflict Spreads in the Andes," *NACLA Report* (July–August 2001): 25.

32. Daniel Lazare, "A Battle against Reason, Democracy, and Drugs," *NACLA Report on the Americas* 35 (July–August 2001): 14–17.

33. James Petras, "The Geopolitics of Plan Colombia," *Monthly Review* 53 (May 2001): 33–41.

34. Nina S. de Friedemann and Jaime Arocha, "Colombia," in *Invisible No More: Afro-Latin Americans Today*, ed. Minority Rights Group (London: Minority Rights Publications, 1995): 67.

35. Peter Wade, "Colombia," *Africana: The Encyclopedia of the African and African American Experience*, ed. Kwame Anthony Appiah and Henry Louis Gates, Jr. (New York: Basic Civitas Books, 1999): 477–78; also see the discussion on legal reform and collective land rights in Peter Wade, *Race and Ethnicity in Latin America* (London: Pluto, 1997): 99.

36. Petras, "The Geopolitics of Plan Colombia," 42.

37. Maria Mies, Veronika Bennholdt-Thomsen, and Claudia von Werlhof, *The Last Colony* (London: Zed Books, 1988).

38. Jacqui M. Alexander and Chandra Talpade Mohanty, "Introduction: Genealogies, Legacies, Movements," in *Feminist Genealogies, Colonial Legacies, Democratic Futures*, ed. Jacqui M. Alexander and Chandra Talpade Mohanty (New York: Routledge, 1997), 5.

39. Ong, "The Gender and Labor Politics of Postmodernity," 289; Gayatri Spivak, "Subaltern Studies: Deconstructing Historiography," in *Selected Subaltern Studies*, ed. R. Guha and Gayatri Spivak (New York: Oxford University Press, 1988): 3–44.

40. National Labor Committee, *Mickey Mouse Goes to Haiti*, video recording (1996).

41. Chandra Talpade Mohanty, "Women Workers and Capitalist Scripts: Ideologies of Domination, Common Interests, and the Politics of Solidarity," in *Feminist Genealogies*, ed. Jacqui M. Alexander and Chandra Talpade Mohanty, 28.

42. Vivienne Wee and Noeleen Heyzer, *Gender, Poverty and Sustainable Development: Towards a Holistic Framework of Understanding and Action* (Singapore: ENGENDER, Centre for Environment, Gender and Development, 1995), xv; Pamela Sparr, ed., *Mortgaging Women's Lives: Feminist Critiques of Structural Adjustment* (London: Zed Books, 1994); Enloe, *Bananas, Beaches and Bases*, 1–18. For an analysis of how gender ideologies informed structural adjustment programs in the Caribbean, see Peggy Antrobus, "Crisis, Challenge and the Experience of Caribbean Women," *Caribbean Quarterly* 35 (1989): 17–28; UNICEF report quoted in Sparr, "How We Got into This Mess and Ways to Get Out," *Ms.* (March–April 1992): 33–34.

43. June Nash, "Global Integration and Subsistence Insecurity," *American Anthropologist* 96 (January 1994): 7–30; Harrison, "The Gendered Politics and Violence of Structural Adjustment."

44. Ong, *Spirits of Resistance and Capitalist Discipline*; Ong, "Gender and Labor Politics of Postmodernity," 289.

45. See Faye V. Harrison, "Negotiating Everyday Neoliberalism in Jamaica and Cuba," in *Outsider Within: Reworking Anthropology in the Global Age*, ed. Harrison (Urbana: University of Illinois Press, 2008).

46. Carollee Bengelsdorf, "[Re]Considering Cuban Women in a Time of Troubles," in *Daughters of Caliban: Caribbean Women in the Twentieth Century*, ed. Consuelo Lopez Springfield (Bloomington: Indiana University Press, 1997).

47. "Five Family Members Killed," *The Weekly Gleaner* (December 27, 2001–January 2, 2002): 12.

48. Ibid. See also Helen I. Safa, *The Myth of the Male Breadwinner* (Boulder, CO: Westview Press, 1995), 139. The issue of these racially differential effects and the unequal distribution of remittances from abroad is discussed in Pedro Pérez Sarduy and Jean Stubbs, eds., *Afro-Cuban Voices: On Race and Identity in Contemporary Cuba* (Gainesville: University Press of Florida, 2000), xv, 5, 34.

49. Julia O'Connell Davidson, "Sex Tourism in Cuba," *Race and Class* 38 (1996): 39–48; Coco Fusco, "Hustling for Dollars," *Ms.* (September–October 1996): 62–70.; Gayle McGarrity and Osvaldo Cárdenas, "Cuba," in *No Longer Invisible: Afro-Latin Americans Today* (London: Minority Rights Publications, 1995), 100; Georgina Herrera, "Poetry, Prostitution, and Gender Esteem," in *Afro-Cuban Voices*, ed. Perez Sarduy and Stubbs, 123–24; Nadine Fernandez, "Back to the Future: Women, Race, and Tourism in Cuba," in *Sun, Sex, and Gold: Tourism and Sex Work in the Caribbean*, ed. Kamala Kempadoo (Lanham, MD.: Rowman and Littlefield, 1999), 81–89.

50. Fusco, "Hustling for Dollars," 67.

51. Fernandez, "Back to the Future," 88.

52. Bengelsdorf, "[Re]Considering Cuban Women," 245.

53. The colonization and market configuration of Cuban bodies through sex work does not target women alone. There are also *pingueros* (male sex workers). Derived from *pinga* ("dick"), *pinguero* was coined in the wake of the legalization of the U.S. dollar. Derrick G. Hodge, "Colonization of the Cuban Body: The Growth of Male Sex Work in Havana," *NACLA Report on the Americas* 34 (March–April 2001): 20–28.

54. The idea of an escape hatch comes from historian Carl N. Degler's classic study, *Neither Black nor White: Slavery and Race Relations in Brazil and the United States* (New York: Macmillan, 1971).

CHAPTER 3

THE MODERN WORLD RACIAL SYSTEM

HOWARD WINANT

AS THE WORLD LURCHES FORWARD INTO THE TWENTY-FIRST CENTURY, there is widespread confusion and anxiety about the political significance, and even the meaning, of race. In this chapter I argue that, far from becoming less politically central, race defines and organizes the world and its future, as it has done for centuries. I challenge the idea that the world, as reflected by the national societies I compare, is moving "beyond race." I suggest that the future of democracy itself depends on the outcomes of racial politics and policies as they develop in various national societies and in the world at large. This means that the future of democracy also depends on the *concept* of race, that is, the meaning that is attached to it. Contemporary threats to human rights and social well-being—including the resurgent dangers of fascism, increasing impoverishment, and massive social polarization—cannot be managed or even understood without paying new and better attention to issues of race. This chapter attempts to provide a set of conceptual tools that can facilitate this task.

The present moment is unique in the history of race. Starting after World War II and culminating in the 1960s, there was a global shift, a "break," in the worldwide racial system that had endured for centuries. The shift occurred because, after the war, many challenges to the old forms of racial hierarchy converged: anticolonialism, antiapartheid, worldwide rejection of fascism, and perhaps most important, the U.S. civil rights movement and U.S.-U.S.S.R competition in the world's southern hemispheric nations all called into question white supremacy to an extent unparalleled in modern history. These events and onflicts linked antiracism to democratic political development more strongly than ever before.

All around the world a centuries-old pattern of white supremacy is more fiercely contested, more thoroughly challenged, *in our lifetimes* than ever before. As a result, for the first time in modern history, there is widespread, indeed worldwide, support for what until recently was a "dream"—Dr. Martin Luther King Jr.'s dream, let us say, of racial equality.

The rise of a worldwide trend toward antiracism and democratization start-ing in the late 1940s was only the first phase, that is, the initiation of the shift or break in the *old* world racial system. A second phase, the containment of the antiracist challenge, came after several decades of fierce struggles. Thus by about 1970, despite all the political reforms and cultural transformations wrought by social movements and democratic politics around the world and despite the real amelioration of the most oppressive features of the old world racial system, the centuries-old and deeply entrenched system of racial inequality and injustice was hardly eliminated. Rather, in a postwar social order faced with an unprecedented set of democratic and egalitarian demands, racism had to be adapted. A new racial politics developed, which was a reformed variety that was able to concede much to racially based democratic and egalitarian movements, yet it maintained a certain continuity with the legacies of imperial rule, conquest, enslavement, and so forth.

Thus, white supremacy has proven itself capable of absorbing and adapting much of Dr. King's dream, repackaging itself as "color-blind," nonracial, and mer-itocratic. Paradoxically, in this reformed version racial inequality can thrive, still battening on stereotypes and fears; still resorting to exclusionism and scapegoat-ing when politically necessary; still invoking the supposed superiority of so-called mainstream (i.e., white) values, and still cheerfully maintaining that equality has been largely achieved. It is ironic that this new, "color-blind" racial system may be more effective in containing antiracist challenges than any intransigent, overtly racist backlash could possibly have been.

Although the officially nonracial version of white supremacy has succeeded in curtailing progress toward the dream in many battles—immigration and citizen-ship, income redistribution and poverty, and above all compensatory programs commonly called "affirmative action"—nonracialism has hardly won the day. It has certainly not eliminated the movement for racial justice that spawned it. Rather, the racial politics that result from this synthesis of challenge and incorpo-ration, racial conflict and racial reform, has proven neither stable nor certain. It is a strange brew, often appearing more inclusive and pluralistic than ever yet filled with threats—of "ethnic cleansing," resurgent neofascism, and perhaps equally insidious, a renewed racial complacency.

The global racial situation, then, is fluid, contradictory, and contentious. No longer unabashedly white supremacist, much of the world is, so to speak, abash-edly white supremacist. The conflicts generated by the powerful movements for racial justice that succeeded World War II have been contained but not resolved. Thus, no new world racial system has yet been created; instead, the problems of the old system have come to a head, and the outlines of what will succeed it can at least be glimpsed, if not clearly foreseen. What does such a glimpse, however preliminary, reveal? The new world racial system will struggle to adapt the rhetoric of egalitarian social movements to the exigencies of a postimperial, post–cold war, and postapartheid reality. To some extent, this system succeeded in reinventing itself along nonracist lines; in fact, its capacity to redefine itself as "beyond race" is in many ways a crucial index of its intransigence. Yet there is also a widespread recognition that the reforms undertaken in the 1950s and '60s have ossified and been derailed far short of their goals. Indeed, they may be providing a kind of cover for a reassertion of white privilege, white rule, and "northern" cultural norms—all under the banner of "postracial" societies, which are now officially considered color-blind and pluralist.

Today's global racial system is obviously not the world's first. The racial dimensions of modernity itself have been widely acknowledged. The Enlightenment's recognition of a unified, intelligible world, the construction of an international economy, the rise of democracy and popular sovereignty, and the emergence of a global culture all were deeply racialized processes. To understand how race was fundamental to the construction of modernity is of more than historical interest; it also explains much about the present. Notably it undermines the commonly held belief that racism is largely a thing of the past and the idea that, after the bad old days of white supremacy and colonial rule, the "race problem" today is resolved.

Before addressing the present, let us recall that past. What are the origins of the world racial system? How have the enigmatic specters of racial difference and racial inequality been loosed on the world?

Examining early modern history, racial patterns present ample precedents for the horrors of our own age. The tension between slavery, on the one hand, and nascent democracy, on the other, structured the lengthy transition to the modern world. Resistance against slavery was pivotal to crafting the broader redefinition of political rights for which early advocates of democracy yearned and fought. Indeed the violence and genocide of earlier racial phenomena prefigured contemporary atrocities, such as the Holocaust, "ethnic cleansing," and totalitarianism.

How racial was nascent capitalism? Were the politics and cultural groundwork of modernity premised on racial distinctions? Did the generally limited democracy of the developed world consist, in part, of an application of the principles of colonial rule to the "mother countries"? In what ways did early forms of resistance to racialized forms of rule—as seen in abolitionism and slave revolts for example—dynamize the worldwide impetus toward democratization? In what ways did antiracism itself become an archetypal democratic movement? Did the resistance to slavery, which grew into antiracism, ultimately accomplish more than fighting for the human, social, cultural, and political rights of racially subordinated groups? Was it not also crucial in permitting the acquisition of those same rights by whites? In other words, is the modern, inclusive form of democracy to which we have become accustomed itself the product of global struggles against racism?

The abolition of African slavery was the great rehearsal for the "break" with white supremacy that took place in our own time. Abolition was made possible by three momentous social changes: the triumph of industrial capitalism, the upsurge of democratic movements, and the mobilization of slaves themselves in search of freedom. Abolition was not completed with the triumph of the Union army in the American Civil War or the passage of the Reconstruction amendments to the Constitution. Only in 1888, when Brazil became the last country to free its slaves, did the first crucial battle in the world's centuries-long war against white supremacy draw to a close. And slavery persists in many forms even today.

But abolition left many emancipatory tasks unfinished as new forms of racial inequality were implemented: Democracy was still partial. Racialization continued to fine the mechanisms of authoritarian rule and the distribution of global resources. Racial thought and practices defined distinct types of human bodies as subordinate. This ranking of human society by race continued to enable and justify world-systemic rule. Generalized processes of racial stratification continued to support enormous and oppressive systems of commercial agriculture and mining. Thus until the mid-twentieth century, the unfulfilled dreams of human rights and equality were still tied up with the logic of race.

Although there was always resistance to racist rule, it was only in the period after World War II that opposition to racial stratification and racial exclusion once again provoked widespread political conflicts. Civil rights and antiracist movements, as well as nationalist and indigenous ones, fiercely contested the racial limitations on democracy. These movements challenged the conditions under which black and brown people's labor was exploited in the former colonies and the metropoles. They drew upon the antifascist push of the war and the geopolitical conflicts of the cold war. They rendered old forms of political exclusion problematic and revealed a panoply of mainstream cultural icons—artistic, linguistic, scientific, and even philosophical—to be deeply conflicted. They drew on the experience of millions who returned from military service to face a segregated or colonized homeland. Such movements recognized anew their international character, as massive postwar labor demand sparked international migration from the world's southern to its northern hemisphere, from areas of peasant agriculture to industrial areas. These enormous transformations manifested themselves in a vast demand to complete the work that began a century before with slavery's abolition. They sparked the worldwide break with the tradition of white supremacy.

As the tumultuous 1960s drew to a close, the descendants of slaves and ex-colonials forced at least the partial dismantling of most official forms of discrimination. But with these developments—the enactment of a new series of civil rights laws, decolonization, and the adoption of cultural policies of a universalistic character—the global racial system entered a new period of instability and tension. The immediate result of this break with past practices was an uneven series of racial reforms that had the general effect of ameliorating racial injustice and inequality. But they also worked to contain social protest. Thus, the widespread demands of the racially subordinated and their supporters were at best answered in a limited fashion. A new period of racial instability and uncertainty was inaugurated.

This shift in racial attitudes and practices was a worldwide phenomenon, but it obviously took very different forms in particular national settings. Racial conditions are generally understood to vary dramatically in distinct political, economic, and cultural contexts. In this chapter I comment, necessarily briefly, on four national case studies: the United States, South Africa, Brazil, and the European Union (considered as a whole). In these comparisons, I argue that the post–World War II break is a global backdrop, that is, an economic, political, and cultural context in which national racial conflicts are being worked out.

THE UNITED STATES

How permanent is the United States' color line? The activities of the civil rights movement and related antiracist initiatives achieved substantial, if partial, democratic reforms in postwar decades. Today, however, these innovations coexist with a weighty legacy of white supremacy that originates in the colonial and slavery era. How do these two currents combine and conflict today?

Massive internal and international migration has reshaped the U.S. population numerically and geographically. A multipolar racial pattern has largely supplanted the old racial system, which is often (and somewhat erroneously) viewed as a bipolar white-black hierarchy. In the contemporary United States, new varieties of interminority competition and new awareness of the international dimensions of racial identity have greater prominence. Although racial stratification varies substantially by class, region, and indeed among groups, comprehensive racial inequality certainly endures. Racial reform policies are under attack in many spheres of social policy and law as opponents of the policies forcefully claim that the demands of the civil rights movement have largely been met and that the United States has entered a "postracial" stage of its history.

The racial break in the United States was a partial democratization, produced by the moderate coalition that dominated the political landscape in the post–World War II years. The partial victories of the civil rights movement were won through mass mobilization and a tactical alliance with U.S. national interests. This alliance was brokered by so-called racial moderates: political centrists largely affiliated with the Democratic Party who perceived the need to ameliorate racial conflict and end outright racial dictatorship but who also feared the radical potential of the black freedom movement.

There was a price to be paid for civil rights reform. It could take place only in a suitably deradicalized fashion, only if its key provisions were articulated (legislatively and juridically) in terms compatible with the core values of U.S. politics and culture: individualism, equality, competition, opportunity, and the fair accessibility of the American dream. The movement's radicals—revolutionaries, socialists, and political nationalists (black, brown, red, yellow, and white)—paid this price, foregoing their vision of major social transformation under threat of marginalization, repression, or death.

The radical vision was, as Dr. King's dream allowed us to call it, an alternative "dream" in which racial justice was central. To be "free at last" meant something deeper than symbolic reforms and palliation of the worst excesses of white supremacy. It meant substantive social reorganization that would be manifested in egalitarian economic and democratizing political consequences. It meant something like social democracy, human rights, and full social citizenship for blacks and other so-called minorities.

But it was precisely here that the moderate custodians of racial reform drew their boundaries, both practically and theoretically.[1] To strike down officially sanctioned racial inequality was permissible, but to create racial equality through positive state action was not. The danger of redistribution—of compensating for the unjustified expropriation and restriction of black economic and political resources, both historically and in the present—was to be avoided at all costs.

Civil rights reform thus became the agenda of the political center, which moved "from domination to hegemony."[2] The key component of modern political rule, of *hegemony* as theorized most profoundly by Marxist writer Antonio Gramsci, is the capacity to *incorporate opposition*. By adopting many of the movement's demands, by developing a comprehensive and coherent program of racial democracy that hewed to a centrist political logic and reinforced key dimensions of U.S. nationalist ideology, racial moderates were able to define a new racial common sense. Thus, they divided the movement, reasserted some stability, and defused a great deal of political opposition. This was accomplished gradually from about the mid-1960s to the mid-1980s.

This partial reconfiguration of the U.S. racial order was based on real concessions but left major issues unresolved, notably the endurance of significant patterns of inequality and discrimination. The reform that did occur, however, was sufficient to reduce the political challenge posed by antiracist movements in the United States. Certainly it was more successful than the intransigent strategy of diehard segregationists—whose tactic was encapsulated in the slogan "massive resistance" to even minimal integration—would have been. Yet the fundamental problems of racial injustice, inequality, and white supremacy, of course remained; they were moderated, perhaps, but hardly resolved.

So in the Unites States race not only retains its significance as a phenomenon for structuring society but also continues to define North American identities and life chances decades after the supposed triumph of the so-called civil rights revolution. Indeed "the American dilemma" may be more problematic than ever as the twenty-first century commences because achieving this moderated agenda requires that the civil rights project be raised. This process, which began in the late 1960s, achieved greater success in the following two decades.

This tugging and hauling, this escalating contestation over the meaning of race, resulted in ever more disrupted and contradictory notions of racial identity. The significance of race ("declining," as one author has suggested, or increasing?), the interpretation of racial equality (color-blind or color-conscious?), the institutionalization of racial justice (reverse discrimination or affirmative action?), and the categories of race—black, white, Latino/Hispanic, Asian American, and Native American—all these were called into question as they emerged from the moderate civil rights gains of the mid-1960s.

The argument is now made that the demands of the civil rights movement have largely been met and that the United States has entered a postracial stage of its history. Some claim that this means racial injustice is largely eradicated. In some circles, such views have become the new national "common sense" in respect to race, gaining not only elite and academic adherents but also widespread support, especially among whites. As a result, the already limited racial reform policies such as affirmative action and the relatively powerless state agencies charged with enforcing civil rights laws (i.e., the U.S. Equal Employment Opportunity Commission) developed in the 1960s are undergoing a new and severe attack. The postracialist advocates of such trends—usually classified as *neoconservative* but sometimes also found on the left—ceaselessly instruct racially defined minorities to rely on their own resources rather than government support to succeed. In a callous distortion of Martin Luther King Jr.'s message, neoconservatives exhort minorities to rely on the "content of their character" rather than "the color of their skin" to propel them to success—a social value of self-reliance that discards claims that racial prejudice still impedes opportunity for some Americans.[3]

In an egregious example, the Supreme Court in April 2001 decided on the *Alexander v. Sandoval* case—in which a Spanish-speaking resident of Alabama challenged the state's policy to administer driving tests in English only—to repeal even the inadequate civil rights reforms of the 1960s. The Court decided that states may offer licensing examinations in English and no other languages—and it said this provision does not violate civil rights laws. Here as elsewhere, by adding a purpose or "intent to discriminate" requirement to anti-discrimination law, the Court makes it almost impossible to get legal relief from discrimination. As critical legal theorist David Kairys has argued, this amounts to creating different equality rules for whites and nonwhites because, in situations in which whites suffer harm, the Court does not require proof of intent or purpose.

As the dust settled from the titanic confrontation between the movement's radical propensities and the establishment's tremendous capacity to incorporate moderate reform, a great deal remained unresolved. The ambiguous and contradictory racial conditions in the United States today result from decades-long attempts simultaneously to ameliorate racial opposition and to placate the *ancien regime raciale*. The unending reiteration of these conflicting and contradictory practices testifies to the limitations of American democracy and the continuing significance of race in the United States.

SOUTH AFRICA

In the mid-1990s, South Africa—the most explicitly race-structured society in the late twentieth century—entered a difficult but promising transition. The apartheid state was committed to a race-based framework of citizenship, civic inclusion, and law in general; the postapartheid constitution incorporates the principle of nonracialism originally articulated in the African National Congress (ANC)-based Freedom Charter of 1955. Yet the country still bears the terrible burden of apartheid's secondary consequence: persistent racial inequality persists across every level of society. The legacy of segregated residential areas, combined with a highly racialized distribution of resources of every sort, force political leadership to take a moderated middle road toward reform. Because whites continue to control positions throughout the economy, white fears must be placated in order to sustain the country's economic base and minimize capital flight. The handful of blacks who have penetrated the corporate and state elites understand very well the price the country would pay for a radical turn in policy.

Yet South Africa is officially committed to racial equality and to promoting both individual and collective black advancement. Can the postapartheid state stabilize the process of political, social, and economic integration of the black majority? Can it maintain an official nonracialism in the face of such comprehensive racial inequality? How can the vast majority of citizens—excluded until recently from access to land, education, clean water, and decent shelter, debarred from Africa's wealthiest economy, and denied the most elementary civic and political rights—garner the economic access they so desperately need without reinforcing white paranoia and fear?

How can the postapartheid state facilitate the reform of racial attitudes and practices, challenging inequality, white supremacy, and the legacy of racial separatism without engendering white flight and subversion?

Both the antiapartheid movement and the new government's policies were shaped in part by global concerns. Internal political debates reflect changing global discussions around race and politics. Just as the South African Black Consciousness Movement drew on the speeches of Malcolm X and Aimé Césaire to understand racial oppression, just as the antiapartheid movement used international antiracist sentiment to build momentum for sanctions on the old regime so, too, is the current government both guided and constrained by international pressures.

Moreover, internal politics also bring international resources to bear: Through the postwar era, the antiapartheid movement drew a great deal of support from an international antiracist movement, which was largely linked to an international trend to support decolonization. Since the 1994 election, however, international constraints have limited the sphere of action of the new democratic government. Critics of affirmative action policies, for example, emphasize the danger of undermining efficiency in the name of redistribution in the same way critics of redistributive policies deploy neoliberal economic arguments to reject nationalization; in each case, they invoke international discourses that are nonracial in form yet have racial implications in practice. The South African state continues to face considerable challenges from the political left and right: Will it be possible to reconstruct the nation by building not only a democracy but also a greater degree of consensus, citizenship, and belonging? To what degree can a policy of "class compromise" (the politically negotiated pace of reform) forestall the dangers of social upheaval and capital flight.[4]

Understanding these processes requires viewing South African racial debates in global perspective and exploring options for local actors who seek to change the terms of engagement as they restructure national politics. Although many white civil servants remain in place, the 1994 elections changed the racial character of the state; affirmative action policies, to which the ANC-led government is committed, could reorganize racial distribution of incomes, if not wealth. Yet in the context of a global debate over affirmative action, and in the face of the threat of the flight of white capital and skills, the process of reform is far slower than many South Africans, black and white, expected. This dilemma remains unresolved: how can democratic, nonracial institutions be constructed in a society where most attributes of socioeconomic position and identity remain highly racialized?

BRAZIL

Brazil presents significant parallels, both historical and contemporary, to other American nations, including the United States. These similarities include Brazil's history of slavery and black inequality, its displacement and neglect of a large indigenous population, its intermittent and ambiguous commitment to immigration, its inconsistent democracy, and its vast and increasingly urban underclass (disproportionately black). Brazilian racial dynamics have traditionally received little attention from scholars or policy makers despite the fact that the country has the second largest black population in the world (after Nigeria). Its postemancipation adoption of a policy of national "whitening," which was to be achieved by concerte0d recruitment of European immigrants, owed much to the U.S. example and also drew on nineteenth-century French racial theorizing.[5]

Amazingly, what has often been called the "myth of racial democracy" still flourishes in Brazil, even though it has been amply demonstrated to be little more than a fig leaf covering widespread racial inequality, injustice, and prejudice.[6] The Brazilian racial system, with its "color continuum" (as opposed to the more familiar "color line" of North America), tends to dilute democratic demands. Indeed, Brazilian racial dynamics make it difficult to promote policies that might address racial inequality. Public discourse resolutely discourages any attempt to define inequality along racial lines; only under the presidency of Fernando Henrique Cardoso (1995–2003) was the subject of racial inequality first officially raised. Although Brazil has now instituted certain racial reforms—such as affirmative action in education and regularization of land titles based in *quilombos* (maroon communities established by runaway slaves under slavery)—serious obstacles still confront efforts to challenge racial injustice. For example, politicians who do point out racial inequalities, and thereby challenge the myth of racial democracy, are subject to charges that they are themselves provoking racial discrimination by stressing difference.

Reliable research on racial stratification and racial attitudes in Brazil has only become available over the past few years.[7] A range of political questions thus remains mysterious. Consider the example of voting rights: Although illiterates were barred by law from voting until 1985, there is no reliable data on the proportion of illiterates who were black—and thus the extent to which black Brazilians have been disenfranchised through this century, though undoubtedly large, remains uncertain.

The emergence of the *Movimento Negro Unificado* (MNU) as a force to be reckoned with—though by no means as strong as the 1960s U.S. civil rights movement—represents a new development. The MNU used the 1988 centennial of the abolition of Brazilian slavery, as well as the 1990–91 census, to dramatize persistent racial inequalities. As in South Africa, this phase of the black movement takes its reference points partly from international antiracist struggles, often drawing on examples, symbols, and images from the civil rights and antiapartheid movements.

In the 1990s, a range of racial reforms were proposed in Brazil—largely in response to the increasingly visible *movimento negro* (black movement). To strike down officially sanctioned racial equality was permissible; to create racial equality through positive state action was not permissible with these reforms; to prompt the state to adopt antiracist policies, however, will require far greater support for change than presently exists. The political dilemma is familiar: Blacks need organized allies in the party system, among other impoverished and disenfranchised groups, and on the international scene. Yet in order to mobilize, they must also begin to assert a racialized political identity, or there will be little collective support for racial reforms. How can blacks address this dualistic, if not contradictory, situation? How can Afro-Brazilians assert claims on the basis of group solidarity without simultaneously undermining the fragile democratic consensus that is emerging across many constituencies? How can democratic institutions be built alongside policies designed to address racial inequalities without undermining a vision of common citizenship and equality?

THE EUROPEAN UNION

The last few decades have established that indeed, "the empire strikes back." Racially plural societies are in place throughout Europe, especially in former imperial powers, such as the United Kingdom, the Netherlands, France, and Spain but also in Germany, Italy, the Scandinavian countries and to some extent in the East. The influx into these countries of substantial numbers of nonwhites during the postcolonial period has deeply altered a dynamic in which the racial system and the imperial order had been one—in which outsiders were mostly kept outside the walls of the "mother country." As a stroll around London, Frankfurt, Paris, or Madrid quickly reveals that those days are now gone forever. Yet the response to the new situation often takes repressive and antidemocratic forms, focusing attention on the so-called immigrant problem (or the "Islamic problem"), seeking not only to shut the gates to Maghrebines or sub-Saharan Africans, Turks, or Slavs (including Balkan refugees), but often also to define the "others" who are already present as enemies of the national culture and threats to the "ordinary German" (or English, or French, etc.) way of life. This rationale for racial exclusion and restriction in Europe is termed "differentialist." It is distinct from the meritocratic logic of discrimination in the United States, a reflection that Europeans value the integrity of national cultures more highly than they value individual equality.[8] Thus, the particular racial issue that must be confronted in Europe is the newly heterogeneous situation, that is, the multiplication of group identities. Currently, antidemocratic tendencies such as new right and neofascist groups are widely visible, widespread, and growing. At the state and regional levels, restrictive policies that jeopardize mobility of employment or residence, and sometimes stigmatize religious or other cultural practices, are gaining popularity. Conflicts over immigration and citizenship have taken on new intensity, with crucial implications for the character of democracy.

The dynamics of integration raise a wide range of questions about future European racial logics. Conflicting principles of citizenship—*jus sanguinis* (where citizenship is determined by ancestry—"blood") versus *jus soli* (where citizenship is determined by birthplace—"soil")—are deeply imbedded in the distinct European national makeups, and their resolution in a common cultural or political framework has not come easily.[9] Nations' relations with their former colonies vary, posing serious questions of immigrant access and of economic ties between the old empires and the new Europe, but also raising anxieties about security and terrorism. Particularly in the early 1980s, popular antiracist sentiments stimulated the formation of many multiculturalist and pluralist organizations. Over the past decade, however, many have ceased to function as mass mobilization initiatives in support of democracy. So, although the slogan "Touche pas mon pôte" ("Hands off my buddy") no longer summons tens of thousands of French citizens into the street in defense of the democratic rights of racially defined minorities, the transition to racial pluralism is still very much underway.

THE TENACITY OF RACE

To understand the changing significance of race in the aftermath of the twentieth century, the century whose central malady was diagnosed by W. E. B. Du Bois as "the problem of the color-line," requires us to reconsider where the racialized world came from and where it is going. In the settings studied, the break that began with movement activity after World War II and that was contained from the late 1960s onward by political reform has not been consolidated. Just as earlier stages of modern racial history failed to resolve many issues, so too does the present epoch. In the first years of the twenty-first century the world as a whole, as evidenced by the above national cases, is far from overcoming the tenacious legacies of colonial rule, apartheid, and segregation. All still experience continuing confusion, anxiety, and contention about race. Yet the legacies of epochal struggles for freedom, democracy, and human rights also persist. To evaluate the transition to a new world racial system in comparative and historical perspective requires keeping in view the continuing tension that characterizes the present. Despite the enormous vicissitudes that demarcate and distinguish national conditions, historical developments, roles in the international market, political tendencies, and cultural norms, racial dynamics often operate as they did in centuries past: as a way of restricting the political influence of racially, socially, and economically subordinated groups. In the contemporary era, racial beliefs and practices have become far more contradictory and complex. The "old world racial system" has not disappeared, but it has been seriously challenged and changed. The legacy of democratic, racially oriented movements such as the U.S. civil rights movement, anti-apartheid struggles, *SOS-Racisme* in France, the *Movimento Negro Unificado* in Brazil, and anticolonialist initiatives throughout the world's southern nations, is thus a force to reckon with. It is impossible to address worldwide dilemmas of race and racism by ignoring or transcending these themes, for example, by adopting so-called colorblind policies. In the past, the centrality of race determined the economic, political, and cultural configuration of the modern world. Although in recent decades movements for racial equality and justice have blossomed, the legacies of centuries of racial oppression remain—nor has a vision of racial justice been fully worked out. Certainly the idea that such justice has already been largely achieved—as seen in the colorblind paradigm in the United States, the nonracialist rhetoric of the South African Freedom Charter, the Brazilian rhetoric of racial democracy, and the emerging racial differentialism of the European Union—remains problematic.

What would a more credible vision entail? The pressing task today is not to jettison the concept of race but instead to come to terms with it as a form of flexible human variety. What does this mean in respect to racism? Racism has been a crucial component of modernity, a key pillar of the global capitalist system, for five hundred years. So it remains today. Yet it has been changed, damaged, and forced to reorganize by the massive global social movements that have taken place in recent decades. In the past, these movements were international in scope and influence. They were deeply linked to democratizing and egalitarian trends, such as labor politics and feminism. They were able both to mobilize around the injustices and exclusion experienced by racially subordinated groups and sustain alliances across racial lines. This is background; such experiences cannot simply recur. Yet surely the massive mobilizations that created the global break following World War II were not fated to be the last popular upsurges, the last egalitarian challenges to white supremacy, to racial hierarchy. In the countries I have discussed and in transnational antiracist networks, these earlier precedents still wield their influence. They still spark new attempts to challenge racism.

At the same time, new political and intellectual leaders have come onto various national stages in recent years, arguing that the worst racial injustices (of the United States, Brazil, South Africa, and so on) are now firmly relegated to the past and that the problem of racism can now be viewed as essentially solved. Why, then, should we maintain affirmative action policies? Why direct resources toward immigrants, victims of segregation and apartheid, or the (disproportionately dark-skinned) poor? Don't we already have equality now?

Will race ever be transcended? Will the world ever get beyond race? Probably not. But the entire world still has a chance of overcoming the stratification, the hierarchy, the taken-for-granted injustice and inhumanity that so often accompanies the race concept.

Like religion or language, race can be accepted as part of the spectrum of the human condition, while it is simultaneously and categorically resisted as a means of stratifying national or global societies. As we enter a new millennium, nothing is more essential in the effort to reinforce democratic commitments, not to mention global survival and prosperity.

NOTES

1. Nikhil Pal Singh, "Culture/Wars: Recoding Empire in an Age of Democracy," *American Quarterly* 50, no. 3 (September, 1998); Stephen Steinberg, *Turning Back: The Retreat from Racial Justice in American Thought and Policy* (Boston: Beacon Press, 1995).

2. Howard Winant, *Racial Conditions: Politics, Theory, Comparisons* (Minneapolis: University of Minnesota Press, 1994).

3. Shelby Steele, *The Content of Our Character: A New Vision of Race in America* (New York: St. Martin's, 1990); Stephan Thernstrom and Abigail Thernstrom, *America in Black and White: One Nation, Indivisible* (New York: Simon & Schuster, 1997).

4. Edward Webster and Glenn Adler, "Toward a Class Compromise in South Africa's 'Double Transition': Bargained Liberalization and the Consolidation of Democracy," *Politics and Society* 27, no. 3 (September 1999): 347–85.

5. Thomas E. Skidmore, *Black into White: Race and Nationality in Brazilian Thought* (1974; repr. Durham, NC: Duke University Press, 1993).

6. Carlos A. Hasenbalg and Nelson do Valle Silva, "Raça e Oportunididades Educacionais no Brasil," in *Relações Raciais no Brasi l Contemporâneo*, ed. Hasenbalg and Silva (Rio de Janeiro: Rio Fundo/IUPERJ, 1992); Michael George Hanchard, *Orpheus and Power: The Movimento Negro of Rio de Janeiro and Sao Paulo, Brazi l, 1945–1988* (Princeton, NJ: Princeton University Press, 1994); George Reid Andrews, *Blacks and Whites in São Paulo, Brazil, 1888–1988* (Madison: University of Wisconsin Press, 1991).

7. Edward Telles, "Segregation by Skin Color in Brazil," *American Sociological Review* 57 (1992): 186–97; Telles, "Industrialization and Racial Inequality in Employment: The Brazilian Example," *American Sociological Review* 59, no. 1 (1994): 46–63; France Winddance Twine, *Racism in a Racial Democracy: The Maintenance of White Supremacy in Brazil* (New Brunswick, NJ: Rutgers University Press, 1997); Datafolha, *300 Anos de Zumbi: Os Brasileiros e o Preconceito de Côr* [*300 Years After Zumbi: Brazilians and Color Prejudice*] (São Paulo: Datafolha, 1995).

8. Pierre-André Taguieff, "The New Cultural Racism in France," *Telos* 83 (Spring 1990) and excerpted in *Racism*, ed. Martin Bulmer and John Solomos (London: Oxford University Press, 1999); Michel Wieviorka, *The Arena of Racism*, trans. Chris Turner (Thousand Oaks, CA: Sage, 1995).

9. Rogers Brubaker, *Citizenship and Nationhood in France and Germany* (Cambridge, MA: Harvard University Press, 1992).

CHAPTER 4

THE ONGOING CONTESTATION OVER NATIONHOOD

ANTHONY W. MARX

THE NATION IS NEVER A FIXED SUBJECT, but always in flux and a major focus of contestation. Elements of the state and factions of civil society project and dispute varying images of who is to be included as a member of the polity, so specifying obligations, allocation of benefits, and rights. The result at any one time is a set of images of solidarity and rules of citizenship that remains fluid. Precisely because this process is central to the defining, demarcating, and implementing of the community and social identities, it remains a primary subject of politics.[1]

As a country torn by political conflict during much of the twentieth century and having recently enjoyed a transition to majority rule, South Africa has experienced a particularly vibrant contestation over nationhood. That contestation remains today in the aftermath of the second popular national election in June 1999. What remains at stake is the definition of who is included as the central constituency of the polity, serving and served by the state. Thinking about the possible future paths of nationhood in South Africa, we might begin with an overview of the past contestation.

The South African state was formed in the aftermath of a major violent struggle between Dutch-descendant Afrikaners claiming national self-determination and a British imperial power pursuing colonial rule. This struggle reached its crescendo with the Anglo-Boer War of the turn of the twentieth century. Outmanned and outgunned by the world's leading imperial power, the Afrikaners valiantly held off a British victory for three years. The British resorted to the first use of machine guns and concentration camps, in which many Boer women and children perished. One in six Afrikaners died in the conflict.[2]

To make peace and impose unitary rule over the newly created South African state, the British sought to appease their former Afrikaner adversaries. As I have argued elsewhere, that peace was built on British concessions to Afrikaner demands to continue the political exclusion and subordination of the majority African population. White racism, common to the Afrikaners and the English (despite the latter's denials) became the basis for an emergent accommodation among whites assuaging their ethnic antagonism.[3] White loyalty to the state was thus built on the basis of exclusion of blacks.

With the creation of the South African state after 1910, two distinct images of nationalism emerged among whites. British rulers joined with more pragmatic Afrikaners to encourage the idea of a white South African nationalism defined, demarcated, and reinforced by continued black exclusion. This ideology remained the dominant image of the first half of the twentieth century and was forged and refined under the governments of Louis Botha and Jan Smuts. This racist nationalism unifying whites as such was challenged by an ethnic nationalism of more recalcitrant Afrikaners, unified in the prior violent conflict with the British and resentful of imperial rule. Rather than embrace a unified white nationalism, the Afrikaner nationalists pressed for the self-determination of their ethnic constituency as the slim majority among whites. Their National Party came to power under General Hertzog in 1924 and held power until 1936, when a coalition emerged.

The contestation between unified white and Afrikaner nationalism was effectively ended with the 1948 victory of the National Party. Under D. F. Malan and his successors, National Party members ("Nats") solidified their constituency with encouragement of Afrikaner culture and language while using public policy to advance the economic interests of Afrikaner workers against richer English-descendant business owners.[4] The Nats also greatly reinforced and systematized the exclusion and domination of blacks under the tremendously costly, pervasive, and draconian policies of apartheid. Many Afrikaners were employed by the state to enforce these rules, thereby both containing the black majority and employing Afrikaners. By the 1960s and 1970s, Afrikaner nationalism was clearly dominant, by force, in South Africa.

As an important example of the unexpected consequences of a particular form of nationalism, the Nats' policies were so effective that they gradually undermined their own imperative. As Afrikaners were empowered, culturally solidified, and economically advanced by state policy, their antagonism against the richer English diminished. And again the combined white interest in racial domination reinforced an English-Afrikaner alliance, despite the self-proclaimed liberal denials of the English. By the 1980s, many English were supporting the National Party, thereby diluting its ethnic separatism.[5] Meanwhile, the consolidation of white power and wealth provided assurance to whites of their continued privilege and allowed for some exploration of reforming apartheid. Of course such reforms were overwhelmed in the 1980s by rising black protests pressing for more pervasive change.

The role of resistance forces the narrative back to a related but distinct form of black nationalist contestation, challenging white nationalism from its very start. Although black opposition predates even the formation of the South African state, majority opposition was consolidated by the formation of that state and its imposition of rules excluding (and thereby unifying) all blacks. Not surprising, the coming enactments of those rules of exclusion prompted the organization of the first black national political coalition, what would become the African National Congress (ANC), formed in 1909.

The ANC's image of nationalism was, of course, directly contrary to the dominant white view. Instead of a nation of whites, solidified by exclusion of blacks, the ANC sought black rights, at first through peaceful petition by elites and, when that failed, with mass organization and protest. Eventually, the ANC adopted a particularly inclusive form of nationalism, defined as uniting all those who opposed the system of apartheid, including white liberals. The inclusiveness of the ANC's nonracial nationalism was its hallmark, purposefully designed to attract the largest number of followers to oppose the government. As long as apartheid remained, it provided the demarcation of the alternative image of a nation, defined as all those who opposed apartheid.

But this definition of inclusive nationhood was also contested among blacks, although the history of this dispute has been largely eclipsed of late by the predominance of the ANC. For instance, after the 1948 victory of the National Party, a section of the youth within the ANC became critical of the ANC's liberal inclusiveness. Arguing that the ANC's stance failed to prevent the consolidation of white domination, this section broke away from the ANC in 1959 to form the Pan-Africanist Congress (PAC). This group advocated a different form of nationhood, which was open only to those who claimed primary allegiance to Africa and thereby effectively excluded whites as well as "mixed-race coloureds" and Indian descendants. The PAC advocated mass insurrection, and when protests increased in 1960, the PAC was banned and forced into exile, as was the ANC.[6]

Constrained by state oppression and exile, the PAC and ANC continued their efforts at opposition, but within the vacuum of internal organization a new group and ideology emerged. Founded in 1969 by Steve Biko, the Black Consciousness (BC) movement sought to rebuild domestic assertiveness and resistance with yet another form of distinctive nationalism based on black separatism. Positive black identity was asserted by organizing political and service organizations that excluded whites, which was in direct ideological opposition to white nationalism. Unlike with the PAC, officially designated coloureds and Asians were included in BC as fellow victims of apartheid, thereby avoiding an essentialist demarcation of blacks as African descendants only. The power of this image built black assertiveness and exploded with the Soweto Uprising of 1976, which was sparked by violent state oppression against a peaceful student march.[7]

As a further complication, the late 1970s saw the blossoming of at least two other distinct forms of protonationalism. Concentrated in Natal, rural Zulus in particular were organized by Chief Mangosuthu Buthelezi into the Inkatha

Freedom Party. This group enjoyed the status of state client, benefiting from the funding and arms provided to the KwaZulu homeland authority (also headed by Buthelezi). But Inkatha resisted some aspects of apartheid, notably rejecting Kwa-Zulu independence, although it distanced itself from the ANC. Despite claims of being nonethnic, Inkatha was generally understood as asserting an ethnically distinctive Zulu nationalism. And at the same time, the trade union movement among blacks was becoming a major social force, with more than a million members by the mid-1980s.[8] Although less clearly nationalist in its framework and organization, the trade union movement (influenced in ways by the exiled Communist Party) projected an image of national rights leading to advancement of working-class interests. This astonishing mix of actors all converged in pressing for the end of apartheid.

The BC movement helped rebuild domestic protest, which was then inspired by continued ANC exile activism to resurface after the 1976–77 crackdown. The result was the United Democratic Front (UDF), which adopted an ANC-inspired image of inclusive nationalism opposed to apartheid. This mass protest effort of the mid-to-late 1980s was reinforced by the activism of the trade unions, although coming into increased conflict with Inkatha and remnants of BC.

By the late 1980s, opposition was largely consolidated under the UDF and its allied unions, bringing South Africa to a state of near-civil war and economic collapse. By then the National Party, encouraged by the consolidation of white power and privilege and discouraged about the prospects of retaining minority rule, social peace, and economic progress, was more open to pursuing reforms. The way was open to a compromise, further encouraged by the end of the cold war. White fears of a Communist-aligned black takeover diminished, even as international sanctions increased whites' interests in finding an accommodation that would end the economic pressures.

The negotiations of the early 1990s largely pitted the ANC image of an inclusive nationalism against the prior rulers of the National Party. Black separatists, Afrikaner nationalists, and Zulu nationalists were marginalized by ANC efforts to appease Inkatha with an alliance. The largest trade union federation, the Congress of South African Trade Unions (COSATU), effectively subordinated itself politically to the ANC, consistent with the long-held Communist Party view of the need for achieving majority rule before the more divisive issues of class power could be addressed. This accommodation by the organized black working class proved crucial to the negotiated transition, for it allowed the ANC to reach out to whites with promises of property protection for accumulated wealth and preservation of white civil service jobs. Whites thereby felt sufficiently confident of their prospects to support a negotiated settlement, which brought majority rule elections in 1994.

During the subsequent five years, at least for the most part, the contestation over South African nationhood finally came to rest upon the images propounded and reinforced by the ANC-led government. The world has been greatly inspired by South Africa's relatively peaceful emergence from being the quintessentially racist state to a state committed to nonracial inclusion. All South Africans are promised equal citizenship rights by the new constitution, with minority rights of culture and language also asserted. Under the inspiring leadership of Nelson Mandela, blacks advanced through state-provided services, some redistribution of land, and limited forms of affirmative action in private employment. Whites retained most of their previously acquired wealth and largely kept their civil service jobs. Business was reassured by neoliberal economic policies and limits on state intrusion. South Africa would appear to have emerged as a successful compromise of minority interests and majority demands for redress within an inclusive nation.

For all that is to be celebrated about South Africa's success at forging an inclusive nation thus far, tensions remaining under the surface continue to bubble up. In the political realm there remain small groupings advocating Afrikaner, Zulu, or black separatism, although these have relatively little electoral support at the moment. The real tensions are more evident in related economic issues. Many blacks enjoy the new provisions of water, housing, electricity, and improved schools and health care, although such redress has come slower than many expected. A relatively small black middle and upper class has profited greatly since the end of apartheid, but urban black unemployment remains at approximately 40 percent, there has been little advancement particularly in the rural areas, and crime is rampant.[9] Meanwhile, despite their continued enjoyment of relative privilege, many whites feel uncertain. Some have migrated, and white-dominated business has made little new domestic investment and has shifted major holdings overseas.

The current economic situation impinges upon the new ANC-led government headed by Thabo Mbeki. Economic policy remains firmly neoliberal, seeking to appease domestic and international capital by limiting efforts at redress or redistribution in order to encourage growth. But amid record low prices for gold exports, world capital flows against the Third World, and lower labor cost competitors elsewhere, South Africa's growth has been disappointing. The government postponed more redistribution to achieve growth but has largely ended up with less of either.

Pressure is brought against this stasis from competing quarters. Particularly the left camp within the ANC, aligned with the unions and Communist Party, would like to see more redistribution. Whites and business, to put it simply, press for less. The ANC's inclusive nationalism and nonracialism inspire continued efforts to finesse these pressures and retain all camps within the "great tent" of a political party, the inclusiveness of which was forged in the anti-apartheid struggle.

Efforts to maintain a compromise and retain coherence within such an inclusive ANC will come under increased pressure if economic dislocation continues. The inclusiveness that worked to bolster the ranks of the ANC in opposition to apartheid is less easily maintained in government, where actual and divisive policy choices must be made. Of course, the ANC would prefer to hold on to its extensive constituency and retain majority power. This challenge presents the ANC and President Mbeki in particular, with at least two temptations:

Given the salience of resentment against continued white privilege, the ANC might be tempted to solidify its support with an antiwhite rhetoric. This approach might be seen as a way to maintain more or less unified black support, thereby avoiding a split of that constituency along the lines of a growing economic divide among blacks since the end of apartheid. There have already been at least three indications of this temptation. In late 1997, even the great statesman Mandela lashed out in a major speech against white selfishness,[10] perhaps not coincidentally during the lead-up to the 1999 election when blacks again overwhelmingly supported the ANC. Mandela and his successor, Mbeki, also sought to consolidate an alliance with Buthelezi's Inkatha by ignoring Buthelezi's past use of violence and rewarding him with a post in the cabinet. The prospect emerged of a Nguni peoples coalition, unifying the largest black ethnic groups, the Xhosa and Zulu, and excluding whites. More vaguely, Mbeki's trademark advocacy of himself and his country as part of an "African renaissance" hints at a cultural exclusion of whites, despite claims that whites can be included as fellow Africans.

In whatever form, such appeasement to hard or soft images of black separatism may be described as South Africa's "Zimbabwe option," consolidating black majority rule and white marginalization. Zimbabwe's experience suggests that this approach may require increased authoritarian efforts with antagonism of whites' interests bringing economic costs and provoking popular unrest being forcefully contained. Although South Africa's history, larger white population, and higher level of development set it apart from Zimbabwe, white South African political parties stoked fears of the prospect of similarity.

The other temptation is the "Brazil option," that is, a coalition of middle- and upper-class blacks and whites that serve their interests while leaving unaddressed the needs of a marginalized poor population, particularly the rural blacks. This temptation is also already evident in nascent form with macro statistics demonstrating that limited redress thus far has indeed created a greater class disparity in South Africa since apartheid.[11]

Even the trade unions may be described as supporting this outcome, for their organized membership tends to benefit from a rising middle class status, even as the unemployed and informal sector remain outside the union ranks. Neoliberal policies and international economic pressures reinforce this outcome, which at least in Brazil has brought economic advance for the top half of the population with little advance and only limited protest and mobilization by the bottom half.

It is worth noting that South African elites may be tempted to pursue both of these seemingly contradictory temptations. They can use either explicitly antiwhite rhetoric or softer forms of "Africanness" to unify black support. And at the same time, they can pursue economic policies that reassure whites and better-off blacks but do not reach out to poorer blacks. Ironically, this combined approach would be consistent with the ANC tradition of trying to be strategically inclusive for pragmatic electoral benefit.

Still, the ANC ideology of inclusion runs against these less liberal strategic temptations. The ANC leadership rightfully prides itself on its nonracial and inclusive nationalism, which provides an inspiring alternative to the long history of more exclusive forms of nationalism in South Africa and elsewhere. This inclusion is the founding image of the new South Africa, which many would strongly resist abandoning. The actual experiences of continued deprivation and specified social exclusion in Zimbabwe or Brazil do not make the temptations of their approaches all that attractive, at least in the long run. Of course, the pressures pushing South Africa in either or both of these directions remain real, for it is difficult in this "new world order" to meet the needs of a poor Third World majority and the interests of a First World minority and its foreign allies. South Africa's situation is emblematic of many countries that include both of these competing constituencies, although every country faces its own unique pressures.

In the absence of sustained growth and foreign support for such growth, South Africa will be further subjected to the pressures that forced Zimbabwe and Brazil down their unfortunate paths of racial populism, authoritarianism, and inequality. When a balance of all social interests cannot be met, choices must be made that forge new coalitions and nationalisms of selective inclusion and exclusion. Thus far the world shows little inclination to support South Africa's efforts to avoid the alternatives of ethnic, racial, or class exclusion. Indeed, the world is now dominated by global economic market forces that do not have the collective agency to even make such a choice of support. Instead, private capital flows heighten domestic tension while overwhelming more limited global public policies. South Africa thus remains poised at a crossroads of alternative images of its nationhood. The ideological highroad of inclusion is challenged by the pragmatic low road of social and economic choices that exclude some from the real benefits of citizenship. The contestation over these alternatives continues. South Africa's future remains open and uncertain. And with it the world will be making a choice, even if not consciously so. For South Africa has emerged as symbolic of the new world in which we live, inspiring many around the globe with its peaceful transition and efforts to redress past injustice while surviving in the global market. If South Africa fails to maintain this trajectory, it will be a signal that the great transformation of the post–cold war era brings not an "end to history" but rather a new phase of contestation and possible conflict.

NOTES

1. See Rogers Brubaker, *Nationalism Reframed* (Cambridge: Cambridge University Press, 1996).
2. See Leonard Thompson, *A History of South Africa* (New Haven, CT: Yale University Press, 1990).
3. See Anthony W. Marx, *Making Race and Nation* (New York: Cambridge University Press, 1998).
4. See Anthony W. Marx, "Class Discord and Racial Order," *Politikon* 26, no. 1 (May 1999): 81–102.

5. Tim Sisk, *Democratization in South Africa* (Princeton, NJ: Princeton University Press, 1995).

6. See Tom Lodge, *Black Politics in South Africa Since 1945* (Johannesburg, South Africa: Ravan, 1983); Gail Gerhart, *Black Power in South Africa* (Berkeley: University of California Press, 1978).

7. See Anthony W. Marx, *Lessons of Struggle* (New York: Oxford University Press, 1992).

8. See Steven Friedman, *Building Tomorrow Today* (Johannesburg, South Africa: Ravan, 1987).

9. Michael Bratton, "After Mandela's Miracle in South Africa," *Current History* (May 1998): 214–19.

10. Report of the president of the ANC, Mafeking, December 16, 1997.

11. *World Development Indicators* (Washington, DC: World Bank, 1988).

PART II

INTERROGATING RACE AND RACISM IN THE AMERICAS

CHAPTER 5

A TALE OF TWO BARRIOS

PUERTO RICAN YOUTH AND THE POLITICS OF BELONGING

GINA M. PÉREZ

ACCORDING TO DAVID HARVEY, "the perpetual reshaping of the geographic land-scape of capitalism is a process of violence and pain." This process intersects with race and gender, forcing new definitions and criteria of belonging in different communities. Since beginning my fieldwork among poor and working-class Puerto Ricans in Chicago and San Sebastián, Puerto Rico in 1995, I have returned to Harvey's description of capitalism's impact on place and space to help me make sense of the shifting realities of thousands of *puertorriqueños* (Puerto Ricans) living on the island and the mainland. When I began my research project, I initially thought I would construct an ethnographic study of circular migration between Chicago and San Sebastián—a small town in the northwestern region of the island—focusing on particular households and family members as they moved between both communities over time. After a few months I began to panic because I had identified only one family that fit this profile. But as I continued to work at one of Chicago's Puerto Rican cultural centers and continued to hear stories of displacement, gentrification, migration, and confusion over varying meanings of community, I redirected my research focus. I began to consider the changing meanings of migration among first- and second-generation Puerto Rican migrants in Chicago to examine what impact this movement had on both communities. Both places have been connected by transnational flows of people, goods, money, and ideologies for more than fifty years.

When migrants seek economic mobility and security for themselves and their children, migration is to be expected. Unexpectedly, however, migration between Chicago and Puerto Rico was an important strategy used by older migrants to protect their offspring from the influences of American culture. As young girls became teenagers, families worried they might develop relationships with "the wrong kind of guy" and eventually get pregnant. For many young boys, adolescence meant gangs, drugs, and violence. Families would often send children to Puerto Rico as teenagers once they felt their children were getting into trouble. But for most part, the mere *possibility* of sending children to Puerto Rico to live with family members always existed as both a resource and a disciplinary threat. Thus, the links Chicago-based families maintained with relatives in Puerto Rico become critical as their U.S.-reared children grow older. This is one way in which adult migrants can activate and deactivate transnational connections at different points in the life cycle of households.

In addition to parents' justified fears of gang involvement, they also worry about the real problems of police harassment and brutality against Latino youth. This dynamic highlights the racial politics of place and belonging. On one hand, the neighborhoods of Humboldt Park, Logan Square, and West Town—which are on Chicago's Near Northwest Side and house the largest concentration of Puerto Rican families—are popularly regarded as dangerous places. English- and Spanish-language print and television media have labeled these places as gang- and drug-infested: Several years ago, one *Chicago Tribune* writer called Humboldt Park "hell's living room," reflecting attitudes that help justify a zero-tolerance law enforcement policy with regard to so-called suspicious youth. This policy, which was at the heart of the city's controversial anti-loitering ordinance, allowed law enforcement officials to detain people for unlawfully congregating on the street with "no apparent purpose."[1] These neighborhoods are also on the western border of some of the city's most rapidly gentrifying neighborhoods, such as Wicker Park and Buck Town, and are even experiencing the beginnings of gentrification themselves. Although many Puerto Rican youth are lifelong residents of these gentrifying areas, they were suddenly considered outsiders. They were constantly watched, harassed, verbally abused, and humiliated by powerful interests, such as new businesses in the area and law enforcement. They are often reminded that they no longer belong. Just as gentrification brings the destruction, gutting, and rebuilding of old buildings, so too does gentrification reconfigure a neighborhood's racial and social landscape by redefining membership in a community and clearly labeling as outsiders those who transgress imposed norms of class, dress, race, and ethnicity.

These questions of membership, community, race, globalization, and transnationalism also emerged in my research in San Sebastián. Since the late 1960s, Puerto Rico has been engaged in an ongoing debate about who properly belongs to the "Puerto Rican nation." This struggle has been fueled in part by a mass return migration from the United States beginning in the 1960s and by heightened immigration from the Dominican Republic. The return migrants' social remittances and globalization introduce new ways of being at the local level.[2] Together, these processes prompted an ongoing national debate on "Puerto Ricanness" and the detrimental effects that *los que vienen de afuera*—that is, outsiders or, literally, those who come from the outside—have on Puerto Rican culture, values, and local communities. In what follows, I will discuss how this *los de afuera* discourse is used in San Sebastián to define membership in the Puerto Rican nation and the local community and how it is also deployed by local politicians to resist the unexpected consequences of globalization in the town.

During my field research in San Sebastián in 1998, discussions about Puerto Rico's political status as a nation and as an American territory were, understandably, near-constant topics of conversation. Almost as often, residents lamented and hotly debated issues concerning *la juventud de hoy* (today's youth) and los de afuera. People argued passionately about contemporary social problems and attempted to explain the cultural changes happening around them. They considered drugs, crime, sexual promiscuity, and a disdain for work, family, and community to be the unwelcome consequences of the transnationalism enabling los de afuera to dwell among them.

One writer from San Sebastián calls it "the thing we don't want to talk about": the Diaspora at home, *los de afuera*. Narrowly defined, the term refers to Puerto Ricans who currently reside on the island after many years living in the United States. More broadly, the label popularly applies to Dominicans—who are almost always racialized as black—*pillos* (criminals), *asesinos* (murderers), or anyone else who does not properly belong in the imagined "circle of the we" of the Puerto Rican nation.[3] Like the term *Nuyorican*, "los de afuera" is usually used pejoratively, connoting a culturally and racially distinct group whose values, behaviors, language, and dress directly challenge so-called authentic Puerto Rican culture and racial identity. The label "los de afuera" is used not only to define membership in the immediate and national community but also to resist the transnational flow of ideas, people, and capital that many believe destabilize traditional understandings of customs, community, and identity in local community politics.

Like many municipalities in Puerto Rico, San Sebastián witnessed unprecedented growth beginning in the 1970s. Between 1970 and 1990, its population increased by more than 20 percent to 38,799, which is remarkable considering the town's population as a whole shrunk by nearly 10 percent in the 1950s and 1960s.[4] These demographic changes, however, are not surprising and, in fact, reflect population increases throughout the island: Although return migration to the island began in earnest in the 1960s, it reached its peak in the 1970s and continued at a steady, although diminished, rate through the 1980s.[5] The life histories and stories of the residents of barrio Saltos—an outlying sector of San Sebastián—reveal myriad economic and noneconomic factors informing migration decisions. They also paint a complicated portrait of life in the town and of the ways in which migration, identity, and the politics of place remain emotionally charged issues in local residents' lives. In what follows, I identify two distinct but related ways in which people make sense of los de afuera and the changes brought about by return migration.

First, young Puerto Rican men and women who do not fit expected norms of behavior, dress, and linguistic performance are often immediately labeled as de afuera regardless of whether or not they have actually lived outside of Puerto Rico. Multiple piercings, hip-hop styles of dress for men, and provocative dress for women are commonly regarded as markers of one's outside status and usually

provoke pointed remarks by older barrio residents. During my field research and
tenure working in a bodega (small grocery store) in barrio Saltos, for example, I
witnessed the heated intergenerational conflicts between young and old men who
consistently demonstrated their disdain for one another. Because the bodega was
located directly across the street from local basketball courts, the old men drink-
ing and playing dominoes would often interact with the young male ball players
and others around the courts. The old men, who frequented the store consistently, complained about the trendy clothing of the young men, their multiple
piercings, and their music, which included rap and heavy metal. They attributed
these behaviors to the negative influences of life in the United States. One evening
as I worked in the store, I teasingly told one young man with pierced ears, eye-
brows, and a lip ring that he should not smoke because he will die young. One of
the older regulars looked up from his beer and sneered, "Qué se muera, no mas!
Mira como es! Déjalo que se muera!" ("Then let him die! Just look at him! Just
let him die!") The young man looked surprised at the outburst, said something
under his breath, and left the store. Other older men playing dominoes witnessed
the exchange and laughed. One commented, "La juventud de hoy está perdida.
Vienen de afuera, vienen para aca, y no hacen na'." (Today's youth is lost. They
come from the outside, they come here, and they don't do anything.)

The idea of "la juventud perdida" was a common lament. Older community
residents would pray for them—"para los jovenes, tan metidos en el vicio, en la
droga, en la prostitución" (for the youth, so involved in bad habits, drugs, and
prostitution)—and stress how vastly different the young are today compared to
how they were when they were younger—"Son perdidos, y no saben vivir la vida."
(They're lost, and they don't know how to live life.) They would accuse the young
people of being lazy, unmotivated, and corrupted by life *afuera*. One day over a
game of dominoes, two old men tried to convince me and another woman, Beba,
that today's youth were worthless. Severino commented,

> 'Chacho! Estos jovenes—yo diría que solo un 25 percentage de la juventud hoy en
> día son buenos. Jovenes que valen la pena y que no están metidos en problemas—en
> las gangas, en la droga.
>
> [Those young people—I'm saying that only 25 percent of today's youth are good.
> Young people who are worthwhile and not in trouble—in gangs, drugs.]

"Sí," Carlos added, "la juventud está perdida" (Yes, the young people are lost).
When Beba protested that more than 25 percent of the youth were worthwhile,
she was corrected by Severino who continued,

> No, no. Mira, los jovenes de hoy no trabajan! No saben trabajar. Cogen el welfare,
> los cupones. No saben trabajar como nosotros . . . por eso los jovenes son bancar-
> rota. Es el modernismo—el mundo moderno. Y es lo mismo que allá [en los Esta-
> dos Unidos].

[Look, the youth of today don't work! They don't know how to work. They get welfare and food stamps. They don't know how to work like us . . . and so the young people are bankrupt. It's modernism—the modern world. And it's the same over there in the United States.]

Young men, according to these narratives, don't know the value of work or the value of making money and living an honest life. Moreover, unlike these older critics who also lived afuera for varying lengths of time, these young men have internalized the cultural values attributed to urban ghetto living and bring them to the island: gangs, drugs, and increased violence in Puerto Rico are blamed, in large part, on los que vienen de afuera.

Notions of "la juventud perdida" also vary widely depending on gender. Although young men are derided for their failure to be productive members of the community, women de afuera are blamed for their domestic failings: They are called immodest, *locas* (crazy girls), *fiesteras* (party girls), and *enamoradas* (boy crazy) and are said to be responsible for failed marriages and the moral deterioration of the town. When I first arrived in San Sebastián, for example, one of my great-aunts took great pride in introducing me to her friends and neighbors. When she introduced me to the men at the *panadería* (bakery) across the street from her home, she pointed out that although I was born and raised afuera, I was not "una de esas locas de afuera. Miren como se viste (one of those crazy girls from over there. See how she dresses), she insisted, as she passed her hand along the front of me, highlighting my appropriately short skirt and conservative blouse. "Es como si fuera criada aquí, como en los tiempos de antes" (It's as if she was born here, like in the old days), she concluded. In contrast, I heard one grandmother complain to her friends that her granddaughter who was born and raised in Puerto Rico, "se viste como si fuera de Nueva York" (looks like she's from New York). Other young women were criticized for socializing away from their homes—away from the supervision of their elders—too much, relying too heavily on relatives to care for their children, and not attending to their husbands properly. Invariably, these problems were attributed to the influence of American gender ideologies that threaten traditional Puerto Rican family values. Thus, the label "los de afuera" not only reproduces a particular understanding of community that is static and smoothes over its true heterogeneity—including class, race, and sexuality—but it also stigmatizes those people whose economic marginalization necessitates migration as a strategy of survival. As other scholars have noted, young people often make convenient scapegoats for larger political and economic problems.[6]

On the level of political discourse, the label "de afuera" is used to resist what some *pepinianos* (residents of San Sebastián) regard as the negative consequences of progress on daily life in the town. Although globalization is routinely viewed as a sweeping, large-scale phenomenon, these statements offer an on-the-ground example of the way globalization—as it informs the concepts of progress and culture—can spark tension and resistance between individuals and within communities.

As one prominent lawyer and *independentista* (one favoring the independence of Puerto Rico) explained:

> The return of those who come and go between here and the United States is a problem. The mayor is a product of the migration. And you can see the impact return migration has had even on the highways. The mayors who migrated, who have been through that, don't make small streets or roads. When people return, they have a larger vision of the world: big projects with big thinking that is metropolitan and even North American. Big cars, big streets. . . . It's a process that affects the vision of life here, how we should do things here. It's a vision that is not from here. It's from over there.[7]

This lawyer interprets "big visions," as a threat to the traditional cultural values and practices that most pepinianos believe characterize San Sebastián and make it unique. Since the 1950s, San Sebastián, the neighboring municipality of Lares, and other towns of the central highland region have been popularly regarded as authentic sites of Puerto Rican culture.[8] And since the early 1970s, local community activists and cultural workers have dedicated themselves to preserving and enhancing these cultural traditions.

For this reason—and many others that are beyond the scope of this chapter—many residents of San Sebastián are extremely critical of their new mayor, Justo Medina, who was elected in 1992. Born in San Sebastián but raised in Perth Amboy, New Jersey (the city *pepinianos* fondly refer to as a mini-San Sebastián on the mainland), don Justo is young, energetic, and committed to bringing "progreso" to the small, largely rural town.

He is also the first New Progressive Party (PNP) mayor in a town deeply loyal to *los populares*, the Popular Democratic Party. An example of his distinct vision was his backing of a new tabloid-size local newspaper, *El Progreso*, which resembled the *New York Daily News* more than the town's tradition of modest newspapers like *El Pepino*, *El Palique*, and *El Maguey*. Many *pepinianos* complained bitterly about these changes, accusing the mayor of using the paper as an unofficial vehicle of his administration and of squeezing out the town's other print media, which frequently criticized him and his new policies. Such tactics, one man assured me, were typical of los de afuera and they underscored rumors of don Justo's corrupt ways, including his alleged financial gains through illegal activities in Perth Amboy (he was repeatedly accused of making his money through *la bolita* (the numbers) in Perth Amboy, although no one could prove this).

One forty-year-old New York-born man told me,

> Mira, este alcalde que tenemos aquí, el vino de afuera. Igual que el alcalde de Aguadilla. Pero él no trabaja nada para su pueblo. Este alcalde vino, trabajo sus cuatro añitos y se tiró para el alcalde. Y ganó. Vino y hizo su residencia y ahora es alcalde.

> [Look, this mayor we have here, he came from the outside. The same with the mayor of Aguadilla. But he doesn't work at all for his town. This mayor came, he

[worked his four years here and then ran for mayor. And he won. He came and established residency, and now he is mayor.]

Not only was the mayor accused of buying the elections—a strategy, according to local populares and independentistas, employed by *los penenpes* (the PNP supporters) on both a local and national level—he is also believed to embody everything that is wrong with los que vienen afuera: arrogance, *vendepatria* (being a sellout), and disloyalty to the Puerto Rican nation.

What is striking about these political struggles is the way in which real fears about the consequences of global capital in a small town are articulated through the discourse of los de afuera, and at the same time residents praise the men and women who migrated in the 1950s and 1960s because they understood *el honor del trabajo* (the honor of work). In fact, the town's migrants occupy an ambiguous ideological and material space. They are celebrated and honored each year during San Sebastián's two festivals, Festival de la Novilla in January and El Festival de la Hamaca in July. And residents continually point out that the town was built not only on the labor of those working in sugar cane but also through migrants' years of struggle and sacrifice working in industrial jobs in North America. But that was, as a young, unemployed return migrant explained to me, in "los tiempos de la migración, en los años '40, '50, y '60" (the time of the migration, in the '40s, '50s, and '60s). Today, the circulation of people between the island and the United States is reviled, pathologized, and blamed for what were and continue to be—that is, the consequences of inequality, domination, and unequal development as a result of Puerto Rico's ongoing colonial relationship with the United States.

Globalization, then, is often about violence, pain, and new understandings of belonging that are viewed through racial prisms. But globalization can be positive and progressive, presenting new opportunities for people to fashion new collective identities. As we look forward to the way migrants push cultural boundaries, we must pay particular attention to the way youth are implicated in these processes. On one hand, they might redirect our attention to liberating possibilities in global cultures and identities. On the other hand, they are often a society's most vulnerable and voiceless citizens and frequent scapegoats for political and economic changes and problems that are not of their own making.

NOTES

1. Although the law was ruled unconstitutional by the U.S. Supreme Court in 1999, many barrio residents believe it still guides police practices in their neighborhoods.
2. Peggy Levitt, *The Transnational Villagers* (Berkeley: University of California Press, 2001).
3. The phrase "circle of the we" comes from David Hollinger, "How Wide the Circle of 'We'?: American Intellectuals and the Problem of the Ethnos since World War II," *American Historical Review* 98 (April 1993): 317–37.
4. According to the census, the population in 1970 was 30,175, the lowest population rate since 1940, when it was around 30,200. By 1980, the population had increased more than

15 percent to 35,690 and grew again during that decade. Although these figures are based on official census data, there is much controversy among local observers who suspect the census seriously undercounted San Sebastián's population.

5. Francisco L. Rivera-Batiz and Carlos E. Santiago, *Island Paradox: Puerto Rico in the 1990s* (New York: Russell Sage Foundation, 1996).

6. Ana Celia Zentella, "Returned Migration, Language and Identity: Puerto Rican Bilinguals in Dos Worlds/Two Mundos," *International Journal of Social Language* 84 (1990): 81–100.

7. "El retorno de los que vienen y van entre aquí y los Estados Unidos es un problema. El alcalde es producto de la migración. Y se ve el impacto que tiene la migración de retorno hasta en las carreteras. Los alcaldes que migraron, que han pasado por eso, no hacen calles pequeñas o callejones. Cuando la gente regresa, tiene una vision más amplia del mundo: proyectos grandes, con mentalidad grande, metropolitana, y hasta norteamericana. Carros grandes, calles grandes. . . . Es un proceso que afecta la vision de la vida por acá, de como se debe de hacer las cosas aquí. . . . Esa vision grande, no es de aquí. Es de allá."

8. In the early 1950s, during the crucial early years of Puerto Rico's industrialization program, *El Mundo* ran a series of articles spotlighting San Sebastián and its long, distinguished cultural history. The most recent example of San Sebastián's importance in the island's cultural imaginary is a recent article in *El Nuevo Día* describing the town's rich cultural traditions evident in this year's Festival de la Novilla. *El Nuevo Día*, January 17, 2000.

CHAPTER 6

REINVENTING THE JAMAICAN POLITICAL SYSTEM*

BRIAN MEEKS

THE NEW YEAR BEGAN WITH A BANG. One Sunday in late January, these were among the leading stories in the *Gleaner*: The first, under the nonchalant headline "Higglers selling prescription drugs on the streets: Open-air pharmacy on the sidewalks," went on to note: "The illegal trade has left the health sector baffled about how so many different kinds of prescription drugs could have found their way on the street side, and concerned about the health risk involved in the abuse of these drugs."

The second headline that appeared under the photograph of a single telephone pole with a spider's web of hundreds of wires leading away from it was entitled "Wired up": "This is one of many illegal wire connections in Majestic Gardens in the Kingston 11 area. The Jamaica Public Service Company estimates that there are 50,000 illegal wire con nections island wide. Majestic Gardens is an inner-city community, but illegal connections are found in various forms in suburban and rural communities."[1]

The third headline had the decidedly more pedestrian headline, "Thousands of drivers using expired licenses." The new computerized system for issuing drivers' licenses had encountered a number of glitches; these, a Mrs. Ferguson of Inland Revenue indicated, would soon be ironed out, but then she raised the more burning problem of the proliferation of illegally issued licenses: "If there is collusion at the depot, there is nothing we can do about that," she explained. "We are talking about illiterate drivers," Mrs. Ferguson admitted. She said drivers who "bought"

---

* This essay was first published in 2001. There have been numerous changes in Jamaican politics since it was first written, but the general framework of Jamaican society remains the same.

their licenses and avoided the standard examinations can hardly be singled out as long as their original documents were signed by legal authorities and they passed the reading test. "If the person can read, the department cannot tell whether the license is bogus."[2]

The fourth headline, on the front page, was the increasingly familiar story: "Gun Battle follows boy's abduction: Seven-year-old Jovayn Miller abducted. Abductors demand $10 million ransom." In this instance, however, the good news was that the boy was later found and returned to his family unharmed.

Finally, and also on the front page, this now traditional, though increasingly antiquarian, report of the political clash. In this case intraparty appeared under the head "Charles, Broderick in heated brawl": "The police confirm that they had to break up a stone and bottle-throwing incident in Summerfield Clarendon Friday night, involving Pearnel Charles, JLP caretaker for the Clarendon north central constituency and Dr Percy Broderick former JLP Member of Parliament for that constituency."[3]

My last visit to the Institute for Research in African-American Studies at Columbia was in 1994, at which time I presented a paper titled: "The Political Moment in Jamaica: The Dimensions of Hegemonic Dissolution."[4] The paper proposed that following the retirement of Michael Manley from public life in 1993 and the reelection of his People's National Party (PNP) under the new leadership of P. J. Patterson, the island was approaching a social and political crisis point. The main suggestion was that Jamaica was in a state of hegemonic dissolution: "The old hegemonic alliance is unable to rule in the accustomed way, but equally, alternative and competitive modes of hegemony from below are unable to decisively place their stamp on the new and fluid situation."[5]

In the absence of a populist organization to take the lead and carry events in a more radical direction and in the context of an international conjuncture that did not seem to provide a permissive opening for radical action, three potential options were mooted: The first was the option of an increasingly authoritarian government "within or outside the constitution"; the second was the possibility of a democratic renewal across the breadth of the Jamaican social and political terrain; and finally, following Marx, was an option described as the "common ruin of the contending classes," or widespread deterioration in the fabric of civil society and the state. In the seven years that have passed since then, this "strange, eventful history"[6] is unveiling an alternative that is altogether none of these, though it appears to include elements of them all.

The Zeeks riots of late 1998, in which thousands of nominally PNP supporters brought Kingston to a standstill in protesting against the detention of their "don," or area leader, and the April 1999 Gas Price riots, during which the entire island was under a state of siege from protests following budgetary increases in the price of gasoline,[7] signaled a new phase in the old situation. The decision by the besieged police officers to ask the detained Donald "Zeeks" Phipps, don of Matthews Lane, to quiet his own crowd indicated their recognition of the relative autonomy of the dons and of the downtown "massive."

Globalization and neoliberalism, in drastically reducing the size and reach of the state, have undermined its critical power, deriving in part from the ability to distribute scarce resources. Simultaneously, new opportunities in the international drugs market provided the formerly loyal political enforcers with independent sources of wealth. Long before the 1998 crisis, then, the dons began to assume quasi-state functions as allocators of social benefits and, in the absence of a reliable and trusted police force, as law enforcers.[8] Yet even now the process is incomplete.

Though following the Zeeks events there was talk of an alliance of dons across traditionally antagonistic party lines and though it is true that a truce of sorts has held in the inner city for three years, the party divisions have not altogether been erased. Thus in the Gas Price riots of the following April, Zeeks's decision not to block the roads in downtown Kingston provided critical relief in its politically most vulnerable zone for Patterson's government. Tactically this was a brilliant move, for it both asserted Zeeks's autonomy and strategic importance and simultaneously sent the message that he was still "PNP" though now more as an important ally than an operative of the governing party. If this were at all in question, then Patterson's visit to Zeeks's stronghold in Matthew's Lane immediately after the riots to thank him for his support should leave no one in doubt as to the unsubtle shift in power relationships. The Jamaican state is still in charge but the cracks are opening up. Governments still rule, but increasingly they require the tactical support of these area leaders.

The complexity of the present moment is evident in all this, for the devolution of power to warlords who gain wealth from illegal drugs and the substitution of the rule of law with the rough retributive justice of the streets are an evident indicator of the collapse of the political system and the approach of the Hobbesian "war of all against all." Yet is there not a democratic kernel in the popular mobilization of people outside of the restrictive confines of middle-class-led political parties? Is there not a certain laudable autonomy that derives from the shrinkage of the pervasive and bloodily divisive patronage networks of the seventies and eighties? None of this potential can fully bloom as long as authoritarian dons run these communities along the lines of feudal fiefdoms, but it would be remiss not to see the implicit potential for a democratic renewal in the collapse of clientele-oriented politics.

The more evident feature, however, is the growth of the authoritarian trend. This is manifest in the police force and in the formation of a new crime fighting group, the Crime Management Unit (CMU) under Senior Superintendent Reneto Adams. Following a new round of particularly heinous crimes, the CMU was set up in the middle of 2000 as the latest special squad to combat violent criminals and Adams's unit has already gained special notoriety for its viciousness and scant regard for procedure. In the most recent incident, seven young men were killed in a shoot-out in a house in Braeton, a lower middle-class suburb southwest of Kingston. The police claimed that the victims were wanted in connection with, among other crimes, the brutal killing of a headmaster of a local primary school. The police assertion that they approached the house and were fired on was hotly denied by citizens from the community. Firsthand reports of grieving neighbors and family members who heard the shoot-out suggest the degree to which notions of the rule of law have been abandoned and hint at a growing groundswell against the formal representatives of the system:

> Residents find cold comfort in relating the blood-curdling reports of screaming youth, mercy pleas and explosions which dominate their conversations in the day, and their dreams at night . . .
> I heard "Gallus" [Andre Virgo] saying the Our Father prayer. He was crying and begging for his life. Then I heard a barrage of shots and he went silent," a neighbor said.
> Mi hear an officer sey "Weh yu a do bwoy? Try run?" and him seh "No, officer, how mi fi run and yu a beat mi," and then mi hear pure explosion and nothing else," another neighbor said.[9]

Any assumption, however, that these acts are carried out by rogue elements in the police force without the support of a significant body of the citizenry would be mistaken. In 1991, Carl Stone's polls showed a 56 percent support for vigilantism in the adult population, and Anthony Harriott in 1994 found that among Jamaican police officers there was a similar support of some 54 percent. If there was any lingering doubt that support for summary justice exists in the very highest echelons of government, it would have been dashed when Justice Minister K. D. Knight, in an intemperate reference to violent gunmen and no doubt spurred on by repeated, highly publicized acts of wanton violence, blurted out in early January that they belong in the morgue. Indeed it is fair to assume that even as there is a growing and vocal constituency for justice and fair play, in the face of repeated incidences of brutal and wanton violence, the constituency demanding law and order by any means necessary is also consolidating.

The weakest tendency in all of this is that of democratic renewal. There is a significant increase in community organization. Trevor Munroe records, for instance, the increase in the number of registered youth clubs from about 596 in 1989–90 to 727 in 1996–97.[10] And there is a laudable return of elements of the middle class to political activism, which is evident in the formation of human rights groups, such as Jamaicans for Justice and the University of the West Indies (UWI) Faculty of Social Science's initiative to educate citizens in the inner-city community of Craig Town. But so far there is no indication that these disparate tendencies might coalesce into some kind of new social movement. Indeed, with the gutted trade union movement struggling for its survival and a new round of emigration undermining nascent community organization, the likelihood of the emergence of new political formations in the short run is limited.

Certainly from the perspective of the initiative of the PNP regime, there has been very little that can be considered supportive of a new democracy. If one thinks about the early years of the Patterson regime, there was an explicit trend to make government more open and to make politics more accessible to ordinary people. Thus, the important Values and Attitudes conference of 1994 brought hundreds of delegates together and sought to tap broad national opinion on the ways to address and overcome the evident deterioration in manners and civic responsibility at the social and individual levels. The prime minister's own "Live and Direct" meetings, in which he established face-to-face contact with people at the community level, were also a part of this period, as was the local government reform initiative of Arnold Bertram's ministry. But these, particularly since the cataclysmic financial sector collapse of 1997,[11] have receded across the policy horizon.[12]

In their place has emerged the silhouette of a largely unimaginative regime, one that is presiding over a decade without any real growth and that, in deciding to underwrite the profligacy of the banking sector, has placed an enormous yoke of debt around its own neck and around the necks of this and future generations of the country's citizens. The structural adjustment policy of high-interest rates, further enhanced to deal with the fallout of the financial sector, has served as a "blunt instrument" to reduce inflation but, more damagingly, it has facilitated, in the context of an already-skewed pattern of distribution, the transfer of wealth from the poor to the rich. Thus, the United Nations Development Program's (UNDP) 2000 Human Development Report sought to count Jamaica, along with Brazil and Guatemala, among the countries with the greatest income inequality, where the top fifth's share in national income is more than twenty-five times that of the bottom fifth's.[13] The outcome of these policies, alongside the decline of any positive democratic initiatives in an already sharply divided community, has resulted in greater alienation and the feeling that the government does not care, that individual ministers are corrupt, and that the ship is rudderless.

Thus, even positive administrative moves, such as Minister Peter Phillips's road repair and rebuilding efforts and the attempt to resuscitate a public urban transport system, have so far had little discernible impact. [14] The peculiarity of the PNP's third term is this: not everything that was done was misguided nor has everything collapsed. The 1999 Survey of Living Conditions asserts that mean consumption has increased, there have been statistical improvements in education, and the health status of the population has remained stable; on the negative side, there has been a noted stagnation in housing development and when the sample was asked whether their personal economic situation had improved from that of five years ago, 55.4 percent of all respondents said it had improved or remained the same, whereas only 44.6 percent said it had worsened.[15] As to purely statistical performance, aside from the daunting failure to grow and the hard inequality indices, it can be credibly argued that the government's performance, in a generally difficult situation, could have been much worse. The regime has certainly not presided over a total demise of the island's infrastructure, which is evident in the previously mentioned road-building program, the upgrading work on the two international airports, and the proposed new trans-island highway.

What, then, accounts for the anomie and drift in which solutions to the endemic violence appear as elusive as ever and the sense of lawlessness and disorder pervades every sphere of life? What accounts for the dissatisfaction when, in

December 2000, the Jamaica Labour Party (JLP) for the first time in seventeen years nosed past the PNP in the national public opinion polls and then defeated the governing party's candidate in a by-election, in one of its safest seats in North East St. Ann?[16]

Many reasons can be put on the table for disenchantment and political dissatisfaction. Ever more blatant income inequality is obviously one reason. The feeling that the government does not care, that it has looked only after itself and has become too fat and operates at a bureaucratic level without reference to the popular base, is another. There is also the simple factor of having been in power for twelve years, an unprecedented length in modem Jamaican politics. An entire cohort of new voters was six years old when the JLP was last in power. For them, there is no memory of the authoritarian "one-man ism" of the Seaga regime of the 1980s. All their frustrations about unemployment, their real experiences of police brutality, and their sense of alienation from an apparently uncaring society are focused on one known reality: the "third-term" PNP government. But to further understand the present moment, we need to return to the notion of hegemonic dissolution, or the disconnection by significant sections of the population from a formal order that they no longer feel any loyalty toward, that they perceive to have disrespected them repeatedly, and that is no longer able to provide many with the modicum of a decent livelihood.

At the economic level, the world built around remittances—the barrel and moneygram—has grown exponentially. Swathes of the island are now completely dependent on this avenue for survival. At the community level, the dependence on the don for social welfare has also grown exponentially. At the level of justice, the disconnection from the formal justice system takes two forms: For those living in the downtown ghettoes, the justice of the dons is increasingly more available and reliable. For those living in the fortress like middle-class townhouse complexes, a similar reliability is to be found in the justice and efficiency of the fast-response guard services. So, as is characteristic of many prerevolutionary situations (though Jamaica is not quite yet at this point), even when there is some apparent statistical social and economic reprieve, the disconnection from the law and from official society, bred on the intensity of the previous downward spiral, intensifies.

The "election" of George Bush in a deeply flawed electoral process in which he received fewer votes than his opponent, Al Gore—and in which he would have lost, by all indications, on a careful recount of the decisive Florida vote[17]—signals a new and dangerous moment in the neoliberal globalization project.[18]

Like every stock market bubble before it, the particularly long ten-year boom, fueled by the fanciful optimism in technology stocks, has recently burst.[19] In the harsh glare of the morning light following the collapse of NASDAQ and then the Dow Jones, the weak fundamentals of the U.S. economy are once again on display for all to see. In an insightful article in *Newsweek*, Fareed Zakaria points out the growing U.S. current account deficit in which the United States spent $435 billion more dollars than it earned in 2000, which is roughly equivalent to 5 percent of its GDP.[20] The United States, he argues, sustains its profligate consumption habit because countries with "spare cash" find it appropriate to invest in U.S. stocks and bonds. If, however, due to falling confidence even a fraction of this infusion of resources were to stop, then, "It could produce a spiral of problems: a falling dollar, which produces rising interest rates, which weakens stock prices and further slows the economy. Practically every time an advanced country has run a large current account deficit this vicious cycle has emerged."

In favor of the possibility of such a vicious cycle is the fact that this "correction" is occurring at a time when the Japanese economy is doing badly, the Asian economies have not yet recovered from the crisis of 1987, and the German economy is slowing. Against the odds, the United States has emerged as the sole surviving world power and the dollar as a world currency. Thus the United States is the only nation that pays its foreign debts in its own currency, significantly enhancing its ability to pay its way out of crisis.[21]

There is, therefore, no certainty that the vicious cycle will be the only option. Zakaria, however (in what, to be fair, is only an op-ed piece), fails to sketch the outlines of a broader and more dangerous secular crisis facing the Organization for Economic Cooperation and Development (OECD) economies. Whereas in the period from 1950 to 1973 the OECD economies as a whole grew at an average annual rate of 4.3 percent, in the period from 1973 to 1995 this average shrank dramatically to 2.4 percent. This was accompanied by a parallel contraction in the rate of growth of private consumption, from 4.3 percent in the earlier period to 2.6 percent. The relative contraction in the rate of increase of final demand has led to a decrease in the propensity to invest in fixed assets. Thus, fixed capital, which grew at an annual rate of 5.7 percent in the 1950–73 period, grew by only 2.1 percent from 1973 to 1995. This secular decline in growth, consumer demand, and the demand for fixed capital has contributed to the exponential growth of investible funds without a clear productive "home." Thus, as Harry Shutt suggests: "Inevitably this coincidence of a continuing steady growth in investible funds with slowing demand for both fixed investment and working capital meant that a significant proportion of such funds were channeled into speculation—that is, into assets that held out greater prospect of gain from capital appreciation than from earnings yield."[22]

If this fundamental reality is used as a point of focus, then much in recent global economic policy can be discerned. The World Trade Organization (WTO) project, for instance, can through this lens be understood as an attempt to find a means of kick-starting the Western world by opening up new markets and somehow stimulating the stalled consumer demand. And yet by shifting wealth from the wage sector to capital, neoliberal economic policies have, if anything, exacerbated the problem. For although more markets have been liberalized and the marginal productivity of capital has improved, failure to stimulate growth and the shift in the distribution of wealth means that there are fewer consumers able to exercise effective demand for the increased production of goods and services.

Therefore, with mountains of idle cash and insufficient productive enterprises to invest in, there is the even greater possibility of speculation to put this cash to work. The result has been the proliferation of a variety of strategies, from the privatization of governmental assets to junk bonds to futures markets to derivatives and the promotion of so-called emerging markets. The outcome of what is essentially a massive speculative bubble based not on production and growth but on the artificial inflation of stock prices without foundation must ultimately, as Shutt bleakly suggests, lead to a correction of epic proportions:

> All the [new financial] devices . . . are ways of artificially boosting the rates of return on investment in response to unrelenting pressures to push them ever higher. Theoretically, of course, this problem might be resolved if these forces were somehow to abate, so that the market rate of return could fall to a more readily attainable level. However, history suggests unambiguously that the only way this can happen under a competitive market system is by means of a destructive "crash" rather than an orderly retreat to lower returns. . . . Hence a sober assessment of these various financial stratagems must surely conclude that, for all the undoubted ingenuity of the financial manipulators, it can only be a matter of time before the forces of gravity reassert themselves and the reality of systemic financial failure must be faced.

The implications for the developing world in such a context are both well-known and well-documented. Countries caught in the quagmire of indebtedness find themselves in a permanent cycle of structural adjustment. Primary goods exporters, traditionally "price takers," have now become "policy takers" as well.[23] The policies of "opening up," export orientation, and privatization that they have been forced to adopt have not, for the most part, led them on to autonomous paths of growth, but rather, these countries must forever "export themselves out of debt," no matter that they are competing with a dozen other countries exporting the same coffee or cocoa—or shoes and shirts. No matter that domestic food production is declining, as export agriculture is favored over food crops and natural resources are pillaged for instant returns, with long-term damage to the environment. The export of commodities, both primary and manufactured, because labor intensive manufactures are the new "commodities," is a way of exporting cheap labor. The increasing volume of these developing country exports have assisted the United States to maintain the long boom of non-inflationary growth in the 1990s. This is the one sense in which "globalization" has increased wealth-in a unidirectional way.

For countries like Jamaica, then, caught in the stare of neoliberalism, the primary policy questions are how to follow the rules of the game more thoroughly, how to make the state "lean and mean," how to find that elusive niche, and how to take full advantage of the purported comparative advantage with the economic wealth and well-being that should naturally follow. The dismal result has been a collapse of the productive economy; the exacerbation of the gap between the rich and the poor; the impoverishment of the countryside; a massive export of talent;

the undermining of the state; and, in some instances, an all-class crisis, or process of hegemonic dissolution.

Above all, there has been a patent failure at the governmental level, by the "loyal opposition" and by most leading intellectuals and representatives of civil society to heed lessons perceived by Michael Manley two decades ago: that capital follows its own logic of accumulation; that powerful nations set the rules and look after themselves; that economically powerful states are always the advocates of open markets even when, like the infant United States, they were premier defenders of protectionism in earlier phases; and that it is therefore absolutely necessary to begin the debate on alternative futures against and beyond the false horizons of neoliberalism.[24]

There is, of course, the present bear market aside, no certainty that a cataclysmic crash will occur. Unprecedented growth (barely understood as a phenomenon), along the lines of that in the 1950s and 1960s, could come again and effectively utilize the capital overhang to undermine a devastating correction. But if there is any substance in the foregoing analysis, and the general unease with which leading Western commentators view the present bear market would tend to support it,[25] then the need for an entire rethinking of the Jamaican agenda is not only necessary but any delay in doing so would verge on the criminal.

The postwar system, based on competitive parties dispensing scarce benefits and led by heroic leaders, is moribund. The populist party, whether in its nationalist, PNP garb or its private sector-oriented JLP form, is dying. The results of the North East St. Ann by-election suggests the degree to which Patterson's PNP, which was elected to power eight years ago with genuine popular support, has lost credibility. In the absence of a genuine alternative, then, popular support will flow in the direction of the default alternative, the JLP, in the Jamaican tradition, though it would not be surprising if the overall turnout at the next general election were the lowest in Jamaican history.

If the present situation can, at least in part, be accounted for by the alienation of people from politics, then what is urgently required to correct the dangerous drift to the rocks of nihilism and authoritarianism is a popular renewal. The ideal alternative would be a new, democratic federation of local and grassroots organizations that would unite a critical mass of Jamaicans around a politics based on participation, inclusion, and transparency. Such pure forms, however, seldom present themselves. An alternative, in keeping with the country's two-party tradition, might well be a critical alliance of well-thinking men and women across party lines that would allow them to maintain their vestigial party loyalties while beginning the conversation that will lead to a new formation.

Such a national alliance might evolve into a government of national unity, but the danger in that possibility is the reassertion of politics from above without the limited safeguard of an opposition in waiting. Such an outcome would not only prolong the politics of the last half century but also possibly entrench an even more dangerous monopoly, when what is desperately required is a new popular politics from below.

The first task on the agenda of such a formation would have to be a constituent assembly of the Jamaican people at home and abroad. The year 2002 marks the fortieth anniversary of Jamaican independence. What more important milestone to begin the debate around a new social contract, this time discussed and ratified by the people, than this important date? The notion of "abroad" is particularly stressed in this formulation, as roughly half of all Jamaicans live overseas. They are "taxed," as it were, by their remittances to their families on the "rock," but they have no representation. The slogan of Jamaicans overseas must be the classical one of "no taxation without representation." Any new contractual arrangements must include the input of the Diaspora and must make constitutional arrangements for its inclusion beyond the conclusion of the initial debates.

The constituent assembly would not be a one-off meeting but would involve a series, perhaps hundreds, of meetings in Browns Town and Above Rocks, Brooklyn and Brixton, to discuss alternative agendas and to ultimately mandate delegates to a national convention to debate and ratify a new constitution. Issues on a possible agenda should not be preempted, but if the experience of the past forty years suggests that there needs to be a thorough democratic reform of the political system, they might include:

1.   The appropriate constitutional arrangements to undermine the constituency-based spoils system and the ruthlessness endemic in a winner-takes-all election. This might include discussions around proportional representation to partially or completely replace the first-past-the-post system, the question of term limits, the matter of the recall of nonperforming representatives, and the entire question of the transparency of elections, with particular emphasis on electoral funding arrangements.

2.   The devolution of power from the center to the communities. This process, in a de facto sense, began as the patronage system receded. To the extent that people have real local control, then the process of reconstructing community can genuinely take hold.

3.   The political inclusion of overseas Jamaicans in any future electoral arrangement, whether by transparent overseas voting procedures or otherwise. The obverse of this is the development and fostering of powerful lobbying groups in the main immigrant centers. The strength of the overseas Jamaican and Caribbean community can only be manifest to the extent that there is active organization at the community level with close ties to natural allies in the African-American community and among environmentalist groups, organized labor, and elsewhere.

4.   The unprecedented extension of democratic procedures into all spheres of government, including a publicly elected commission to oversee the operations of the police and the military; the opening up of the budget debate to the nation by a discussion at the grassroots level months in advance of the preliminary budget and the election of mandated delegates to a national budget debate; and the further extension of the principle of democracy and

accountability to key national institutions, such as the election of direct delegates to the Public Accounts Committee to sit alongside members of the house and similar directly elected representatives to sit on the committee of the Ombudsman and that of the Contractor General.

5. The initiation of a debate around national social and economic imperatives. These must no longer be subject to secret negotiation by international agencies with no responsibility for the survival of the Jamaican people. Thus a national food policy that would ensure a strategic agricultural sector and educational and health policies that would have twenty-year plans, subject to variation only on the basis of democratic decision-making, would all be high on the agenda.

6. The beginning of a conversation on the role of international capital and its character. Such a debate cannot begin after a crash, when it is too late, but must raise some of the critical issues mentioned in this chapter with a full recognition that capital and the market is not only far from benign but also, as we speak, in deep crisis, with profound implications for the future of the world.

7. The proposal for a federation of the Caribbean peoples. The island-by-island sovereignty of the 1960s is as moribund as its accompanying political system. The Caribbean is our natural region, geographically and culturally. It is here that we need to look for the forms of solidarity and cooperation that will give us more meaningful space, allow us to develop while taking better advantage of economies of scale, and allow us to play a more coherent role in world affairs. There is no danger to be found in a federation based on democratic principles, in which power is devolved to the grassroots and cooperation is based on principles of respect and the search for mutual benefit.

To many, these proposals may seem farfetched and tendentious. They are meant to be. For too long we have played it safe by restricting ourselves to thinking only within the box. That box, the stultifying paradigm of neoliberalism, is now in tatters. We need to begin thinking outside of it. It is therefore appropriate to end on a quote from Kari Levitt, who has recently sought, in a similar vein and spirit, to address the elements of a program for the Caribbean and all developing countries beyond the failed neoliberal paradigm:

> You may dismiss this wish list as idealistic. Perhaps so, but it is certainly more realistic than the assumption that the world can continue on its present path without courting disasters more terrible than any we have yet visited upon ourselves. Without dreams, nothing is possible. Without hope, there is no future.[26]

The collapse of Jamaica in its present form, whether by slow burn or more rapid denouement, may be closer than we think. Jamaica, I submit, shall be democratic or not at all.

NOTES

1. *Sunday Gleaner*, January 21, 2001.
2. Ibid., 3. A crude calculation suggests that if there are 50 thousand connections and each leads to a household of 5 people, then there are roughly 250 thousand people, or 10 percent of the entire population, illegally connected to the electricity network
3. *Sunday Gleaner*, January 21, 2001.
4. Ibid.
5. Brian Meeks, *Radical Caribbean: From Black Power to Abu Bakr* (Kingston: University of the West Indies Press, 1996), 134.
6. *Sunday Gleaner*, January 21, 2001.
7. The paper, under the title "The Political Movement in Jamaica"was first published in *Race and Reason* 3 (1996–97), 39–47. It was later republished as a chapter in my collection *Radical Caribbean* and more recently in Manning Marable, ed., *Dispatches from the Ebony Tower: Intellectuals Confront the African American Experience* (New York: Columbia University Press, 2000), 52–74. References are to the version in the UWI collection.
8. Meeks, "The Political Movement in Jamaica," 134.
9. Ibid., 137.
10. William Shakespeare, "Coriolanus," in *The Concise Oxford Dictionary of Quotations* (Oxford: Oxford University Press, 1981), 205.
11. For a critical reading of these events, see the introduction to my *Narratives of Resistance: Jamaica. Trinidad, the Caribbean*, 1–24 (Kingston: University of the West Indies Press, 2000).
12. For an insightful analysis of the social situation in the inner city and the root causes of violence, see Horace Levy, et al., "They Cry Respect: Urban Violence and Poverty in Jamaica," *Centre for Population, Community and Social Change* (Mona, Jamaica: University of the West Indies, 1996).
13. Although police killings per year have declined from the average of about 200 in the 1980s to 130 in the 1990s, the numbers are still unacceptably high as are the numerous reported cases of summary justice, though, as Harriott suggests, very few of these cases were ruled as unjustifiable and the offenders charged with murder. See Anthony Harriott, *Police and Crime Control in Jamaica: Problems of Reforming Ex-Colonial Constabularies* (Kingston: University of the West Indies Press, 2000), 82.
14. *Gleaner*, March 20, 2001.
15. See Carl Stone, "Survey of Public Opinion on the Jamaican Justice System," Unpublished report to U.S. Agency for International Development, Kingston, 1991, p. 30; and Anthony Harriott, "Race, Class and the Political Behaviour of the Jamaican Security forces," Ph.D. dissertation, University of the West Indies, Mona, Jamaica, 324.
16. See Trevor Munroe, *Renewing Democracy into the Millennium: The Jamaican Experience in Perspective* (Kingston: University of the West Indies Press, 1999), 127.
17. In 1997, after a half decade of a freewheeling financial bubble, the banking sector experienced a severe crash. Many smaller banks folded, leaving depositors high and dry. When the momentum threatened to engulf the entire economy, the minister of finance intervened purportedly to rescue the small investors with a sector-wide bailout. The cost of this and the subsequent "restructuring" exercise has even more severely encumbered the government and, by implication, the taxpayers in debt. See, for instance, "Davies Explains Restructuring," *Sunday Gleaner*, April 26, 1998.
18. It should be noted, however, that in the face of the loss of the North East St. Ann seat there seems to be an early attempt to regroup the PNP around the slogan "deepening democracy." The governor general's "throne speech" opening the 2001–2 financial year

in parliament emphasized, for instance, the need to deepen democracy at the local and community levels. See "A Vision for Prosperity," *Gleaner*, April 3, 2001.

19. Jamaica's GDP grew at a rate of 0.1 percent for the period 1990–98. This put it behind all Caribbean territories except Haiti, which experienced negative growth, and its own performance in the previous decade that, although relatively anemic, was at an average of 2.0 percent. See World Bank, World Development Report 1999–2000, *Entering the 21st Century* (New York: Oxford University Press, 2000), 250.

20. United Nations Development Program, *Human Development Report 2000* (New York: Oxford University Press, 2000), 34.

21. For a discussion of the convoluted and sorry history of urban transport in Jamaica in which I laud the minister's efforts but argue for an as-yet-untried popular approach to the management of public transport, see my *Narratives of Resistance.*

22. Planning Institute of Jamaica, 1999 Survey of Living Conditions. Table 7.22, p. 40.

23. In a three-way race on March 8, 2001, the JLP's candidate, Shahine Robinson, defeated the PNP's Carrol Jackson by a margin of 509 votes on first count. The significance of this is profound, as the North East St. Ann seat was one of the ruling party's safest rural seats. In the last general election in 1997, the PNP romped home in this seat by a margin of about 2,000 votes. If such a remarkable swing were to be repeated island wide, the ruling party would be lucky to retain a handful of seats in the next election. See *Jamaica Gleaner*, March 9, 2001, http://www.jamaica-gleaner.com.

24. This is the name of a popular electronic service for the quick remittance of foreign exchange to family and dependents.

25. See OECD National Accounts Statistics and IMF International Financial Statistics, in Harry Shutt, *The Trouble with Capitalism: An Enquiry into the Causes of Global Economic Failure* (London: Zed Books, 1998), 38.

26. *A Jamaican Case Scud* (Washington, DC: Howard University Press, 1987); and Michael Manley, *The Poverty of Nations* (London: Pluto, 1991). The front-page headline of *Newsweek* in the first few days of the stock market decline is as indicative as any. It stated: "Are You Scared Yet? The Sinking U.S. Economy. Will It Take the Rest of the World Down with It?" *Newsweek*, March 26, 2001.

CHAPTER 7

AFRO-COLOMBIA

A CASE FOR PAN-AFRICAN ANALYSIS

JOSEPH JORDAN

AFRO-COLOMBIANS, ONE OF THE MOST RESOURCEFUL AND RESILIENT COMMUNITIES in the Americas, are under siege. They have struggled largely outside of the gaze of most of the world despite suffering massive social upheavals. According to some sources, Afro-Colombians constitute up to 25 percent of the country's total population of about forty million and as a group have a population in excess of ten million. These figures may vary according to the criteria used.[1] Despite such a large population size, Afro-Colombians face constant threats of displacement from both the Colombian state and transnational corporations. In the wake of such attacks, Afro-Colombians have adopted new strategies of resistance to combat repression from the state and its multinational allies. By looking inward or engaging in self-valorization as an initial strategy, Afro-Colombians have determined the character of the struggle. However, self-valorization, a form of identity politics, is not sufficient as a basis for struggle. When connected to a larger movement for political institution-building, self-valorization has the potential to bring Afro-Colombian communities together with other groups of African-descent in Latin America that are currently engaged in mass movements for social change. At this present moment, Latin America is in transition, and popular movements are challenging regional hegemonies. This article examines the changing political landscape in Colombia and offers some preliminary thoughts on analyzing the course of Afro-Colombians' struggle for self-determination from a Pan-Africanist perspective.

The early history of Africans in Colombia begins with the Spanish conquest and settlement of various areas in the early 1500s. The importation of Africans as slaves provided labor for Spanish conquest and slavery remained a key feature of Colombia's history until it was abolished in 1851. The formation of an Afro-Colombian identity built upon a tradition of resistance can be attributed to a continuous history of slave rebellions and to the existence of independent maroon communities called *palenques*, which were formed by escaped African slaves. The term *palenque* is translated, literally, as stockade. Today, Palenque San Basilio, on the Pacific coast of Colombia, maintains an autonomous culture of resistance that recalls the independent spirit of those early maroons. Sustaining a culturally based consciousness of defiance and resistance is a key component for a Pan-African project that endeavors to reclaim Afro-Colombian autonomy, cultural heritage, security, and ancestral land, particularly in regions such as Choco, on the northern end of the Colombian Pacific littoral. In September of 2002 organic intellectuals from all sectors of the Afro-Colombian community as well as other states in the region assembled for the first Afro-Colombian national conference to advocate for a maroon-based perspective on the Afro-Colombian struggle for self-determination. In particular, conference speakers drew up the political actions undertaken by black intellectuals in the English-speaking Caribbean as a model for Afro-Colombians' movement building:

> During times of crisis, the Maroon experience was used to construct an explanation for the situation in the 1920s and 1930s. The methodology and style of thinking that emerged from these efforts was characterized by its capacity to analyze, verify and apply what had been learned within a perspective that had been based on that accumulated experience. Such praxis/research collectives developed a direction and a set of questions independent of the research methods that reproduced the existing order. . . .
>
> Although individuals represented in themselves several different ways of per-ceiving experience—through race, class, gender, nationality, religion and political party—it was the emergence of a shared vocabulary, created through the exchange of experience, that made it possible for groups across the region to make sense of their situation.[2]

The maroon experience thus consisted of a process that begins with a specific form of consciousness that produces a specific critical analysis based on accumu-lated experiences. When considered collectively, the separate aspects of the process produce a maroon complex or group of factors that characterize Afro-Colombian resistance to oppression.

In addition to those models of resistance examined at the 2002 national con-ference, there are other approaches that have examined the primacy of black pop-ular culture as a means of both recovery and as a catalyst for social change and movement building. Scholarship by Arocha, Escobar, Wade, Asher, and others,[3] has addressed the genesis of African and Afro-Colombian identity in Colombia using various analytical frames that include the roles of popular cultural forms, modernity, and transnational cultural flows.

A common critique of these scholars' work suggests that certain practices that signify "blackness" or "African" identity and culture in Colombia, according to the contemporary sociopolitical sense of both designations, should be read as new rather than historically grounded cultural practices. According to these critiques, the signification of blackness or African identity in Colombia replicates the cultural production of other external black populations and, therefore, bears all of the limitations of an imitative culture.[4] Using this logic, youth in Cali, Colombia such as those studied by scholar Peter Wade who use hip-hop to make statements about their sociopolitical conditions could not be said to have *created* a cultural product, but instead are reproducing the cultural frameworks of black youth experiences in the United States. One, however, could contend that hip-hop itself is a recreation in that its producers have borrowed from several cultural traditions within and outside of the U.S. context. Therefore the adaptation of hip-hop by Colombian youth follows the recombinant strategies that have produced the range of black popular cultures throughout the Americas. Wade, who studied this phenomenon, argues that dismissing these strategies, "also raises issues about the analysis of cultural politics and the deconstruction of cultural 'inventions' without thereby invalidating them as a locus of ethnic solidarity."[5]

Francis Njubi sees efforts, such as those of the youth of Cali, Colombia as the counter-penetration of information technology by people of African descent in the struggle against human rights violations produced by slavery, colonialism, apartheid and globalization. Historical antiracist movements like abolitionism, Pan-Africanism, and the antiapartheid and civil rights movements all used communications technology to expose the extent and brutality of white supremacy and to create international networks of resistance movements.[6]

Richard Powell, in "What is This 'Black' in Black Popular Culture?" references Stuart Hall's validation of the reality, veracity, and even strategic necessity of such a concept. Hall says:

> It is this mark of difference inside forms of popular culture—which are by definition contradictory and which therefore appear as impure, threatened by incorporation or exclusion—that is carried by the signifier "black" in the term "black popular culture." It has come to signify the black community, where these traditions were kept, and whose struggles survive the persistence of the black experience (the historical experience of black people in the diaspora), of the black aesthetic (the distinctive cultural repertoires out of which popular representations were made) and of the black counternarratives we have struggled to voice.[7]

Powell identifies five components, implied in Hall's definition of black popular culture, that define black diaspora cultures:

- They constantly struggle against claims, from within and without, of racial quintessence and against dominant (and often hegemonic) cultural and political forces.
- Black diaspora cultures are defined by having the sufficient demographic numbers and critical mass to proclaim common beliefs, value systems, and goals toward building community-based institutions and products.
- Black diaspora cultures have a structural dependence upon an acknowledged collection of life experiences, social encounters, and personal ordeals

the sum of which promotes a solidarity and camaraderie that creates community.

- The demonstration of a distinctive cultural repertoire expressed through a "black aesthetic," or a collection of philosophical theories about the arts of the African Diaspora.
- Black diaspora cultures are characterized by forms that are alternative to their mainstream counterparts but proactive and aggressive in their desire to articulate, testify, and bear witness to that cultural difference.[8]

While acknowledging the general framework these Africanist interpretations provide, this chapter also offers a preliminary examination of the Afro-Colombian struggle to organize for their own advancement as a national community and the means through which they continue to rationalize a group identity in relation to the Colombian nation-state. The dynamics observed in the Afro-Colombian struggle also appear in other Latin American nations where African-descended populations, having suffered exploitation and the threat of psychic annihilation, form political struggles or movements to articulate (or rearticulate) their histories.

A primary contention of this chapter is that Afro-Colombians have articulated a new black cultural identity as a strategy for mobilization within the current historical process of Colombia's nation-definition. Reimagining "blackness," African consciousness, Afro-Colombian ethnicity, and maroon identity provides a means for an extra-national citizenship, and it connects Afro-Colombians to struggles waged by other communities of African-descent throughout the Americas. This cultural connectedness lays the foundation for African-descended populations in general and Afro-Colombians in particular to construct political alliances that acknowledge and build upon cultural and historical commonalities.

Contemporary cross-cultural alliances between black and indigenous communities in Colombia and other Latin American nations recognize both groups' colonial histories of oppression under European settlers and militaries and the formation of resistance struggles by earlier coalitions of blacks and indigenous people to fight against these forces.[9] In forming the current cross-cultural alliances, Afro-Colombian activists worked to stave off the development of fractions between racial and ethnic groups leading up to Colombia's new constitution in 1991. Current work by cultural, political, and intellectual cadres to sustain coalitions between indigenous and African-descended communities throughout the region has a more progressive orientation. In doing so they avoided what Frantz Fanon saw as a failure of national consciousness, at the height of anti-colonial struggles, to articulate a consciousness oriented toward notions of justice beyond parochial interests. Hence, the lack of inclusive representation and participation from all groups during the formation of a self-governed nation-state would reproduce the exploitation that the oppressed had suffered. Fanon also addressed the issue of culture and how it was recovered and expressed:

Culture is not a folklore nor an abstract populism that believes it can discover the people's true nature. It is not made up of the inert dregs of gratuitous actions, that is to say actions which are less and less attached to the ever present reality of the people.

A national culture is the whole body of efforts made by a people in the sphere of thought to describe, justify, and praise the action through which that people has created itself and keeps itself in existence.[10]

For African-descended communities, this process of engaging an extra-national citizenship built around a shared cultural and racial identity and similar diasporic experiences can be placed under the general rubric of Pan-Africanism. Pan-Africanism is a concept used to describe a movement and is sometimes divided into cultural and political manifestations. For the purposes of this argument, the separation of culture and politics seems to be a false dichotomy given that history shows the opposite is true. Pan-Africanism, a concept with a 200-year-old history, theorizes a politico-cultural link between Africa and black people in the African Diaspora contending with similar experiences of colonial oppression based on the racialized notions attached to blackness, and therefore establishes a foundation for a common struggle against colonialism to regain self-governance.

Although mentioned at various moments in the major studies of Afro-Colombians, most analysts examine the deeper implications of Afro-Colombians' self-identification in a diasporic or Pan-African context or engage the theoretical literature on Pan-African identity in their works. Nonetheless, these largely historical and ethnographic studies add to a growing literature on the Afro-Colombian community's development as a distinctive presence in the nation. Even when Pan-African connections are made in the literature, few seem to imagine the possibility of a region-wide or broader international alignment with other black struggles against racialism, imperialism, state and criminal terrorism, and capitalist exploitation.

As Horner notes, historians of Latin America "seem to lose all interest in the Negro as soon as abolition is accomplished. In any case, he disappears almost completely from the historical literature." Clearly, Africana histories must neither begin nor end with the history of slavery in the Americas. Yet, Watson notes:

National states have been built through purposeful racial, ethnic, religious, class, or other internal exclusions such that the nationalization of society has tended to move in tandem with the racialization of society. Many states practice forms of self-racialization, by constructing national identity and defining national belonging via criteria like birth, ancestry, or naturalization, with the effect of using citizenship to transform commensurable diversities into markers of relative difference. . . . Paradoxically, to become civil and national under sovereign states people have had to enter the nation-state as alienated national persons.[11]

What then, provides the means for a recovery and resituating of the Afro-Colombian as a subject rather than as an object in the nation's historical consciousness?

Luis Gilberto Murillo, former governor of Chocó State in Colombia, addressed this issue in his presentation of *El Choco: The African Heart of Colombia* in New York in February 2001. In this monumental presentation, Murillo outlined the plight of Afro-Colombians and points to the growth of an explicit self-defined Afro-Colombian identity, which represents an evolution or break from an implied (or imposed) Afro-Colombian identity:

> We now use the term Afro-Colombians as a way to keep our cultural identity, whether we are rural or urban Afro Colombians. We say Pacifico or Bio-geographical Choco, instead of only Chocó. We say pluriethnic and multicultural, instead of Black, indigenous or mestizo. We have rights (at least on paper) to Pacific coastal collective lands, as well as decision-making power. We defend our right to self-determination and cultural, economic and political development. We have our own Afro Colombian agenda at regional, national and international levels.[12]

Murillo's presentation also confirms the evolution of a distinct and modern Afro-Colombian identity that breaks from (and evolves from) previous notions of Africanness. And he hints at the prospects for an enlightened politics based upon this new outlook both for Afro-Colombians and other Afro-Latin communities facing similar struggles.

Afro-Latin collectives and communities are consolidating new ideological positions and collaborative alliances as a way to resist state hegemony. In the case of Colombia, these alliances also hold the promise of providing a political vehicle for the expression and pursuit of Afro-Colombian and indigenous agendas for participation in the political system. Murillo notes that, in previous generations, government and societal attempts to theorize and valorize *mestizaje* meant black and indigenous lives were devalued.

Mestizaje, and promotion of the notion of a "Cosmic Race," is understood as the state's attempt to fix the hierarchy of racialized identities around the notion of a Spanish-Indian-black hybrid. Where "Blackness" as a distinct presence would be gradually eliminated. As Arocha notes, "Carved into the stone walls of the Spanish Language Academy building in downtown Bogotá is a golden motto: ONE GOD, ONE RACE, ONE TONGUE."[13]

But, in reality, achieving the true equality necessary to realize this motto is impossible in a context where European aesthetics and cultural and social values inform a rigid hierarchy of race and identity. The Colombian state continues to reject the idea of Afro-Colombians as legitimate cultural and political subjects. Furthermore, state forces dismiss the sociocultural structures of Afro-Colombians and indigenous peoples as primitive, anachronistic, and incapable of forming a foundation for life in the modern world. Mestizaje assumes, in the most Darwinist and Manichean manner imaginable, that Afro-Colombians desire and need to shed their identities and whiten themselves in order to participate in Colombian society.

In this atmosphere, blacks are forced to struggle against the social pressure toward *blanqueamento*, or the whitening imperative, as both an act of political resistance and as an affirmation of their own humanity and identities. Afro-Colombians' pursuit of political recognition and ancestral rights to land based upon heritage strikes at the heart of the "one race" proposition of the Colombian state, and supports, instead Murillos's idea of a pluriethnic society and the collective rights for blacks and indigenous peoples. The multiethnic proposition provides for the mestizaje myth and would presume that political and economic rights and benefits are distributed equally among all individuals deemed eligible for those rights and benefits. In this system, group rights are minimized in favor of the capitalist model of individual ownership and exploitation of resources based upon the individual(s) yielding to modernism and state imperatives. In contrast, the pluriethnic model critiques these claims based upon the reality of existing structural inequalities in class status and other hierarchized sectors as a result of centuries of racism, ethnocentrism, and economic exploitation. Recognizing these historical realities, the pluriethnic model identifies the existence of marginalized ethnic groups and their rights to collectively own land and proposes redistribution of national resources based on ancestry, tenure, and other historical grounds. Though not a perfect alternative, it is one that accounts for the inherent discrimination that exists within state articulations of the Colombian individual and the impact on Colombia's communities of color as a collective.

Afro-Colombians engaged in an extraordinary process of political organizing during the constitutional reform period during the 1980s and were preparing to take advantage of new openings promised by the new constitution of 1991 and the passage of Law 70 in 1993. The new constitution acknowledged ethnic diversity in the nation and had provisions for collective land ownership rights for Afro-Colombians and autonomous indigenous territorial units for Indians. Despite their activism, Afro-Colombians' lack of strong organizational structures, lack of access to technical assistance, and the general lack of sympathy for their plight on the part of state officials made it difficult for them to take full advantage of the new political openings. Added to these complications were the state's commitment to mega-development projects that required control of lands historically occupied by Afro-Colombians and indigenous peoples.

In contrast, Arocha documents the effectiveness of the indigenous community's links to external and national Non-Governmental Organizations (NGOs), as well as their own collective organizations during the period of petitioning and filing claims for land in 1993. The comparative weakness of Afro-Colombian political organizations and the relative lack of extra-national support meant they were not able to take advantage of new laws to claim land in the same manner as members of Colombia's indigenous community.[14] Since then Afro-Colombian activists realized that obtaining collective land rights has important implications for Afro-Colombians' reconciliation with the state as well as for their concomitant reconstruction within society. These activists have been reconfiguring their efforts to build coalitions with other African-descended communities throughout Latin America to build a broader political movement to redress their specific problems.

Such initiatives follow a long tradition of Pan-African cooperation and problem solving in the Americas and promises to provide new directions for the Pan-African movement as a whole.[15]

The development of Afro-Colombian's pluriethnic alliances was and is a tedious process, hence this essay draws these connections in broad strokes. This study aims to explore the development of racialized cultural and political practices within the Afro-Colombian community that actively draws upon Pan-African internationalism as a natural vehicle for constructing indigenous struggles. Former Governor Murillo feels:

> Faced with formidable attacks on our cultural, economic and physical well-being, the most important challenge to Afro-Colombians today is to build a strong alternative movement. We support all efforts towards national reconciliation based on the inclusion of all Colombians in a pluriethnic country where all cultures have the same opportunity for self-development, without violence, without discrimination of any sort, and with *territorial autonomy. Black and Indigenous people* need to think and plan our future independently from the rest of the nation. Now, and in the future, despite the deep troubles we are suffering today, we are the African heart of Colombia. Colombia, as well as the rest of Latin America, cannot deny its African roots. To the contrary, we should take pride in these roots![16]

At the same time that Afro-Colombia gazes outward to make those extended Latin American connections, regional developments also echo these sentiments. At the Preparatory Conference of the Americas against Racism, Racial Discrimination, Xenophobia and Related Intolerance held in Santiago, Chile in December of 2000, those gathered issued a statement entitled, the Chile Declaration of African Descendents that opens with these words: "Affirming that the main victims of racism in the Americas are African descendants and indigenous peoples."[17] In the body of the document attendees further "affirm that African descendants have the right to our cultural identities, to recognition of those identities and to the adoption of measures that protect and develop them, as well as educational systems and institutions that respect our history, cultures, and identities."[18] This statement marks both a transgressive and affirming action to articulate an African selfhood by persons from nations where blackness has historically been a mark of backwardness and exclusion. Further pronouncements from the Conference were equally impressive. Those gathered agreed to: "condemn the situation of exclusion and marginalization that leaves our peoples submerged in poverty in all the Americas, a situation aggravated by the implementation of economic policies and development models that do not respect diversity, promote homogeneity, and perpetuate the systematic violation of our economic, political, social and cultural rights. As a consequence, we demand that states, multilateral organizations and private enterprise adopt compensatory measures, including reparations."[19]

Similar dynamics are evident in efforts in Ecuador, Guatemala, Peru, and Brazil. Nascent Africanist and Pan-Africanist projects that build on sporadic but persistent efforts over the last two hundred years are developing in these countries as well as in Uruguay, Costa Rica, Mexico, Honduras, and Bolivia. In other countries such as Panama, Puerto Rico, and Cuba a strong Africanist and Pan-Africanist formation has existed for many years but retains a legacy of racialized thinking reified in the rigid color hierarchies still evident in the political, economic, and aesthetic sectors of these societies.

While traveling in Suriname in 1972, S. Allen Counter and David Evans made contact with a maroon community. The Suriname chief of the maroon community asked both men the following question: "are you still the white man's slaves?" To that Counter and Evans replied: "We are not slaves in our land. Well not really in the true sense of the word." His people having fought for their freedom and won, the chief asked another profound question: "Well, have you won your fight?" To that they responded: "The battle is still being fought." The conversation continued and produced this telling inquiry from the Bush chiefs: "Well, have you found the road home?"[20]

In this chapter I examined the situation of Afro-Colombians as an oppressed community within Colombia in relation to the exploitation of African descendants in the Americas. The intent was to propose validation of the Pan-African perspective both as an analytical frame for understanding the evolution of Afro-Colombian identities and as a strategy for internationalizing their struggle for rights and recognition. This holds important implications for the Pan-African movement as a whole, which has experienced a theoretical and practical stasis since the fall of the apartheid regime in South Africa and the emergence of modern reactionary regimes on the African continent. The emerging Afro-Colombian rights movement is an important historical development for Pan-African solidarity, because it has critically articulated and theorized Africana identities, histories, and futures. In addition, Afro-Colombians acknowledge the need for commitment to theory and practice and for critical reflection upon both as a way to nurture the developing Afro-Latin movement that aims to further institutionalize its gains.

Returning to the history of maroon traditions will enable theorists and activists of the Afro-Colombian struggle to gain further insight on ways to shape campaigns of resistance in the Americas. These traditions of resistance serve as crucial reservoirs for consciousness and appear as significant touchstones when attempting to forge a national and international sense of unity.

The maroon complex and the independent settlements it spawned have important symbolic meanings since these runaway groups provided a structure for black and Indian cooperation in opposition to state repression. Antonio Benitez-Rojo has reminded us that, "Some day, when the holistic investigations on this matter are undertaken, the Caribbean itself will be surprised to learn how close it came to being a confederation of Maroon states. . . . It would certainly be necessary to examine the participation of the Maroons in the social struggles and independence movements of the region."[21]

All of these imperatives pose new questions. How should theorists and activists execute further projects to sustain solidarity and support in a pluriethnic context?

How can the lessons of the past twenty years be incorporated into these projects to create a comprehensive model of action based upon the unique aspects of the Afro-Colombian and Afro-Latin conditions and the special remedies demanded by these conditions? Lastly, how do solidarity projects of this nature provide for the further restructuring of the Pan-African internationalist movement in the United States? Will it help to resolve lingering ideological or philosophical questions about the nature of the movement? Will NGOs, community-based support groups, and government officials find imaginative ways to integrate their work and thus produce a more viable movement?

With the official recognition of African-identified communities in Latin America the Pan-African movement has entered a period wherein the theoretical stagnation of the past twenty years can be addressed. As Afro-Latin communities begin to insist upon a regional integration of their individual political struggles based upon the commonality of their conditions, a more complete critique of globalization and neo-liberal capitalism and imperialism becomes possible. Moreover, when Afro-Latin national and regional groups extend their views and endeavor to situate their analysis within a Pan-African context, the movement for self-determination is profoundly transformed. However, even though a new and empowered identity accompanies the growth in Afro-Colombian political activism, in order to be successful this process must consider the perspectives from other African-descended communities dispersed across many nations who have experienced similar transformations. Lastly, Afro-Colombian activists must continue to struggle with the histories of denial and the often self-imposed identity isolation that has separated their communities from the African Diaspora in general and from the rest of African America, which reaches from the northern to the southernmost regions of the Americas.

NOTES

1. Arturo Escobar, "Displacement, Development and Modernity in the Colombian Pacific," *International Social Science Journal* 175 (March 2003): 157–67.
2. W. F. Santiago-Valles, "The Caribbean Intellectual Tradition that Produced James and Rodney," *Race and Class* 42, no. 2 (October–December 2000): 61–79.
3. For Jaime Arocha, see "Inclusion of Afro-Colombians: Unreachable National Goal?" *Latin American Perspectives* 25, no. 3 (May 1998): 70–90; "Los Negros y la Nueva Constitucion Colombiana de 1991," *America Negra* 3 (1992): 39–54; and "Afrogenesis, Eurogenesis y Convivencia Interetnica," *Pacifico: Desarrollo o Biodiversidad? Estado, Capital y Movimientos Sociales en el Pacifico Colombiano*, ed. Arturo Escobar and Alvaro Pedrosa (Bogota: CEREC: 1996). For Arturo Escobar, see "Displacement, Development and Modernity in the Colombian Pacific," *Moving Targets: International Social Science Journal* 175 (March 2003): 157–67; and "Culture, Economics, and Politics in Latin American Social Movements Theory and Research," in *The Making of Social Movements in Latin America: Identity, Strategy, and Democracy*, ed. Arturo Escobar and Sonia Alvarez, (Boulder, CO: Westview, 1992), 62–85. For Peter Wade, see *Blackness and Race Mixture: The Dynamics of Racial Identity in Colombia* (Baltimore: Johns Hopkins University Press, 1993); "The Cultural Politics of Blackness in Colombia," *American Ethnologist* 22, no. 3 (1995): 341–58; "Blackness, Music, and National Identity: Three Moments in

Colombian History," *Popular Music* 17, no. 1 (1998): 1–19; "Representations of Blackness in Colombian Popular Music," *Representations of Blackness and the Performance of Identities*, ed. Jean M. Rahier (Westport: Bergin and Garvey, 1999); "Current Anthropology, Making Cultural Identities in Cali, Colombia," *August* 40, no. 4 (October 1999): 449–68; and "Racial Identity and Nationalism: A Theoretical View from Latin America," *Ethnic and Racial Studies* 24, no. 5 (September 2001): 845–67. For Kiran Asher, see *Constructing Afro-Colombia: Ethnicity and Territory in the Pacific Lowlands*, PhD dissertation (University of Florida: 1998).

4. See, for example, Eduardo Restrepo, "Afrocolombianos, antropologia y proyecto de modernidad en Colombia," in *Antropologia en la Modernidad: Identidades, Etnicidades y Movimientos Sociales en Colombia*, ed. Maria Victoria Uriba and Eduardo Restrepo (Instituto Colombiano de Antropologia, Bogota: 1997).

5. Wade, "The Cultural Politics of Blackness in Colombia," 341–58.

6. Francis Njubi, "New Media, Old Struggles: Pan Africanism, Anti-Racism and Information Technology," *Critical Arts* (January 2001): 117–35.

7. Richard J. Powell, *Black Art: A Cultural History* (London: Thames and Hudson, 2003), 13–15.

8. Ibid.

9. Adam Halpern and France Winddance Twine, "Antiracist Activism in Ecuador: Black-Indian Community Alliances," *Race and Class* 42, no. 2 (October–December 2000): 19–32.

10. Frantz Fanon, *Wretched of the Earth* (New York: Grove, 1963), 233.

11. Hilbourne Watson, "Theorizing the Racialization of Global Politics and the Caribbean," *Experience Alternatives: Global, Local, Political* 26, no. 4 (October–December 2001): 449–84.

12. Luis Gilberto Murillo, *El Choco: The African Heart of Colombia*, speech given at The American Museum of Natural History, Sponsored by The Colombia Media Project and The Caribbean Cultural Center, New York, February 23, 2001.

13. Arocha, "Inclusion of Afro-Colombians: Unreachable National Goal?," 72.

14. Ibid, 82

15. See Karen Juanita Carillo, Afro-Latinos Prepare for Third AFROAMERICA XXI Conference, Mundo Afro Latino, February 25, 2004, http://www.mundoafrolatino.com/english/020504.htm (accessed June 2004). See also Chile Declaration of African Descendents, Preparatory Conference of the Americas Against Racism Racial Discrimination, Xenophobia and Related Intolerance, December 5–7, 2000, Santiago, Chile.

16. Murillo, *El Choco*; emphasis added.

17. *Chile Declaration of African Descendents*, Preparatory Conference of the Americas Against Racism Racial Discrimination, Xenophobia and Related Intolerance, December 5–7, 2000, Santiago, Chile, http://academic.udayton.edu/race/06hrights/VictimGroups/AfricanDescendants/AfricanDescendants.htm.

18. Ibid.

19. Ibid.

20. S. Allen Counter and David L. Evans, *I Sought My Brother: An Afro-American Reunion* (Cambridge, MA: MIT Press, 1981).

21. Antonio Benitez-Rojo, *The Island Is Repeated: The Caribbean and the Post-Modern Perspective* (Hanover, NH: Ediciones del Norte, 1989), quoted in W. F. Santiago-Valles, "The Caribbean Intellectual Tradition," 47.

PART III

MUTUAL INSPIRATION

Radicals in Transnational Space

Chapter 8

The Havana Afrocubano Movement and the Harlem Renaissance

The Role of the Intellectual in the Formation of Racial and National Identity

Ricardo Rene Laremont and Lisa Yun

From the writings of Frederick Douglass and Toni Morrison in North America to those of José Martí and Carlos Moore in Cuba, intellectuals and writers in the Americas have played a role in creating political and social identities for the descendants of Africans in this hemisphere. From the 1920s through the 1940s, parallel intellectual movements took place in New York and Havana that attempted to find voice and identity for the descendants of Africans in the United States and Cuba. The Afrocubano movement in Havana not only found a voice for Africans in Cuba but also redefined the definition of what it meant to be Cuban, making it difficult for Cubans to assert Cuban national identity without embracing both European and African cultures. In contrast, New York's Harlem Renaissance embarked on a different intellectual project. Rather than redefining American national identity, the movement constructed an African American identity within American and European culture so that African American culture would become admirable and comparable to Anglo American culture. In the process Harlem Renaissance writers created a culture that was ennobled but also one that would parallel Anglo-American culture rather than fundamentally change the dominant discourse in the decades to come. Writers of the Negritude movement (Aimé Césaire and Leopold Senghor, principally) did the same for Africans in Europe. Highlighting the glorious and romantic aspects of African culture, the Negritude movement emphasized that, in measuring Africanity against whiteness, African culture was worthy of admiration by Europeans and Americans.

The intellectuals in the Havana Afrocubano movement, by attempting to merge African and European cultures not only distinguished themselves from the Harlem Renaissance and the Negritude movement but also did something novel: they redefined the definition of national identity. This is not to say that racism was erased in Cuba. Castro's regime has significantly improved the social and economic conditions of working-class Cubans, most of whom are of African descent. Carlos Moore has pointed out, however, that Afro-Cubans are still excluded from the upper echelons of political power and social prestige. The color-line, what W. E. B. Du Bois called the problem of the twentieth century, still operates in Cuba and is similar in some respects to that which operates in Brazil. Cubans embrace African culture—there is not that aversion to African culture that exists in the United States—yet Cubans themselves often relegate descendants of Africans (especially darker descendants) to poverty and social exclusion. A contradictory paradigm of race and colorism operates. African culture is included in the definition of what it means to be Cuban or Brazilian, yet socioeconomically, the darker African is excluded.

Although racism is operative in Latin America, this chapter will not attempt to address the categories of relative racism in Cuba and the United States. The aim here is instead to clarify that in Cuba from the period from 1920 to 1940 intellectuals gathered to create a new definition of the Cuban nation that would integrate African culture and Spanish culture. During the same period, African American intellectuals embarked on a related but different project, creating a discourse still circumscribed by separate-but-equal paradigms of dominant Anglo-Saxon culture. In the following generations, white Americans could appreciate Harlem Renaissance writers and artists without fundamentally challenging their own images of whiteness as the center of American national identity and without recognizing blackness as part of their own heritage.

The first essential political and social reality to grasp is that a counterrevolution occurred in Cuba after the war for liberation against Spain in 1898. In the war against Spain, Africans and descendants of African slaves, Chinese coolies, and poor European whites, or *criollos*, formed the bulk of the military forces that fought against Spain and the white landowning elite. Their principal military strategist and general was a black man (or more accurately, a mulatto) named Antonio Maceo. The other leader of the revolution, a criollo named José Martí, died in the first year of the war. After the success of the revolution in 1898; two events took place that would frustrate the hoped-for political and social empowerment of Africans in Cuba: the arrival of large numbers of Spanish immigrants and the arrival of Americans with their capital. From 1898 until the Great Depression of 1929, large numbers of Spaniards immigrated from Spain to Cuba. The Cuban government subsidized this immigration as part of an unwritten policy of "whitening" the Cuban nation.[1] By 1929, nine hundred thousand Spaniards had immigrated to Cuba, which is more than had immigrated in the entire period from 1511 to 1899. This immigration significantly changed the demographics and racial politics of the new Cuban nation. The Spanish who arrived in Cuba entered the labor market or set up businesses that directly competed with those run by criollos or phenotypically Mediterranean-looking Cubans. These criollos and Spaniards shared not only points of cultural commonality (both had origins in Spain) but also points of friction (the criollos had just fought a war of independence that was to have liberated them from Spain and Spaniards). Criollos, with Africans and Chinese, had only recently expelled the Spanish from the island in what was a class war shouldered by a multiracial underclass. The arrival of large numbers of white Spaniards in Cuba beginning in 1901 seemed like a reinvasion to many Cubans.

Many of the Spanish who arrived came with views of race informed by the European intellectual tradition rather than by the Cuban racial experience. Those Spaniards who arrived in Cuba in the first two decades of the twentieth century had their views of Africans informed by European schools of racial eugenics that were influential in both Europe and the United States. Such views included those of Arthur Gobineau and Oswald Spengler (nineteenth-century intellectuals later associated with the Nazi Party in Germany), and Enrico Ferri, Cesare Lombroso, and Alfonso Asturaro (Italian intellectuals who would contribute to Francoism and Fascism).[2]

Besides mass immigration by the Spanish, Cuba was effectively invaded by American capital during the period from 1896 to 1911. American financiers began investing heavily on the island, focusing initially upon the purchase of sugar plantations and refineries. Land purchases and other sales to Americans quintupled from $50 to $250 million. The portion of the sugar crop produced by U.S. owned mills increased from 15 percent in 1906 to 48 percent in 1920 to 75 percent by 1928.[3] With its substantial interests on the island, the United States interfered in Cuban politics to obtain results that were compatible with American investment.

To protect American capital, the United States also sent marines in 1899 to assure that Cuba would remain pacific. Many officers injected a new racist perspective that was different from the racism that had been practiced in Cuba in the past. Americans brought views on race informed by the North American experience. Miguel Barnet's *Biography of a Runaway Slave* records a slave's recollection of the new brand of "nigger" racism brought to the island by the Americans. Instilling fear in Cuban whites, the Americans helped in effectively disenfranchising blacks who had gained stature in the revolution.[4]

Because of the arrivals of the Spanish and the Americans, politics in Cuba during 1899 to 1920 underwent a white nationalist period. White racism toward blacks attained a new level of virulence. Borrowing from the sociology of eugenics, some white Cuban essayists called for the absolute elimination of the black race.[5] Criollos within Cuba also objected to blacks taking low-paying jobs. They asserted that blacks accepted substandard wages for agricultural and other industrial jobs (cigar rolling, stevedore work) that pushed them out of these labor markets. This racial pattern was similar to labor rhetoric leveled against blacks, Chinese, and Mexicans in the United States during the same period. Poor criollos became even more disgruntled when, in 1913, President Jose Miguel Gomez relaxed the general ban on black immigration to Cuba that the Cuban government had imposed in 1898. Agricultural laborers flooded into Cuba from Haiti and Jamaica. In the period from 1913 to 1925, 150,000 black workers immigrated to Cuba from Haiti and Jamaica, taking low-paying agricultural jobs that some white criollos felt should have been assigned to them.[6] Because of these developments, Cuban criollos in the labor market were being pushed from three directions: by the new arrivals from Spain, by the newly liberated Africans of Cuba, and by immigrants from Haiti and Jamaica.

The expansion of Haitian and Jamaican migration took place one year after the most bloody incident involving race on the island of Cuba. In 1907, Evaristo Estenoz and Pedro Ivonet organized the Partido Independiente de Color (Independent Party of Color), a party formed as a result of growing frustration among blacks and mulattoes in post–independence Cuba. Blacks and mulattoes found themselves still marginalized socially, economically, and politically. Even after their valiant participation in the war, they were still subjected to extensive economic and political discrimination like blacks in the post–Civil War American South. They were limited in the educational opportunities afforded to them and were relegated to second-class lower-paying agricultural, industrial, and service jobs. The new Cuban legislature was composed overwhelmingly of white delegates who refused governmental positions even to prominent blacks who had participated in the war. Blacks and mulattoes were pushed aside (often with a nudge from white American officials who emerged from paradigms of Jim Crow, anti-miscegenation, and segregation and who were overseeing the island politically, militarily, and economically). Because of their exclusion from government and from skilled-labor jobs, Africans in Cuba began organizing politically, principally within the Partido Independiente de Color.

Although the party had been founded in 1907, by 1910 it had twenty thousand members and posed a significant threat to the conservative and liberal parties in elections. Fearing this threat and claiming that the Partido Independiente de Color was intent on fomenting a race war, the Cuban Congress enacted a law prohibiting the formation of political groups based on race, thereby making the Partido Independiente de Color an illegal party. Unwilling to accept the demise of their party, Estenoz and Ivonet allegedly staged a revolt against the state in Oriente Province in May 1912. Taking advantage of this situation, President Gomez sent troops to crush this rebellion. In the resulting pogrom, the Cuban army killed more than four thousand blacks and mulattoes.[7] This *Guerrita de Doce* (Little War of 1912) renewed Cuban awareness of the role of race in politics and had lasting consequences for race-based politics on the island. The political and ideological developments of 1900–1920, the demographic changes accompanying them (the arrival of Spaniards, Haitians, Jamaicans), and the economic boom from expanded sugar production all affected the discussion in Cuba about race. Cuba was about to embark in two directions regarding race.

One approach emulated American Jim Crow and European Fascist and Nazi views that aggressively promoted the separation of the races and the elimination of the black race as legitimate political objectives. During this period, racist groups including La Liga Blanca de Cuba (The White League of Cuba) and El Orden de los Caballeros (Order of the Knights)—a Cuban chapter of the Ku Klux Klan— were formed with the express purpose of "whitening" Cuba demographically and culturally. Intellectuals including Rafael Conte, Jose M. Campany, Mario Guiral Moreno, and Carlos de Velasco argued in favor of eliminating or diluting the vote for black men (notably, the constitution provided for universal male suffrage regardless of race) or replacing Cuba's social system of race with the system of racial apartheid (Jim Crow) existent in the United States.[8]

Because of the shocking pogrom of the Little War of 1912 and the introduction of American- and European-influenced styles of racism in Cuba, some intellectuals within Cuba began questioning whether these approaches to race relations were relevant to the Cuban racial experience. By the beginning of the twentieth century, most Cubans were genetically mulattoes. Few Cubans could claim to be either pure Spanish or pure African. Marriage and concubinage between white men and black women were common. (Nevertheless, there was still social resistance to marriage between black men and white women.[9]) Cuba, after four hundred years of tolerated miscegenation, was not socially and attitudinally prepared to adopt the brand of legislated racism being practiced or discussed in the United States or Europe. Cuba was, at least genetically if not yet culturally, a mulatto nation.

Despite this tolerance of miscegenation and acceptance of interracial offspring, many Cubans still practiced racial discrimination against the darkest blacks— mulattoes were still more acceptable socially than darker blacks. This structure of social reward based upon color encouraged lighter-complexioned Cuban elites to practice marriage with white women while sustaining concubinage with black women as a means of maintaining their social status. That said, the tolerance of miscegenation in Cuba also led to social practices that allowed blacks and whites to congregate, mingle, socially engage, and marry. This particularly occurred among blacks and whites from the non-elite classes. Elite bars, schools, and restaurants often denied entry to blacks. (Even today, dark blacks are barely visible in the best Cuban restaurants and hotels, such as the Hotel Nacional.) Nevertheless, in most locales, social segregation never approximated the Jim Crow practices of the United States. Although criollos may have attempted to deny blacks access to public spaces and whereas they traditionally excluded blacks from access to better paying jobs, the American taboo and the shame associated with interracial relations were not replicated in this context. Most prominent Cuban "families had mulatto cousins: the better the family, the more probable this was."[10]

Spanish males and criollos in Cuba from the sixteenth through the twentieth century seemed to have preferred sexual relations with black and mulatto women. The extent of interracial sexual contact in Cuba reached levels unprecedented in the United States. Their offspring provide genetic proof of such contacts. Furthermore, in Cuba, paternity across the races was socially acceptable and legally recognized. Social acceptance of interracial progeny reached levels unthinkable in the United States, which legislated interracial marriage as a crime punishable under the law. Because of this socially accepted practice of miscegenation, racial tolerance was much greater in Cuba than in the United States. Cuban criollo attitudes toward African Cubans during these periods were mostly schizophrenic. Although tolerance of interracial relations was acceptable, the appearance of the black man in well-paying jobs was not.

In the final analysis, some intellectuals in Cuba concluded that the political programs and the racial ideologies being offered by both Europe and the United States were connected to imperialist policies that may have played a role in subjugating Cuba to neocolonial status. Because of Cuba's neocolonial dependency, they concluded that Cuba needed to embark on a new set of cultural politics that would liberate it politically and culturally from the United States and Europe. The political developments of the first quarter of the twentieth century and the contradictions at work within Cuban society itself forced some Cubans to create countercultural alternatives regarding the definitions of racial and national identity that would stand out in contrast to those being offered to Cuba by the United States and Europe. Cuba's failure to deal adequately with the question of race in the first quarter of the twentieth century provoked some Cuban intellectuals to examine the question of race and the definition of Cuban identity. This movement eventually redefined Cuban national identity so that it included African religions and Africanity.

This objective within the Havana Afrocubano movement led to a centering of Africanity that was different from the employment of Africanity in the Harlem Renaissance. The Havana Afrocubano movement, led principally by mulattoes and whites, inverted social and psychological preference among some Spanish elites and criollos that preferred and revered hispanidad (Spanishness) and purexa de sangre (blood purity).

The unusual aspect of the Cuban project was that it was led primarily by mulatto and white intellectuals who consciously tried to create a new national identity linked to racial inclusion. The principal leaders of this movement were Gustavo Urrutia, Nicolas Guillen Fernando Ortiz, Ramon Guerra, and Lydia Cabrera. From approximately 1928 to 1940, Cubans engaged in a significant debate within Cuba to define the racial composition of the new Cuban nation. Cuban intellectuals examined European and American paradigms of race relations and rejected them in favor of a paradigm that specifically sought to merge the Spanish and the African cultural traditions. In effect, they created a very different paradigm of race relations.

Gustavo Urrutia began this movement when he initiated an editorial column entitled "Ideales de una raza" in Cuba's largest-selling newspaper, El Diario de la Marina. The editorial series began in 1928 and continued until 1933. In this series, Urrutia and other sympathetic writers wrote articles about the need for cultural proximity between Africans and Europeans, and about the need for improving the economic conditions for recently liberated African slaves. Most important, they wrote about the need to redefine the meaning of what it meant to be Cuban. They concluded that "descendants of Martí" (criollos) and the "descendants of Maceo" (mulattoes or blacks) were to be included in the definition of what it meant to be Cuban.[11] This meant explicitly and implicitly that criollos and Africans were entitled to Cuban nationality. The founding fathers of the Cuban nation (Martí and Maceo) were a white man and a black man. Therefore, their heirs—the whites and blacks born on the island—were more entitled to be Cubans than the Spanish and Americans born abroad. Similarly, the racial views of the Europeans and Spanish, which only served to divide the Cuban nation, were to be rejected because they impeded national unity and contributed to the continued dependency of Cuba, particularly upon the United States. They argued that those recent arrivals—those from Spain and the United States—did not really belong to the definition of the Cuban nation.

The Americans' views regarding the derogation of the black race may have been acceptable in many parts of North America, but from the Cuban writers' viewpoints, they were inconsistent with the Cuban social experience. In 1929, Nicolas Guillen wrote in "El Camino de Harlem" that the North American paradigm of race relations was to be avoided in Cuba:

> We are terrified when we reflect upon the essence of the problem, which is grave and sensitive, we are separating ourselves one from the other when we should be uniting. And as time passes this division will become so profound that there will not be room for a final embrace. This will be the day when every Cuban city and town—all of them—will have a "black ghetto" as exist with our neighbors in North America. And this will be the road that all of us-those of us who have the color of Martí and those of us who have the color of Maceo-must work to avoid. That is the road to Harlem.[12]

Fernando Ortiz was a central figure within the Afrocubano movement. Ironically, this white lawyer and ethnographer initially began studying the African religion in Cuba because, as a criminologist, he believed that there was a link between African religious practices and criminality. His first book, Los negros brujos, amounted to a condemnation of Cuban–African religious traditions. The more he studied African religions, however, the more Ortiz underwent the conversion from critic to disciple of the religion. His immersion in African religion and music led him to adulation of the African tradition both in his personal and his academic life. His studies moved him to create the Society for Afro-Cuban Studies, which served as the locus for intellectuals who began reimagining a new Cuban national identity.

The intellectual objective of the Society for Afro-Cuban Studies was explicit. Jose Antonio Ramos wrote in the first issue in his article entitled "Cubanity and the Mixture of Races":

> Cuba will comprise a mixture of races . . . or it will not be Cuba. Objectively and subjectively, both realistically and spiritually, Cuba is neither white nor black. It is mulata, mestiza. . . . The first obstacle that confronts us is the development of Cubanity, this definite national spirit, within the larger colonial family, we must identify with this, and with our historical destiny that unites us. One must truly be an idiot to ask, again, for another opportunity to create a white Cuba, which would please the bankers in New York who oppose our views on mixed race. Don't they secretly rely upon Hitler's help to create a pure Aryan race![13]

What emerges from the writings of the Havana Afrocubano movement is the belief that both America and its system of race relations were direct threats to the formation of an independent Cuban nation. For Cuba to liberate itself from American neocolonialism and American and European paradigms of race, it had to redefine what it meant to be Cuban. This ideological project implied resistance to the "Colossus of North." The intellectuals within the Afrocubano movement argued for creating an alternate paradigm of race that would unite the Cuban nation against American oppression.[14] For them, to be Cuban was to consciously create a culture that united the African and Spanish cultural traditions. This was the Cuban nationalist and patriotic project.[15] Fernando Ortiz himself had a grander vision. While Guillen, Herminio Portell Vila, Ramos, Guerra, and Cabrera focused on uniting Africans and Spanish into one Cuban nation, Ortiz went further to urge the inclusion of mestizos and Chinese in the new definition of the Cuban nation. His definition of the Cuban nation was dynamic rather than static. It provided for the incorporation of multiple groups into the definition of Cuban identity.[16]

The Havana Afrocubano intellectuals who encouraged the inclusion of Africanity worked on fertile ground. Cuba, like Brazil, was a place where African religious traditions survived. Three factors can explain the vibrancy of African culture in these two countries: (1) the large number or proportion of Africans brought to those countries; (2) the long period of slavery (slavery was abolished in Cuba in 1886 and in Brazil in 1888); and (3) the possibilities for survival of African deities within the pantheon of saints of the Roman Catholic Church. Protestantism (embraced by the black churches in the United States), with its absence of a pantheon of saints, has not provided fertile ground for the survival of African religions in either North America or the British Caribbean. The roles of the Roman Catholic and Protestant churches, their different styles of Christianity, and their different potentials for the syncretism of non-Christian beliefs into Christian beliefs have become crucial to the transcultural survival of African practices.

In Cuba, Yoruba culture is prevalent, perhaps dominant. Yoruba cultural dominance became possible because it was culturally flexible and structurally compatible with the other dominant island religion, Catholicism. Yoruba orishas (West African deities) merged with multiple saints of the Catholic Church, and in some cases the practice included the Chinese gods San-Fan-Con (Gwan Gung in Cantonese) and the Buddha, forming a pan ethnic religion. Yoruba cultural flexibility also incorporated its only serious African religious rival on the island (the religion of the Mayombe ethnic group).[17] Yoruba influence is not only prevalent in Cuba but also readily observable in Brazil's candomble religious tradition and, to a lesser extent, in Trinidad's chango religion and Haiti's voudou. It is in Cuba, however, where it is claimed that the Yoruba tradition has been most faithfully preserved.[18]

Cuban intellectuals and working-class Cubans fashioned and practiced a definition of Cuban national identity that included the African and Spanish traditions. Although racism and colorism exist in Cuba, the definition of what it means to be Cuban synthesizes these traditions. In the United States, a different course of identity formation occurred, due to a long history of entrenched American nationalism that centers on antiblack and anticolor politics.

Definitions of "American," in terms of cultural nationalism and institutional politics were once and still are the discursive bastion of white exclusionary politics. The geographic entities that were first the English colonies of North America and that became the United States have been profoundly racist and exclusionary in their definition of who could become "American." From its inception, the predominant cultural paradigm for emulation in America was Anglo-Saxon Protestant society. Blacks by practice and by law were excluded from the definition of what it meant to be American. Early Asian arrivals were similarly excluded. Labeled "nagurs," a "new type of mulatto," and a "slight removal from the African race," they were denied the rights to citizenship, to own property, to testify in court, and to attend white schools.[19] White non-Anglo-Saxons, including Irish and Italians, were initially marginalized before becoming integrated into the definition of "American." Even Catholics did not really believe that they fully belonged to America. Only the election of President John F. Kennedy granted them the legitimacy they wanted. Most non-Anglo-Saxon groups in the United States have been excluded to some degree from the definition of what it means to be American. The Anglo-Saxon ethnic is at the center of that definition.

With this paradigm already in place, the writers and artists of the Harlem Renaissance movement engaged in cultural and political objectives that were dissimilar to the tactics and strategies undertaken by the Havana Afrocubano movement. Intellectuals of the Afrocubano movement worked on the project of cultural inclusion. The principal leaders in the Harlem Renaissance did not work toward cultural inclusion but rather toward cultural comparability and parallelism. After the Harlem Renaissance transpired (arguably with a denouement by James Baldwin), America underwent a cultural process that empowered and culturally "legitimized" African Americans. Nevertheless, the Harlem Renaissance did not change the systemic realities of a national culture based on profound separatism and exclusion from the seventeenth century onward.

After the Harlem Renaissance, America had two admirable cultures: Anglo-Saxon American and African American. This process, it was hoped, would result in two cultural Americas that would examine each other with respect, which is something that is still in contention today. Principal proponents of the Harlem Renaissance movement wanted, in the words of Alain Locke, to move in a direction of creating a "New Negro" who would be acceptable to the dominant society. To accomplish this, the New Negro's culture had to be high bourgeois culture, mostly divorced from "low" grassroots culture or African folk tradition. The principal leaders of the Harlem Renaissance attenuated aspects of culture that were explicitly African. The New Negroes marginalized those who tended to emphasize African folk culture in their work, like Zora Neale Hurston. In their view, such work was too lowbrow and inconsonant with their aspirations. The leaders of the movement, by ascribing to "high" culture, were reconstructing Africanity selectively and symbolically while avoiding the direct inclusion of African culture in what it meant to be African American.

Mainstream Harlem Renaissance theorist Alain Locke, in his work "The New Negro," emphasized uplifting the African American so that African American culture would become admirable to whites in Europe and the Americas. Locke revealed his rationale for the emphasis on class, stating: "The particular significance in the re-establishment of contact between the more advanced and representative classes is that it promises to offset some of the unfavorable reactions of the past, or at least to re-surface race contacts somewhat for the future."[20] In his work, there was little discussion claiming Africanity. Rather, the New Negro, as a more socially and culturally "advanced" representative and class, was separable from the African. Locke venerated the African tradition while placing it in a category that was different from and inferior to African American culture. Locke described African art as "pure art" and "exotic art" that contributed to modern artistic movements in Europe, including cubism. Yet he rejected connecting African art to African American art, calling African art "stagnant," "decadent," and "dismal."[21] In his article "The Legacy of the Ancestral Arts," Locke also maintained that African American art was severable from African art.

In his "The Criteria of Negro Art," W. E. B. Du Bois claimed that legitimate art is linked to propaganda that uplifts the race, but he did not include African art in his arguments.[22] Given this preference, one can understand how Du Bois's influential journal shaped African American public opinion in the direction of respecting African American art forms that were only indirectly related to African iconography. In contrast, the Havana Afrocubano movement specifically revered African, and usually Yoruba, art forms. Still, Du Bois began the initial deconstruction of Americanism when he wrote of the "twoness" of the black man as being "American" and being "negro" in *Souls of Black Folk*. An Americanism equated with whiteness, as Du Bois pointed out, does not include the darker races. In the United States, the deconstruction of American identity and the "Africanizing" of America remained in question, while in Cuba, the Africanizing of Cuban culture had already begun. American people of color recognized and lived within

Europeanized culture but white Americans had yet to recognize the colored heritage of its national identity and culture. Both Locke and Du Bois believed that they were among a vanguard of intellectuals who would serve as "the advance-guard of the African peoples" (this advance guard embraced being African American while rejecting African culture itself) and lead the way toward the restoration of legitimacy for African American culture and people. This movement, however, could be devoid of a specific African cultural content. The New Negroes spoke of pan-Africanism, yet they admired African culture without fully embracing it. Locke, Du Bois, and Richard Wright also rejected Zora Neale Hurston's alleged lowbrow "minstrelsy." To varying degrees, Hurston and others (such as Langston Hughes, Wallace Thurman, Jean Toomer, Aaron Douglas, Claude McKay, and John P. Davis) integrated folk culture and, in some cases, explicit references to Africa and Africanity into the experience of African Americans. In their "Negro quarterly of the arts" called *Fire!*, they tried to distinguish themselves from the genteel New Negroes who eschewed "lowbrow" folk culture.[23]

The New Negroes, Du Bois in particular, disliked the publication and roasted it because of its allegedly crude depictions of the African American working class.[24] This literary and artistic vision of working-class black America was one that the New Negroes would have preferred had never been published. In his essay "The Negro Artist and the Racial Mountain," Langston Hughes criticized "Nordicized Negro intelligentsia," praised Toomer's *Cane*, and urged the creation of a true Negro art in the United States that would provide expression for "truly racial art" and "the low-down folks" who were being excluded from the emerging definition of the African-American identity.[25] Hughes wrote of "the man in the ditch."[26] Hughes's close colleague Hurston went further to incorporate folk culture and the vernacular into her writings, to make linkages between African-American culture and African culture, and to observe Caribbean culture.[27] She linked African American images of angularity to African sculpture, claimed African shields as the only "true Negro painting," and observed that African American folklore was rooted in West African culture.[28]

The dancer and choreographer Kathleen Dunham also linked Africanity to her work. She researched African religious and cultural traditions in the Caribbean and included these traditions in her choreography, creating forms that embraced African culture. By so doing, she incorporated Africanity into the meaning of what it meant to be African American. Perhaps no one focused as much as Dunham did on the inclusion of African traditions. Both Hurston and Dunham had anthropological training and were willing to use those techniques to reinvigorate their art with respect to African culture. However, the politics of sexism and class within the Harlem Renaissance marginalized both women. "Any discourse that contradicted the ideology of the New Negro was excoriated, silenced, or otherwise dismissed."[29] Hurston's rural, radical voice that invoked a possibly African past was incompatible with the projection of an African America as bourgeois. Langston Hughes suffered criticisms from traditionalists but ultimately did not suffer as much.

Still, neither the "radicals" such as Hughes, Hurston, and Dunham nor the "mainstreamers" such as Locke, Du Bois, or Countee Cullen really examined the long-range implications and possibilities of incorporating Africanity into what it meant to be African American or what it meant to be "American." Here is a consistent point of departure between the Havana Afrocubano movement and the Harlem Renaissance. The activists in the Afrocubano movement embraced cultural, religious, and racial boundary crossings involving African and European cultural and social endeavors. The intellectuals in the Harlem Renaissance, because they were placed in a radically different social, political, and cultural milieu, struggled with issues involving class and race, "highbrow" and "lowbrow" art and culture, and "American" and "African" identity. In the end, the Harlem Renaissance intellectuals created an African American culture that became comparable to Anglo-Saxon culture while maintaining its separateness.

The success of the "New Negroes" had significant implications for future identity movements in the United States. Because of the success of those in the Harlem Renaissance and their descendants (principally Ralph Ellison, Richard Wright, and James Baldwin), the definition of American nationality remains grounded in racial and cultural separateness. The African American identity movement's progeny—the Latino movement and Asian American movement—have similarly established identity and cultural discourses that have been implicitly based upon racial and cultural separation and the politics of "authenticity."

These movements—because of the operation of the law, the development of history and politics, and the role of the intellectual—have not created an intercultural paradigm for the definition of national identity. They operate under the paradigm of cultural separateness and even conflict. The cultural project of racial separateness is reified by dominant structures of power and image-making. This paradigm is understood under the code word "multiculturalism," a term of derision for conservatives and celebration for liberals. We need to critically examine how the multicultural movement continues to disempower people of color and racialize the discussion of what it means to be American. The Cuban approach of cultural inclusion has not been discussed as an option, because it has not been understood by most Americans as an alternate cultural and identity model. Furthermore, the Cuban model may not be replicated due to the obvious factors, such as American demographics, size, and so on. The Cuban model may not even be desirable to some. Nevertheless, it can be critiqued for future discussion of racial and national identity in the new millennium.

For African Americans, emerging from the legacy of North American slavery with the systemic destruction of the African family and traditions, the African American path toward its own cultural nationalism has been substantially severed from Africanity, at least from the perspective of Cuban and Brazilian culture. Now, more than 130 years after slavery and a half century after the Harlem Renaissance, with changes in racial politics, economic and cultural globalization, and the significant impact of diasporic people of African descent from Latin America, the Caribbean, and Africa, we can reassess the relative inflexibility of this political and cultural model.

NOTES

1. Aline Helg, *Our Rightful Share: The Afro-Cuban Struggle for Equality, 1886–1912* (Chapel Hill: University of North Carolina Press, 1995), 56.
2. Julio Le Riverend, *La republica: Dependencia y revolución* (Havana: Editorial de Ciencias Sociales, 1973).
3. Robin Moore. *Nationalizing Blackness: Afro-Cubanism and Artistic Revolution in Havana* (Pittsburgh, PA: University of Pittsburgh Press, 1997), 27.
4. Miguel Barnet, *Biography of a Runaway Slave* (Willimantic, CT: Curbstone, 1994), 194
5. Helg, *Our Rightful Share*, 234–35.
6. Ibid., 56, 193–226.
7. Rafael Conte and Jose M. Capmany, *Guerra de razas (Negros y blancos en Cuba)* (Havana: Imprenta, 1912).
8. Hugh Thomas, *Cuba: The Pursuit of Freedom* (New York: Da Capo, 1998), 524.
9. Guillen, "El Camino de Harlem," *Diario de la Marina* (April 1929).
10. Jose Antonio Ramos, "Cubanidad y mestizaje," *Estudios Afrocubanos* I, no. 1 (1937):107–8.
11. Herminio Portell Vila, "Los prejuicios raciales y la integracion nacional norteamericana," *Estudios Afrocubanos* 2, no. 1 (1938): 47.
12. Ramiro Guerra, "Nuevas y fecundas orientaciónes," *Diario de la Marina*, January 13, 1929, p. 6.
13. Fernando Ortiz, "La Cubanidad y los Negros," *Estudios Afrocubanos* 3, no. 1 (1939): 3–15.
14. Lydia Cabrera, *El monte* (Havana: Ediciones C. R., 1954).
15. Roger Bastide, *Us Amériques noir* (Paris: Payot, 1967).
16. Ronald Takaki, *A Different Mirror* (New York: Penguin, 1989), 101, 329.
17. Alain Locke, "The New Negro," in *The Norton Anthology of African-American Literature*, ed. Henry Louis Gates (New York: W. W Norton, 1997), 966.
18. Locke, "A Note on African Art," in *The Critical Temper of Alain Locke: A Selection of His Essays on Art and Culture*, ed. Jeffrey C. Stewart (New York: Garland, 1983), 135.
19. Deborah G. Plant, *Every Tub Must Sit on Its Own Bottom: The Philosophy and Politics of Zora Neale Hurston* (Urbana: University of Illinois Press, 1995), 66.
20. Locke, "Enter the New Negro," *Survey Graphic Harlem Number* 6, no. 6 (1925): 632.
21. Ibid, 631.
22. W. E. B. Du Bois, "Criteria of Negro Art," *The Crisis* 32 (1926): 290–97.
23. Thomas, *Cuba*, 524.
24. Locke, "The New Negro," 966.
25. Langston Hughes, "The Negro Artist and the Racial Mountain," *The Nation* 122, no. 3181 (1926): 693.
26. Zora Neale Hurston, "Characteristcs of Negro Expression," in *Negro: An Anthology*, ed. Nancy Cunard (New York: Continuum, 1996), 28.
27. Ibid., 24.
28. Ibid., 26.
29. Plant, *Every Tub*, 66.

CHAPTER 9

ESLANDA GOODE ROBESON'S *AFRICAN JOURNEY*

THE POLITICS OF IDENTIFICATION AND REPRESENTATION IN THE AFRICAN DIASPORA

MAUREEN MAHON

"I WANTED TO GO TO AFRICA." This sentence opens Eslanda Goode Robeson's 1945 travel narrative *African Journey*, an account of her summer excursion to sub-Saharan Africa.[1] A simple statement, it reflects the belief in historical, material, and spiritual connections between people of African descent that motivated Robeson to study, visit, and write about Africa. Although best known as the wife of world-renowned singer, actor, and activist Paul Robeson (1898–1976), Eslanda Goode Robeson (1896–1965) was a scholar, writer, and activist in her own right. In 1934, Robeson began graduate study in anthropology at the London School of Economics and University College, London. Her goal was to learn more about African history and culture.[2] During the 1940s, she emerged as a critic of U.S. racial policy and a proponent of African nations' independence from European colonial powers. She drew on the concepts she learned in anthropology seminars, her experiences traveling in sub-Saharan Africa, and her concerns as a U.S.-born African American woman to argue for the liberation of people of color around the world. Like many Pan-Africanists in her cohort, she viewed European colonialism in Africa and Asia and segregation and racism in the United States as linked forms of oppression. Using her visibility as the wife of one of the most recognized black

Americans in the world, Robeson carved out a role as a progressive public intellectual that lasted until her death in 1965.[3]

Robeson set out to earn a doctorate in anthropology, but she never focused her energy on academic writing. Instead, she communicated her critique through books, pamphlets, articles, and lectures that urged her audiences to shift long-standing racial practices and ideologies. Concerned with changing the ways black and white Americans thought about Africa and their relationship to the continent, Robeson chose to challenge negative mainstream depictions of Africa by producing an image of a "Modern Africa." This was a region involved in international politics and populated by intelligent citizens who maintained cultural traditions in spite of colonialism and were capable of self-governance. In the ensuing discussion of Robeson's writings and lectures about Africa, I show how a Pan-Africanist perspective influenced her intellectually and politically. This Third World viewpoint informed her efforts to address the social and political restrictions placed on black people around the world. Throughout her career she worked with concepts of culture and identity that established bonds among people of African descent and celebrated the vibrancy of African people and culture.

Robeson's decisions about what to study and how to interpret her experiences were shaped by a diasporic politics of identification. This viewpoint recognizes, creates, and extends cultural and political connections among people of African descent. Built on a global perspective and rooted in a politicized black consciousness, a diasporic politics of identification is a self-conscious construction of connections among people of African descent that views their history and fates as linked and considers that link to be politically, culturally, and emotionally meaningful. In this chapter the term "politics of identification" is used to stress the aspects of choice and construction that inform this engagement with Africa. Concerned with black solidarity and black liberation, this perspective is the foundation of intellectual and social movements, such as Pan-Africanism and Afrocentric thought. It also animates contemporary academic interest in African Diaspora studies.[4] This expanding field has developed alongside a general interest in Diaspora and growing scholarly attention to globalization, migration, and transnational cultures. Much of this turn-of-the-millennium scholarship highlights differences within the African Diaspora and the impossibility of defining what constitutes authentic or representative blackness. Often, these contemporary scholars struggle to articulate the relevance of racially based political connections and social identities in the face of these differences (and in the face of acknowledging the spuriousness of biological race as a meaningful category alongside recognition of the continued political, social, and economic marginalization of black people that make race feel quite real). This approach contrasts with that of Robeson and others in her cohort.

In the early twentieth century, a critical mass of black artists, intellectuals, and activists were working to reconstitute linkages among people of African descent that historical processes had severed.[5] These Pan-Africanists were concerned with the political and social progress of black people who, whether in the United States, the United Kingdom, the Caribbean, or Africa, were denied political rights, economic access, and social mobility. Pan-Africanist intellectuals and activists sought to consolidate a sense of connection rooted in shared African ancestry and enhanced by similar structural conditions—subordination by Europeans and racialized marginalization. Some Pan-Africanists extended their concept to all nonwhites struggling against colonialism and imperialism. Privileging their similar structural position over national, ethnic, and class differences, they used formations like "the colored world" and "the color line" to establish connections between blacks in the New World and people in sub-Saharan Africa, India, and China, all of which were battlegrounds for the liberation of people of color from the dominance of Western Europeans.

During this period there was no single form of Pan-Africanism, but whether Garveyite or Du Boisian, pan-Africanist activists worked to mobilize shifts in consciousness. Widely circulated colonial discourses taught that Africa was the "Dark Continent." For Robeson and other Pan-Africanists it was a repository of black history and a positive source of identity. Pan-Africanist activists, then, were fighting for both political and psychological independence from Western hegemony. This led to political activism and to the development of cultural productions that demonstrated pride in African heritage and forged connections across the Diaspora. African Journey was part of this project to build Pan-African consciousness. In her narrative, Robeson reimagined Africa, a much-maligned region, as a site of cultural development. She encouraged black Americans to recognize a connection with Africa, form politically meaningful connections with contemporary Africans, and eschew their negative views of sub-Saharan Africa. Her approach was built on a sincere embrace of African cultures and a belief in the significance of African connections, but it meant expanding, not rejecting, an American identity. It was also a strategic move that anticipated that increased political influence and positive black consciousness would develop if people of African descent recognized what they had in common.

Robeson's efforts to produce alternative representations of Africa and Africans are also part of the tradition of "racial vindication" that developed among African American intellectuals and activists in response to the ways black people are depicted in scholarship and public discourse. As Faye V. Harrison and Ira E. Harrison explain in their history of African American anthropologists, vindicationist writing and scholarship "emerged in reaction to racist assertions that Africans are degraded savages, that Africans and African Americans have no culture, that blacks are inherently inferior, and that miscegenation is degenerative to whites and white culture."[6] In order to vindicate and uplift the race, these black intellectuals overturned the old meanings associated with blackness. Typically, they argued for the validity of black culture while developing a socially and economically grounded critique of the power relations that marginalized black people. As a vindicationist and public intellectual, Robeson used her writing to provide information, comment on political conditions, and assert a particular type of black subjectivity. These tendencies were present in *African Journey*, her most widely circulated publication.

African Journey is an account of Robeson's 1936 trip to South Africa and Uganda. Her prominence as the wife of a famous man and the book's unusual subject matter combined to secure *African Journey* media attention. The book was reviewed in the New York Herald Tribune, the New York Times, the Chicago Tribune, and the Christian Science Monitor. *African Journey* sold out of its first printing and was selected by the American Library Association as one of the fifty outstanding books of 1945.[7] Also in 1945, the National Council of Negro Women named Robeson "one of twelve outstanding women in American life."[8] She began to lecture on Africa and civil rights and published on these issues in left and black presses. By the late 1940s, she was involved in electoral politics, co-founding and chairing the Connecticut Progressive Party. She ran (unsuccessfully) as the Progressive Party candidate for Connecticut Secretary of State in 1948 and for Congress in 1952. She was also active in the Council on African Affairs (CAA)—a New York organization whose members campaigned for decolonization in Africa.[9] Paul Robeson was one of the organization's founding members and its chairman. The CAA published E. G. Robeson's pamphlet on contemporary conditions in sub-Saharan Africa and through her affiliation with the organization, she became an accredited observer at the United Nations.[10]

It was a long-standing concern with the condition of black people and curiosity about Africa that led Eslanda to travel to the continent. Born in 1896 in Washington, DC, she was the maternal granddaughter of Francis Lewis Cardozo, who had been "a pioneer in Negro education and in the fight for Negro rights" and Secretary of State of South Carolina during the Reconstruction era.[11] Her father, John Goode, a clerk in the War Department, died when Eslanda was four years old. Her mother, Eslanda Cardozo Goode, supported Eslanda and her two brothers by working as a beautician catering to a wealthy black and white clientele.[12] By 1905 the family had moved to Manhattan where Robeson distinguished herself as a student. At sixteen years old, she graduated from high school and won a four-year scholarship to the University of Illinois. In her senior year, she transferred to Teachers College, Columbia University and earned a B.S. in chemistry in 1923. From 1918 to 1925, she worked at New York City's Presbyterian Hospital as a surgical technician and chemist in the Surgical Pathology Department.[13] The first African American to hold such a high position at the hospital, Robeson planned to become a doctor.[14] These plans changed after she married Paul, a star college athlete and Columbia University Law School student, in 1921. She gave birth to their only child, Paul, Jr., in 1927.

Reasoning that American racism would limit Paul's prospects as a lawyer, Robeson encouraged her husband to pursue his interest in acting and became his business manager. To capitalize on performance opportunities in England and Europe, the Robesons moved to London where they lived from 1928 to 1939. During this period, Robeson's interest in the history and culture of African Diaspora people flourished. She researched the African Diaspora community in

Paris and published a series of profiles about them.[15] Her study of Africa intensified as she met African students who lived in London and followed the extensive
coverage of African politics in the British press. Both Paul and Eslanda availed
themselves of opportunities to learn about African culture in British universities: Paul studied African languages and Eslanda studied social anthropology. In a
1934 letter, she explained: "I am working in anthropology at the London School
of Economics and University College, both connected with London University,
and will in due time take a Ph.D. in the subject. I am specializing in African cultures, and am more interested in them than I have ever been in anything. When
we get through, we will know something about 'our people.'"[16]

Eslanda began her graduate work at a fortuitous moment. In the mid-1930s
when she started attending classes at the London School of Economics, Bronis-
law Malinowski, one of anthropology's towering figures, was offering his weekly
seminars in social anthropology and establishing fieldwork and his theory of
functionalism as central to anthropological research.[17] In a social, political, and
intellectual context that devalued non-Western people and their practices, British
social anthropology sought to identify and understand the logic of non-Western
cultures and thereby claim validity for them. The field offered an unusual academic opportunity to take nonwhites and their cultures seriously, even to present
them in a positive light. Still, this attractive quality was sometimes undercut,
Robeson learned, by a conservative streak in the discipline. The quest for knowledge about the other that drove African studies in Great Britain was shaped by
the colonial context out of which the field developed.[18] As a "colored" woman
attending Malinowski's seminars, Robeson encountered firsthand the ways British anthropologists labeled nonwhites and the Eurocentricism that shaped their
attitudes. In *African Journey* she recalled, "After more than a year of very wide
reading and intensive study I began to get my intellectual feet wet. I am afraid I
began to be obstreperous in seminars. I soon became fed up with white students
and teachers 'interpreting' the Negro mind and character to me. Especially when
I felt, as I did very often, that their interpretation was wrong."[19] Much to her
chagrin, they also interpreted her mind. Her fellow students and the professors
insisted that because of her class and education, she was more like the British
anthropologists than the Africans they were discussing.[20] Although she followed
many Euro-American cultural norms, she refused to abandon the connection she
felt with African people. "I'm Negro. I'm African myself," she recalled asserting.
"I'm what you call primitive. I have studied my mind, our minds. How dare
you call me European!"[21] She held to her position in part because she knew that
her blackness prevented her from enjoying the rights guaranteed to people of
European descent. In London, Paul and Eslanda avoided the daily indignities of
de facto and de jure racism that prevailed in the United States. Still, neither had
lost sight of the ways their race impeded their access to full equality or the way it
shaped them as individuals. Affirming her Africanness was also a way to challenge
the tendency of whites to make negative generalizations about black people and
view successful blacks as exceptions who were more white than black.[22]

Robeson recognized that part of the reason her colleagues so confidently made their proclamations was that they had conducted research in Africa and used their experience of fieldwork to claim the authority to speak about the other.[23] Robeson decided she could establish authority of her own—beyond her being black—by conducting research in Africa. That Robeson could start her African fieldwork while living in Europe is a result of early twentieth-century transnational movements of people. She and Paul began to "seek out all the Africans we could find, everywhere we went: in England, Scotland, Ireland, France; in the universities, on the docks, in the slums."[24] In these encounters, Paul and Eslanda put aside obvious differences in language, culture, ethnicity, skin color, religion, and class between themselves and their African acquaintances; instead, they focused on what they shared. In *African Journey*, she reported: "The more we talked with them, the more we came to know them, the more convinced we were that we are the same people: They know us, we know them; we understand their spoken and unspoken word, we have the same kind of ideas, the same ambitions, the same kind of humor, many of the same values."[25] Her book's dedication, "For the brothers and sisters, who will know whom I mean," underscores this identification and connection with black Africans.

While Western European anthropologists debated which theoretical models and field methods would best contribute to the development of their discipline, Robeson approached anthropology as a tool that would enable her to contribute to the project of black liberation. Given this imperative, it is not surprising that she did not write a standard ethnographic monograph. Like Zora Neale Hurston, a black anthropologist who had voiced her enthusiasm about Eslanda's foray into the discipline, Robeson constructed a text that reached beyond an academic audience while making an argument about the integrity of black culture.[26] Where Hurston focused on black Americans in the South, Robeson focused on the Africans she met in South Africa, Uganda, and the Congo. As with Hurston, Robeson wrote in an accessible style and placed herself in the text using first person narrative. She also incorporated the insights of her travel partner, her nine-year-old son, Paul, Jr. (Paul, Sr., did not go on the trip). This positioning allows the reader to get a sense of Robseon's personality—her sense of humor, her outrage over European colonialism—through her interpretations of African culture and politics. As with Hurston, her status as a black woman conducting research with black people upset the traditional, usually unquestioned, disciplinary dynamic in which a (usually male) person of European descent "studied" non-Europeans. The form and content of *African Journey* challenge this binary and the scientific distancing typically found in ethnographies and probably contributed to distancing Robeson from the mainstream of anthropology, both in her era and today. There has been some recent attention to Robeson's work as an anthropologist, but for the most part Robeson is not talked about in the field of anthropology.[27] This is the case even though she held a Master's degree in anthropology, did doctoral studies in anthropology, and identified herself as an anthropologist throughout her career. Robeson completed the course requirements for her PhD at the Hartford

Theological Seminary but, like Zora Neale Hurston who is similarly on the margins of the discipline, she did not complete her doctorate.

Robeson organized *African Journey* as a series of journal entries. They take the reader through her trip, describe the African and European people she encountered, and comment on the incidents that shaped her travels. Many of the entries are amplified with quotations from primary and secondary sources. Aware that racial commonality and the identification she openly felt with Africans might lead readers to accuse her of bias, Robeson cited the work of European writers to indicate that her views were grounded in a reality that others had also observed. The book's opening chapter situates her journey in a desire to learn about people with whom African Americans shared a historical bond that slavery had ruptured. The body of *African Journey* documents the everyday life of black Africans. Two broad questions inform the text: "What are African people like?" and "What is happening in Africa now?" The book is divided into two sections, the first is about South Africa and the second about Uganda. We follow the two travelers, "Essie" and "Pauli," as they take a sea journey from London to South Africa; make excursions into South African cities and townships; travel to Uganda for a four-week stay with the Toro, a cattle-herding people; and undertake side trips to Dar-Es-Salaam, Zanzibar, Mombassa, the Congo, and Cairo. This would be quite a trip now and it was a serious adventure in 1936. Robeson documented the rough sea voyage, car breakdowns, and bouts with tropical fever, but she also commented on the warmth and hospitality of the African people that more than offset the difficulties.

The section on South Africa details the harsh conditions that black Africans contended with as people living under minority rule. Robeson traced the impact of European appropriation of land, the institution of the pass system, and the segregation that shaped every aspect of daily existence.[28] To Robeson's mind, South Africa recalled the United States. At one point, she observed, "This traveling about Africa reminds me of traveling through the Deep South in America: You are passed from friend to friend, from car to car, from home to home, often covering thousands of miles without enduring the inconveniences and humiliations of the incredibly bad Jim Crow train accommodations and lack of hotel facilities for Negroes" (*African Journey*, 65). She learned the term kaffir, a South African word with the same connotations as "nigger" (59); had confrontations with white South Africans whose racist attitudes reminded her of "crackers" back home (71); and noted the ways housing segregation and the legal system stripped black Africans of rights and mobility in their own homeland (37, 73).[29] Her descriptions of the impact of poverty and racism detailed the uncivilized aspects of European colonialism and argued for freeing South Africa from an abusive system. In her section on Uganda, where British colonial officials administered the local government, Robeson depicted everyday life as less oppressive than that in South Africa. The Toro, for example, were able to maintain many traditional practices, particularly cattle-herding, and their precolonial tribal structure and landholding patterns did not seem to have been as disrupted as those of blacks under South African rule.

Robeson spent time with the Mukama (king) of Toro, the cousin of one of her London-based African friends. Here she conducted her fieldwork, focusing on work related to the care of cattle and the processes of milk production.

Robeson's presentation of the life and landscape of southern and eastern Africa in *African Journey* demonstrated that black Africans were intelligent and engaged with the world beyond their borders. At one point, she recounted a discussion between South Africans and black Americans that covered the political situation in Africa, the United States, Italy, the Soviet Union, India, Spain, Germany, and Japan. She then stated:

> And I thought as I talked and listened, These Africans, these "primitives," make me feel humble and respectful. I blush with shame for the mental picture my fellow Negroes in America have of our African brothers: wild black savages in leopard skins, waving spears and eating raw meat. And we, with films like *Sanders of the River*, unwittingly helping to perpetuate this misconception.[30]

In this passage, Eslanda acknowledged the role of media in perpetuating American ignorance about Africa. Released in 1935, *Sanders* was a bitter experience for Paul who had agreed to play the role of an African chief with the hope that the movie would present a realistic depiction of African culture. The resulting film, however, was a reaffirmation of British superiority and African inferiority.[31] Her husband's *Sanders* experience underscored for Eslanda the fact that when Europeans were in control of representation, Africans were all too often shown in a negative light. She chose to battle stereotypes by making her own representation of Africa and Africans.

The connection Robeson recognized between black Americans and Africa encouraged her to take a nontraditional approach to the classic anthropological encounter. Instead of seeking the difference of a non-Western other and contrasting it with a western self, she documented what people of African descent in Africa and the United States had in common and described Africa and the United States as contemporary.[32] She presented Africa as a region that had made important contributions to world civilization and that continued to play a role in world affairs. Similarly, she represented Africans as interested in the contemporary world. Inverting the image of Africa and Africans as primitive, Robeson insisted that this modern continent and these modern people were due respect and independence.

The choices Robeson made as a writer creating a representation of sub-Saharan Africa in the 1940s anticipate the critiques of the practice of fieldwork and the production of anthropological knowledge that have emerged from within the discipline over the last thirty years.[33] Robeson's perspective contrasts with the traditional anthropological approach in which, as anthropologist Lila Abu-Lughod observes, "otherness and difference may have assumed . . . 'talismanic qualities.' " Whether its goal is to engage in cultural self-critique or to assert enlightened tolerance through relativism, anthropology needs others that are different from the self."[34] Abu-Lughod emphasizes that the anthropological concept of culture often has the effect of creating, exaggerating, and fixing difference between the (usually) Western anthropologist and the (usually) non-Western subject.[35] She further observes that ethnographic writing has "trafficked in generalizations," using "details and the particulars of individual lives to produce typifications" that can make people living in other societies "seem simultaneously more coherent, self-contained, and different from ourselves than they might be."[36]

Robeson challenged the presumed distance between the researcher and the researched. She looked for—and found—connections, commonality, and "modern Africans." *African Journey* introduced the reader to people who were knowledgeable about issues that also concerned her American readers. Robeson reported, for example, that South African women were curious about women's organizations in the United States and that the Toro women asked questions about the lifestyle of American women.[37] Throughout, she focused on contemporary conditions, particularly the impact of British colonialism. She had a network in Africa because of her contacts with African students based in London, the colonial metropole. She was able to conduct her research because many of her male and some of her female interlocutors spoke English, a consequence of prolonged contact with Europeans. Colonialism and the accompanying forced entry into modernity were inescapable. Domination by European powers was an experience that black Africans and black Americans had in common and this link could help establish Pan-African consciousness.

Robeson's race, nationality, gender, and class mediated her encounters. Her husband's success as an entertainer gave her the financial means and leisure time to pursue the nonremunerative projects of writing, graduate study, and research travel. Like many female anthropologists, she took advantage of "honorary male" status. She was introduced to and interacted with men; indeed, she had to make it a point to ask to meet women.[38] Furthermore, although she was a person of African descent, she was a foreigner and a scholar; as a result, she was also accorded a kind of "honorary white" status that helped her navigate South Africa and Uganda and allowed her access to arenas from which local blacks were excluded. Still Robeson identified with black Africans. This sense of connection led her to create a representation of a place where people were, in many ways, much "like us." Her efforts to highlight the "normalness" of the Africans are evident in her textual descriptions and also in the visual images that supplement the text. *African Journey* features over sixty black and white photographs that Robeson took during her trip and they provide a striking example of Robeson's commitment to de-exoticizing Africans and Africa.

Anthropologist Deborah Poole observed that photographs of non-European people held a "scientific and voyeuristic fascination" for nineteenth-century Europeans partly because they were "image objects [that] lent support to an emerging idea of race as a material, historical, and biological fact."[39] In the 1940s, race was still being used uncritically as a way to classify people. The visual images Robeson incorporated into her text aimed to counter the dominant emphasis on racial ranking and difference. Here she followed the anthropological process of providing information to make "the exotic" seem familiar, selecting photographs that attacked dominant representations of "primitive Africa." There are pictures of a wedding party, family groups, and neatly attired children standing outside their school. These are not wild-eyed natives but people who have homes and families. Their hair is carefully groomed and their clothes—whether Western or African—are tidy. Many of the photo captions identify people by name or occupation, individualizing them and creating a feel of informal snapshots from a vacation rather than of scientific data. Robeson also included photos of traditional housing, goods for sale at a marketplace, and the physical terrain. Together, these pictures gave a sense of what Africa looked like and showed the reader that, while different, Africa was not a fundamentally incomprehensible place. Discussing the culturally and historically specific factors that inform the ways people "experience the pleasure of the visual," Poole suggests that "the spell photography weaves around us is multiplied in images of people from the colonial or non-European world who appear both like and not like us."[40] Robeson was likely fascinated by the phenotypic similarities she saw when comparing Africans to African Americans, including herself and Paul. For her, the pleasure of looking and her interest in showing stemmed from a desire to mark similarities and dismantle barriers. Her photographs amplified the text's emphasis on connecting with "brothers and sisters" in Africa and advancing the then revolutionary idea that "Africans are people."[41]

Robeson found that the Africans she met were in some respects more advanced than westerners. After spending time with herdswomen in Uganda, she reported:

> I have learned a great deal about the very important business of living, and as a result have rearranged my sense of values to some considerable extent. The leisurely approach, the calm facing of circumstances and making the most of them, is very different from the European hustle and hurry and drive, and worry and frustration when things don't go well. The African gets things done, gets a great deal done, but gets it done without the furious wear and tear on the nervous system.[42]

Here and throughout *African Journey*, Robeson used a culturally relativist approach to shift the terms of evaluation and reveal the logic of African practices and the persistence of European biases.

As Robeson documented Africa's complexity and modernity, she succumbed to some exaggeration. It was only the Europeans who were loutish, unattractive, and ignorant. The Africans she met were dignified and wise, generous and patient. The continent would be in good hands, Robeson was suggesting, if the Europeans would return it to its rightful stewards. As with any representation, the one Robeson produced of Africa was selective and based on choices she believed would convey her larger points. In her pamphlet on Africa, she commented on this process, "Now, I don't want to give you a misleading picture of Africa. There is frightful, ruthless exploitation, segregation, discrimination, oppression. The examples I have given are the exceptions, to show what the African can do, has done, against terrific odds."[43] With the negative images of Africa and Africans so dominant, Robeson highlighted the positive. Robeson took from anthropology certain assumptions, like the importance of approaching culture from a relativistic point of view and the value of pursuing knowledge about non-Western cultures. She rejected, however, an emphasis on the difference of the "primitive other" and a synchronic analysis that ignored history, process, and change. With the publication of *African Journey*, she disseminated her representation of "Modern Africa" and was soon discussing it in public forums.

With the her 1941 return to the United States, Robeson continued her study of anthropology and embarked on a career as a public intellectual. In speeches and articles she attacked racism in the United States and European colonialism in the non-Western world. Working from a global perspective, she recognized the interconnections among the world's nations. In the early 1940s she began to publish in black newspapers like the *California Eagle*, the *New York Amsterdam News*, and the *Baltimore Afro-American*, and was syndicated by the Associated Negro Press news service. She also wrote for left publications like *The Daily Worker* and *New World Review* and saw her articles reprinted in papers in England, India, and the Soviet Union. She began to give lectures about her travel in Africa in the early 1940s. In 1947, for example, she went on a seventeen-city lecture tour that took her from her home in Connecticut to Massachusetts, New York, Ohio, Indiana, Arizona, and California. She spoke before audiences of librarians, teachers, women's club members, social workers, church groups, high school and college students, missionaries, PTAs, and interracial groups. Whether she was discussing anticolonial movements in Africa, the politics of the cold war era, or race relations in the United States, a Pan-Africanist perspective informed her approach. Furthermore, she situated her comments within the anthropological frame of culture.

For Robeson, an anthropological perspective enabled social analysis that revealed and critiqued the logic that supported social inequality. She emphasized the context of human behavior and custom, particularly the practices and ideologies that kept black Americans subordinated to and segregated from white Americans. In a lecture that she delivered on a number of occasions in the mid-1940s, Robeson outlined her definition of snthropology:

> I like to think of Anthropology as the study of man and his behavior in relation to his fellow man, and to his changing environment. Thus it includes the study of primitive man under primitive conditions, of modern man under modern conditions, of human relations, race relations, of education, of social institutions. Thus it includes the study of the migrations of man, of wars, of the Industrial Revolution. And it includes some religion, economics, sociology, science, politics, literature, music and art.[44]

This "dynamic interpretation," as she called it, allowed her to encompass the issues that interested her: the changing position of American women, the battle for African American civil rights, the situation of war refugees, juvenile delinquency, education, and the impact of the World War II on social life.[45] All of these matters, she explained, "definitely have to do with man's attempt to adjust himself to his fellow man and his changing environment and so are clearly within the field of anthropology."[46] For Robeson, anthropology went beyond British social anthropology's study of "primitive sociology." She saw anthropology as a means for understanding contemporary social organization and as a valuable tool for rebuilding ruptured social relations in the postwar era. "Maybe you think, as I once did," she would tell her audiences, "that anthropology is a high-brow, intellectual, highly specialized subject, and therefore you are not interested in it. I thought I wasn't interested in it. Now I can scarcely talk, think, live—without plunging smack into Anthropology."[47] Robeson used anthropology to frame her presentation of information and, like Boasian anthropologists who were also publishing in the 1940s, to attack racism.[48]

As Mrs. Paul Robeson, she was able to present her views to white audiences, an extraordinary situation for a black American woman in the post–World War II era. Robeson exhorted white Americans to think of African Americans as full American citizens and to recognize the responsibility the United States had on the world stage. In her estimation, the primary postwar issue was interracial and intercultural relations—that is, how people from different backgrounds and cultures were going to get along in the world. Robeson stressed that if the United States was to become a truly democratic nation, whites would have to learn to coexist with nonwhites without depriving them of rights and opportunity. In speaking to black audiences, she noted that for black Americans to take their rightful place as American citizens, they would have to embrace their African heritage. She explained the reasons that this process was necessary:

> The conditions under which we Negroes live in America have made us sensitive about Africa. Most of us feel somehow insulted if we are called African. . . . We do not want to look back to slavery, and therefore we do not look further back to Africa, fearing it might be even worse than slavery. I think it is now time for us, as a people, the Negro people, to look back; to find out about our background, our roots, our "old country." No people have a future without a past. In the past there may be information and inspiration for the future. And when we look back now we can become very proud of our history and of our race.[49]

Like anthropologists Franz Boas and Melville Herskovits whom she echoed here, Robeson described African accomplishments to counter the view—held by black and white Americans alike—that Africans were without culture.[50] She provided evidence of a tradition of African law, using the example of the Ashanti and Hausa legal codes and courts.[51] She noted that the Hausa had a publicly supported system of policing long before Europeans did and explained that although "most people think of the so-called witch doctor and medicine-men as primitive ignorant men, playing upon the superstition of their people," they in fact have "a considerable knowledge of materia medica, and treat diseases on purely scientific principles." She credited that Africans "were among the first people to raise cattle and use their milk" and noted the influence West African masks and carved figures had on celebrated European artists like Picasso, Matisse, and Brancusi.[52] She also indicated that Africans took what Euro-Americans might view as the "civilized" path before Europeans did so, citing Boas to make her case: "It seems likely that at a time when the European was still satisfied with crude stone tools, the African had invented the art of smelting iron. Neither ancient Europe nor ancient Western Asia, nor ancient China knew iron, and everything points to its introduction from Africa."[53]

While Boas and Herskovits recuperated Africa through a focus on the past, Robeson brought Africa into the present and attacked the common sense notion that "Africa is a long way away, and we as Americans haven't really very much to do with Africa."[54] Refusing the sharp distinction between "the field" and "home" that informed anthropological research of her era, Robeson showed that the west and the non-west were interdependent:[55]

Every time you eat a chocolate bar, use soap, or ride in a car on rubber tires, you come into not-so-indirect contact with Africa and African labor, because most of the cocoa (chocolate), palm oil (soap) and a lot of the rubber come from West and Central Africa; a lot of the copper in your electrical wires, a lot of uranium for the atomic bomb, most of the gold come from Central and South Africa.[56]

By focusing on contemporary African trade relations, Robeson identified the phenomena of globalization and transnational economic flows long before they became familiar academic concepts. She pointed out that the wealth that contributed, in the last four centuries, to the emergence of Western European nations and the United States as world powers was partly the result of extracting material goods and human labor—notably slave labor—from Africa. Her critique of European exploitation of and dependence on Africa presages the argument that Walter Rodney explored in his seminal text on European colonialism in Africa.[57] Robeson also dispelled Eurocentric myths about the colonization of Africa that painted Europeans as the bearers of civilization, countering the European colonialists' claim that their presence benefited the region. She argued that exploitation, not development, was their motive, and stressed that Africans had resisted colonization "to the utmost of their ability."[58] Similarly, in an effort to decenter European hegemony, she noted, "This really isn't a White World at all. It is actually a predominantly Colored world with a White minority."[59] With these observations, Robeson reframed debates about race relations and civil rights.

In the 1940s, few Americans of any race had traveled in Africa and there was a growing interest in the region. Robeson became known for her knowledge of Africa and her insight into race relations. This was also a time when public discussion of African American civil rights was increasing. During the 1950s and early 1960s, Robeson wrote and spoke about these issues and also covered the United Nations and the peace movement. Throughout, she called on the anthropologist's trope of rendering the unusual understandable and worked from a Pan-Africanist commitment to African independence and African Diaspora interdependence.

Robeson's work demonstrates the ways an African American public intellectual drew on a Pan-Africanist perspective and developed a politics of diasporic identification. The resulting practices of representation sought to reconfigure the public image of black people and contribute to their ideological and political liberation. Reflecting on her time in Africa, she observed, "Finding out about my new relations, about my African brothers and sisters, has been a thrilling and gratifying adventure. It answered vague questions and fulfilled vague yearnings which had faced me as a young American."[60] This statement is a testament to the process of construction and the power of imagination that are a part of the formulation of identities. In particular, it indicates the ways "Africa" could be collapsed and manipulated—as much an evocative idea as a physical place. Europeans may have "invented Africa," but people of African descent living in the United States, the Caribbean, and Europe have also invented an Africa that is meaningful to their political and ideological projects.[61] While rooted in the historical facts of common geographical roots and similar historical experiences, the views that sustained early Pan-Africanists like Marcus Garvey and W. E. B. Du Bois and that inspire contemporary scholarly attention to the African Diaspora depend upon the elaboration of these imaginative constructions.

Sub-Saharan Africa, a region where blacks were the majority, has been a place that blacks who were marginalized in the West could imagine as a welcoming home. Although she did not frame her trip as "going back to Africa," Robeson understood herself to be making a pilgrimage to her "old country."[62] This was true even though she was traveling to South and East Africa, regions that were not involved in the transatlantic slave trade and to which black Americans do not have the ancestral connection that exists with West Africa. Nevertheless, because the areas she visited were part of black Africa and because they were subject to European rule, she felt a meaningful identification with the region. Her apparent delight in describing instances where she and her son were accepted as African is telling.[63] If she and Pauli had the experience of not being viewed as "really African" that other black American travelers report, she does not mention them. This process illustrates the passionate and not wholly rational feelings that are part of diasporic identification. The desire to connect and a longing for home can trump clear cultural differences, historical ruptures, and, indeed, the lack of actual historical connections. The motivations behind Eslanda and Paul's trip to Africa exemplify the seductive contradictions of Diaspora: one can have an investment in a place where one does not literally belong and one can return to a home where one has never lived.

Unlike many of the Harlem Renaissance artists who embraced Africa, Robeson did not dwell on an African past. Nor did she seek African survivals as her contemporary Melville Herskovits did. These historical ties may have been of interest to her, but for her purposes, the common present-day experiences of racism, exploitation, and marginalization were of greater significance. The Diaspora connections she formed and described to the reader were the consequence of her experiences of travel in the contemporary period and led her to emphasize contemporaneity and modernity. Arguably, this focus was facilitated by the fact that, in her travels, many of the Africans she met were local elites and many had been educated in Europe. Beyond shared continental ancestry, similarity in status and education provided an immediate sense of connection. Robeson portrayed the less-privileged Africans she met as sympathetic figures who were subjected to the whims of the white ruling class. Their economic situation was dire and she gave money to a number of the people she met. When discussing these encounters, she emphasized the dignity with which poor and working-class Africans accepted her donations; she did not, however, dwell on the class differences that so clearly separated her from them (and also from the working-class black Americans who could not afford to make their own African journeys).[64] Robeson negotiated the potentially treacherous differences in power and access through a stated and enacted political commitment to black unity and liberation that she hoped would bridge a vast gap.

Because Pan-Africanist activists confronted a context in which black people had not internalized a sense of connection across ethnic, religious, and cultural differences, a key part of their project was to produce a feeling of common cause among physically dispersed and culturally diverse people. In contrast to the Pan-Africanist emphasis on unity, much contemporary African Diaspora scholarship makes attention to diversity within the diaspora a central theme. These contemporary scholars are interrogating the ways differences of ethnicity, gender, class, and nationality undermine often idealized claims about race-based commonality. It is dissimilarity and misunderstanding, they argue, that are central to and constitutive of diaspora.[65] They remark on the coexistence of the desire for connections and the difficulties of achieving them, and their studies describe a condition of separation, displacement, dislocation, and movement.

Here the emphasis is on an analysis of the dialectics of similarity and difference, relations of connection and conflict, bridgeable and unbridgeable differences, gaps and translation, and varying degrees of power and powerlessness. In contrast, the Pan-Africanism of Robeson's era grew out of a politics based on roots, unity, and cohesion as activists sought ways to heal what they considered a painful rupture and separation. Most surprising to earlier generations of Pan-Africanists, perhaps, would be the fact that Africa is not at the center of contemporary African Diaspora scholarship; indeed, in some cases, it is marginal to, even absent from, the discussion.[66]

The distinct but related projects of Pan-Africanism and African Diaspora studies reflect and contribute to debates about the formation of black identities globally and the forms of intellectual and political activity that shape and are shaped by these formations. Robeson's work, like the work of contemporary Diaspora scholars, addresses the social, cultural, political, and intellectual project of creating ideological ties among Africans and people of African descent. Robeson's visit to Africa was a transformative personal journey and a path into politicized practices of representation. By wanting "to go to Africa," thinking in terms of Pan-Africanism, and seeking Diaspora connections, Robeson worked to effect the political and ideological changes that would promote greater equality for citizens of "the colored world."

NOTES

1. Eslanda G. Robeson, *African Journey* (New York: John Day Company, 1945), 13.
2. For biographical information used throughout the article, I have drawn from materials in Eslanda Robeson's Biographical Sketches File, Personal Papers Box; Eslanda G. Robeson Papers in the Paul and Eslanda Robeson Collection, Moorland-Spingarn Research Center, Howard University, Washington, DC; Pearl S. Buck and Eslanda Goode Robeson, *An American Argument* (New York: John Day Company, 1949); Martin Bauml Duberman, *Paul Robeson* (London: Pan Books, 1989); and Paul Robeson, Jr., *The Undiscovered Paul Robeson: An Artist's Journey, 1898–1939* (New York: John Wiley & Sons, 2001).
3. For example, see multiple authors, "Tribute to Eslanda Robeson," *Freedomways* 6, no. 4 (1966): 327–57; Faye V. Harrison and Ira E. Harrison, "Introduction: Anthropology, African Americans, and the Emancipation of a Subjugated Knowledge," in *African-American Pioneers in Anthropology*, ed. I. E. Harrison and F. V. Harrison (Urbana: University of Illinois Press. 1999); Barbara Ransby, "Eslanda Goode Robeson, Pan-Africanist," *Sage 3*, no. 2 (1986): 22–26; and Robert Shaffer, "Out of the Shadows: The Political Writings of Eslanda Goode Robeson," in *Paul Robeson: Essays on His Life and Legacy*, ed. J. Dorinson and W. Pencak (Jefferson, NC: McFarland & Co. 2002). Shaffer provides a concise overview of Robeson's writings and career as journalist and lecturer.
4. Examples include Jacqueline Nassy Brown, *Dropping Anchor, Setting Sail: Geographies of Race in Black Liverpool* (Princeton, NJ: Princeton University Press, 2005); Tina Campt, *Black Germans and the Politics of Race, Gender, and Memory in the Third Reich* (Ann Arbor: University of Michigan Press, 2004); Kamari Clarke, *Mapping Yoruba Networks: Power and Agency in the Making of Transnational Communities* (Durham, NC: Duke University Press, 2004); St. Clair Drake, *Black Folk Here and There: An Essay in History and Anthropology* (Los Angeles: Center for Afro-American Studies, UCLA, 1987); Brent Hayes Edwards, *The Practice of Diaspora: Literature, Translation, and the Rise of Black Internationalism* (Cambridge, MA: Harvard University Press, 2003); Paul Gilroy, *The Black Atlantic: Modernity and Double Consciousness* (Cambridge, MA: Harvard University Press, 1993); Faye V. Harrison, "Introduction: An African Diaspora Perspective for Urban Anthropology," *Urban Anthropology* 17, no. 2–3 (1988): 111–41; and Sidney J. Lemelle and Robin D. G. Kelley, eds., *Imagining Home: Class, Culture, and Nationalism in the African Diaspora* (London: Verso, 1994).
5. For discussions of the impact of Pan-Africanism on black scholars and activists in the United States, see St. Clair Drake, "Diaspora Studies and Pan-Africanism," in *Global Dimensions of the African Diaspora, Second Edition*, ed. J. E. Harris (Washington, DC: Howard University Press); Brent Hayes Edwards, "The Uses of Diaspora," *Social Text* 66, no. 1 (2001): 45–73; Joseph E. Harris, ed., *Global Dimensions of the African Diaspora*, 2nd ed. (Washington, DC: Howard University Press, 1993); Sidney Lemelle and Robin D. G. Kelley, "Imagining Home: Class, Culture, and Nationalism in the African Diaspora" (New York: Verso, 1994); Elliott P. Skinner, "Afro-Americans in Search of Africa: The Scholars' Dilemma," in *Transformation and Resiliency in Africa As Seen by Afro-American Scholars*, ed. P. T. Robinson and E. P. Skinner (Washington, DC: Howard University Press, 1983); and Penny M. Von Eschen, *Race Against Empire: Black Americans and Anticolonialism, 1937–1957* (Ithaca, NY: Cornell University Press, 1997).

6. Harrison and Harrison, "Anthropology, African Americans," 12; cf. St. Clair Drake, "Anthropology and the Black Experience," *The Black Scholar* (September–October 1980): 2, 10; Skinner, "Afro-Americans in Search of Africa," 6.

7. Mildred C. Ludecke to Eslanda Robeson, February 15, 1946, Eslanda Robeson Correspondence File, Eslanda G. Robeson Papers, Paul and Eslanda Robeson Collection.

8. Duberman, *Paul Robeson*, 293.

9. Hollis R. Lynch, *Black American Radicals and the Liberation of Africa: The Council on African Affairs, 1937–1955* (Ithaca, NY: Africana Studies and Research Center, Cornell University, 1978), 19.

10. Eslanda G. Robeson, *What Do the People of Africa Want?* (New York: Council on African Affairs, 1945); Lynch, *Black American Radicals*, 29, 34.

11. E. G. Robeson, *African Journey*, 13.

12. P. Robeson, Jr., *The Undiscovered*, 47.

13. Buck and Robeson, *American Argument*, 15.

14. P. Robeson, *The Undiscovered*, 47.

15. E. G. Robeson, "Black Paris," *Challenge: A Literary Quarterly* 1, no. 4 (1936): 12–18, and "Black Paris, II," *Challenge: A Literary Quarterly* 1, no. 5 (1936): 9–12.

16. E. G. Robeson to Carl and Fania Van Vechten, April 5, 1934, Carl Van Vechten Collection: Box Rj-Rz, Folder 1934–39. Yale Collection of American Literature, Beinecke Rare Book and Manuscript Library, Yale University, New Haven, CT. 116 Souls Summer 2006

17. Adam Kuper, *Anthropology and Anthropologists: The Modern British School*, 3rd ed. (London: Routledge, 1996). In addition to training a generation of anthropologists, Malinowski taught African Diaspora leaders like Ralph Bunche, who became a civil rights activist and Nobel Prize winner, and Jomo Kenyatta, who became the first prime minister of independent Kenya.

18. Talal Asad, ed., *Anthropology and the Colonial Encounter* (London: Ithaca Press, 1973).

19. E. G. Robeson, *African Journey*, 14.

20. Ibid., 15.

21. Ibid.

22. Ibid., 16.

23. James Clifford, "On Ethnographic Authority," in T*he Predicament of Culture: Twentieth-Century Ethnography, Literature, and Art* (Cambridge, MA: Harvard University Press, 1988).

24. James Clifford, "On Ethnographic Authority," 15.

25. Ibid., 15.

26. Zora Neale Hurston to Eslanda Robeson, April 18, 1934, Eslanda Robeson Correspondence File, Eslanda G. Robeson Papers, Paul and Eslanda Robeson Collection.

27. Zora Neale Hurston, *Mules and Men* (New York: Perennial Library, 1978).

28. For example, Harrison and Harrison, "Anthropology, African Americans," 23–25; Faye V. Harrison, "Black Women Putting Anthropology into Action from Corporate Academia to the United Nations: The Unexpected Consequences and Unfulfilled Expectations of a Reclamation Project." Paper presented at the Annual Meeting of the American Anthropological Association, 2002.

29. Apartheid became official South African policy in 1948, but the country was highly segregated at the time of Robeson's visit in 1936.

30. E. G. Robeson, *African Journey*, 65.

31. Ibid., 47–48.

32. For details about the filming, see Duberman, *Paul Robeson*, 178–82.

33. For discussion and critique of these tropes of the other in anthropology, see James Clifford and George E. Marcus, eds., *Writing Culture: The Poetics and Politics of Ethnography* (Berkeley: University of California Press, 1986); Johannes Fabian, *Time and the Other: How Anthropology Makes Its Object* (New York: Columbia University Press, 1983); Renato Rosaldo, *Culture and Truth: The Remaking of Social Analysis* (Boston: Beacon, 1993); and Mariana Torgovnick, *Gone Primitive: Savage Intellects, Modern Lives* (Chicago: University of Chicago Press, 1990).

34. For examples, see Asad, *Colonial Encounter*; Clifford, "Ethnographic Authority"; Fabian, *Time and the Other*; Akhil Gupta and James Ferguson, "Discipline and Practice: 'The Field' as Site, Method, and Location in Anthropology," in *Anthropological Locations: Boundaries and Grounds of a Field Science*, ed. A. Gupta and J. Ferguson (Berkeley: University of California Press, 1997); George E. Marcus and Michael M. J. Fischer, *Anthropology as Cultural Critique: An Experimental Moment in the Human Sciences* (Chicago: University of Chicago Press, 1986), and Rosaldo, *Culture and Truth*.

35. Lila Abu-Lughod, *Writing Women's Worlds* (Berkeley: University of California Press, 1993), 12–13.

36. Ibid., 7, 10–11.

37. Ibid., 7; cf. Rosaldo, *Culture and Truth*, 201.

38. E. G. Robeson, *African Journey*, 103.

39. Ibid., 93.

40. Deborah Poole, *Vision, Race, and Modernity: A Visual Economy of the Andean Image World* (Princeton, NJ: Princeton University Press, 1997), 15.

41. Ibid., 17.

42. E. G. Robeson, *African Journey*, 154, emphasis original.

43. Ibid., 109.

44. E. G. Robeson, *What do the People of Africa Want*, 15, emphasis original. *What Do the People of Africa Want?* (New York: Council on African Affairs, 1945).

45. Typescripts of these lectures with places and dates of lectures in Eslanda G. Robeson Papers: Box 12. Paul and Eslanda Robeson Collection.

46. E. G. Robeson, "Inter-cultural and Inter-racial Relations," April 7, 1944, 9. Typescript, Eslanda G. Robeson Papers: Box 10. Paul and Eslanda Robeson Collection.

47. Ibid.

48. Ibid., 9–10.

49. Ibid., 1.

50. Some representative examples include Ruth Benedict and Gene Weltfish, *The Races of Mankind. Public Affairs* Pamphlet 85 (New York: Public Affairs Committee, 1943); Melville J. Herskovits, *The Myth of the Negro Past* (1941; repr. Boston: Beacon, 1990); and M. F. Ashley Montagu, *Man's Most Dangerous Myth: The Fallacy of Race* (New York: Columbia University Press, 1942). For an analysis of the role American anthropologists played in the construction and deconstruction of the concept of race, see Lee D. Baker, *From Savage to Negro: Anthropology and the Construction of Race, 1896–1954* (Berkeley: University of California Press, 1998).

51. E. G. Robeson, "A Negro Looks at Africa," *Asia and the Americas* November (1944): 501.

52. Franz Boas, "The Outlook for the American Negro," in *A Franz Boas Reader: The Shaping of American Anthropology, 1883–1911*, ed. G. W. Stocking, Jr. (Chicago: University of Chicago Press, 1989); and Herskovits, *Myth of the Negro Past*.

53. E. G. Robeson, "A Negro Looks at Africa," 501.

54. Ibid., 502.

55. Ibid. Boas made these comments in a speech delivered to the 1906 graduating class of Atlanta University at the invitation of W. E. B. Du Bois; Boas, "The Outlook," 311.

56. E. G. Robeson, "Africa: Its Cultural Heritage and Present Problems," 1949, 10. Typescript, Eslanda G. Robeson Papers: Box 12. Paul and Eslanda Robeson Collection.

57. Gupta and Ferguson, "Discipline and Practice," 12–15.

58. E. G. Robeson, "Africa: Its Cultural Heritage," 10.

59. Walter Rodney, *How Europe Underdeveloped Africa* (Washington, DC: Howard University Press, 1974).

60. E. G. Robeson, "The Negro Pattern of World Affairs," 1947, 22–23, Typescript, Eslanda G. Robeson Papers: Box 12. Paul and Eslanda Robeson Collection.

61. E. G. Robeson, "Race Relations Lecture," 1944, 12. Typescript, Eslanda G. Robeson Papers: Box 10. Paul and Eslanda Robeson Collection.

62. E. G. Robeson, "A Negro Looks at Africa," 501.

63. V. Y. Mudimbe, *The Invention of Africa: Gnosis, Philosophy, and the Order of Knowledge* (Bloomington: Indiana University Press, 1988).

64. E. G. Robeson, *African Journey*, 13.

65. Ibid., 133.

66. See, for example, discussions in Brown, *Dropping Anchor*; Campt, *Black Germans*; and Edwards, *Practice of Diaspora*.

CHAPTER 10

Du Bois's Double Consciousness versus Latin American Exceptionalism

Joe Arroyo, Salsa, and Négritude

Mark Q. Sawyer

It has been tempting for observers to use W. E. B. Du Bois's ideas of double consciousness to contrast U.S. race relations with those in Brazil and Cuba. In those cases, Du Bois's theory becomes a paradigmatic representation of the inability of the United States to accept U.S. blacks' basic humanity, while the obvious patriotism of blacks in places like Cuba and Brazil becomes a clear sign that racial politics there differ in important ways that contradict Du Bois's proposition (Glasco, 1992). However, I will contend in this essay that both the positions of blacks in Latin America and the United States are more nuanced both on the ground and as they relate to Du Bois's theory of consciousness than most readings would allow. By not applying Du Bois to racial politics in Latin America we obscure the struggles of Afro-Latinos for political, cultural, social, and economic equality and misread the legacy of black U.S. politics. I propose that the elegance of the construction of double consciousness is capable of capturing the differing dimensions of inclusion and exclusion in Latin America *and* the United States. By invoking Du Bois's double consciousness in relation to Latin American racial politics, we are able to reconnect Latin America to the African Diaspora and build a bridge for critical engagement of black politics in Latin America. Consequently,

I also argue that denying the existence of double consciousness in Latin America obscures the struggles of African descended people in Latin America and leaves little possibility for understanding assertions of black identity and challenges to racial oppression and inequality.

While this chapter is motivated by a substantive interest in racial politics in Latin America, I hope to equally provide a framework to consider comparative racial politics and theoretical issues of comparison more broadly. I argue that one can apply what political scientists David Collier and James Mahon, Jr., call a "radial category" to understand racial politics comparatively. There are a set of core concepts that define racial inequality—especially antiblack racism. These core concepts are: a history of oppression and unequal incorporation in social, political, and economic life of the nation; a negative and limiting set of stereotypes that operate to define the group; formal legal and informal barriers to achievement; and an ideology that justifies the domination and oppression of the group. As Collier notes, radial categories have a set of core attributes to identify cases but are frequently delineated by the combination through which these concepts appear in any given case.[1] This paper argues that scholars have foreclosed opportunities for comparison by not understanding that racial inequality is a radial category that has several elements. These scholars discard opportunities for comparison by not engaging these core concepts. Thus, the absence of one part of the core concepts causes them to prematurely reject the category altogether. Further, Du Bois's theoretical conception of double consciousness defines a key idea in the antiracist struggle that crosses national boundaries. Thus, double consciousness applies not only to the U.S. context but also to the situation of African descended peoples in Latin America.

Blacks in Latin America are patriotic and a critical part of the national symbol.[2] At the same time, they have historically had unequal access to social, political and economic power. It is this duality that developed in a growing literature on race and Latin America and that can be described as "inclusionary discrimination."[3] Hanchard describes this when he explains, "African elements of Brazilian culture were selectively integrated into the discourses of national identity. With the ascendance of the ideologies of racial democracy and whitening, Afro-Brazilians came to be considered part of the cultural economy, in which their women and men embodied sexual desire and lascivious pleasure. At the same time, Afro-Brazilians were denied access to virtually all institutions of civil society that would have given them equal footing with the middle classes of modernizing Brazil."[4] While Hanchard is writing specifically about Brazil, similar processes occurred throughout Latin America. Despite miscegenation and a lack of formal segregation, a growing literature has pointed to manifestations of racial inequality in a variety of social structures in Latin America. However, it remains to be seen whether it is possible for comparisons to be drawn with the United States by using Du Bois's construction of double consciousness as a pivot point.

Du Bois defined double consciousness with this statement:

After the Egyptian and Indian, the Greek and Roman, the Teuton and Mongolian, the Negro is a sort of seventh son, born with a veil, and gifted with second-sight in this American world . . . a world which yields him no true self-consciousness, but only lets him see himself through the revelation of the other world. It is a peculiar sensation, this double-consciousness, this sense of always looking at one's self though the eyes of others, of measuring one's soul by the tape of a world that looks on in amused contempt and pity. One ever feels this twoness,—an American, a Negro; two souls, two thoughts, two unreconciled strivings; two warring ideals in one dark body, who dogged strength alone keeps it from being torn asunder.[5]

Du Bois, through the notion of double consciousness, grounded the struggle over the politics of representation and the unequal incorporation of blacks in the national project as well as in world systems. Yet Du Bois was not a separatist and saw room for a political system that recognized the specificity of the black experience while challenging, from a black perspective, injustice in national and world systems. The struggle over culture in Du Bois's theory is paramount and can be seen in his concern for black culture and organizations. As Manning Marable notes, "he also fully identified with the cultures, heritage, and political resistance of people of color throughout the Third World, particularly in the Caribbean and sub-Saharan Africa. His academic research focused largely on the cultural and political role of Africa in world civilization."[6]

Thus Du Bois's concept of double consciousness was in its inception an international construct that challenged the politics of representation. Representation was not confined to a brute political, but was also invoked in order to question the unequal inclusion of blacks, first as slaves or colonial subjects and later as subjects without basic human rights or the ability for social mobility or development. These ideas were applied to Latin America. Du Bois wrote specifically about Latin America and challenged the myth of racial democracy and miscegenation as the road to racial paradise in his work:

We cannot allow the West Indies and Central America to be made deliberate slums for the profit and vacation activities of the whites. In South America we have long pretended to see a possible solution in the gradual amalgamation of whites, Indians and Blacks. But this amalgamation does not envisage any decrease of power and prestige among whites as compared with Indians, Negroes, and mixed bloods; but rather an inclusion within the so called white group of a considerable infiltration of dark blood, while at the same time maintaining the social bar, economic exploitation and political disfranchisement of dark blood as such. We have thus the spectacle of Santo Domingo, Cuba, Puerto Rico and even Jamaica trying desperately and doggedly to be "white" in spite of the fact that the majority of the white group is of Negro or Indian descent. And despite facts, no Brazilian nor Venezuelan dare boast of his Black fathers. Thus, racial amalgamation in Latin-America does not always or even usually carry with it social uplift and planned effort to raise the mulatto and mestizos to freedom in a democratic polity.[7]

However, the challenge to Du Bois's ideas derive from a growing literature that, while acknowledging racial inequality in Latin America, seeks to challenge race as a frame for understanding social inequality there and discounts the use of race as a potential frame for social movements. The literature takes the lack of formal segregation to extend the idea that there are few similarities or possibilities for comparison between the United States and Latin America. Theoretically, the United States is used as paradigmatic of racial oppression, and Du Bois is used as the predominate intellectual voice of a specifically black American perspective.

By localizing Du Bois to the United States, the ability to compare and draw parallels with the United States and construct diasporic politics is attacked. As philosopher Anthony Bogues explains, the struggle of blacks has consistently been to construct a tradition that is recognized for its broadness and depth. However, he notes more often than not, "At best this tradition continues to be viewed as particularistic, mired in fossilized, irrational conceptions and myths not worthy of serious study."[8] Leaders in this literature include Laurence Glasco, Peter Fry, Alejandro de la Fuente, Pierre Bourdieu, and Loïc Wacquant. At issue has been the work of Michael Hanchard. His study of the Movmiento Negro Unificado (Unified Black Movement) in Brazil has been the center of a growing controversy about the study of race in Latin America. Peter Fry, Pierre Bourdieu, and Loïc Wacquant have attacked Hanchard's work as imposing an African American racial perspective on the politics of Brazil: "Instead of dissecting the constitution of the Brazilian ethnoracial order according to its own logic, such inquiries are most often content to replace wholesale the national myth of 'racial democracy' (as expressed for instance in the works of Gilberto Freire, e.g., 1978) by the myth according to which all societies are 'racist,' including those within which 'race' relations seem at first sight to be less distant and hostile."[9]

Bourdieu and Wacquant (1999) go on to attack the work of philanthropic organizations and other Non-Governmental Organizations (NGOs) that, they argue, have imposed an African American perspective on racial politics in Brazil. Specifically they say that U.S. foundations have wrongly forced affirmative action upon Brazil. Further, they argue that the modes and paradigms of U.S. racial politics and resistance stemming from the civil rights movement are being imposed on Brazilians and other blacks in Latin America by U.S. researchers and foundations. In his article, "Race, National Discourse, and Politics in Cuba: An Overview," Alejandro de la Fuente (1998) also expresses concern over the work of Aline Helg and Vera Kutzinski that follows a similar path to the work of Michael Hanchard. De la Fuente writes,

> Recent scholarship has stressed, however, that this foundational discourse, frequently referred to as "the myth of racial equality," was an ideological construction of the elite that masked the objective structural subordination of Afro-Cubans in society. These researchers recognize that, once established and accepted, these myths become, as Viotti da Costa (1985: 235) puts it, an integral part of social reality, but they tend to see their effects in only one direction: That of the subordination and demobilization of Blacks. This is an interpretation that minimizes the capacity of subordinate groups to appropriate these Inclusionary ideologies and use them to their advantage.[10]

While Hanchard, Helg, and Kutzinski use a Gramscian model to discuss the struggle to overcome the myth of racial democracy, a philosophy that denies the existence of racism, de la Fuente sees that myth not as a constraint for black activism but as a discourse that prevented further discrimination and allowed blacks to be incorporated into the national project. De la Fuente questions the emphasis on constraint and suggests that black movements for racial equality in Latin America are not necessary because of the ability of existing discourses and institutions to accommodate black concerns. While recognizing some racial oppression, these authors emphasize difference from the United States and in so doing attack the idea of race as a principle category for analysis or political struggle in Latin America.[11]

However, while de la Fuente has emphasized the inclusiveness of Latin American nations, historian Laurence Glasco (1992) has specifically taken up the issue of Du Bois. In his essay entitled "National Versus Racial Identity: Juan Gualberto Gomez of Cuba and W. E. B. Du Bois of the United States" he attempts to draw a contrast between the Cuban independence figure Juan Gualberto Gomez and Du Bois. Glasco asserts that, despite the challenging racial exclusion and sad realities of his day, he can still draw a contrast between Du Bois and Gomez and, in so doing, contrast the racial situations in Cuba and the United States. However, Gomez's sense of black organizations was similar to that of Du Bois: "Gomez also founded two race-oriented newspapers in Cuba. *La Fraternidad (Fraternity)* and *La Igualidad (Equality)*. Through their pages he helped to convince Black Cubans that the struggle for independence was intimately linked with that for abolition and Black advancement. He also worked to unite the island's colored societies, establishing the first Central Directory of the Societies of the Colored Race in order to protest discrimination and support the independence struggle."[12] Du Bois had a similar record. He helped found *The Crisis* and the NAACP, and he supported American national causes both for patriotic reasons and as instrumental avenues toward full citizenship for blacks. Similarly, in 1918, Du Bois wrote a call to African Americans to close ranks behind the U.S. war effort in World War I: "Let us, while this war lasts, forget our special grievances and close ranks with our own fellow white citizens and the allied nations that are fighting for democracy. We make no ordinary sacrifice, but we make it gladly and willingly with our eyes lifted to the hills."[13] His statement is directly parallel to Gomez's call. Thus, Glasco selectively reads the record or ignores these parallels in order to construct a difference between the United States and Cuba.

Glasco's exploration of Du Bois becomes a pivot to argue that race relations in Cuba were far gentler and did not require racial identification or organization among blacks. Like Bourdieu and Wacquant, he argues that any suggestion of comparison or similarity with the United States smacks of Black Nationalism and imposes an improper frame on Latin American racial politics. These authors also suggest that slavery in Cuba and Brazil were kinder and gentler than U.S. slavery. This is a topic de la Fuente takes up directly in his paper entitled, "Slave Law and Claims-Making in Cuba: The Tannenbaum Debate Revisited." De la Fuente concludes that "one need not romanticize the experience of slavery in the

Spanish colonies to realize that under Spanish law slaves, depending on their location in the productive structure and the specific phase of development of the slave system, were able to claim some rights and to create some avenues for advancement" (2003). Similarly in the postslavery era, Glasco comes to the conclusion that "Cuba was a country that discriminated, but whites were more accepting of blacks."[14] In drawing this conclusion, Glasco obscures points in his own narrative where Gomez himself employs what Du Bois describes as "second sight" to establish a critical position on the development of the Cuban national project. Glasco fails to understand that double consciousness does not imply separatism, but an uneven pattern of inclusion and exclusion that allow African Americans to complicate their relationship with the national project. Glasco ignores Du Bois's radical integrationism in *Souls of Black Folk* where he wrote:

> He would not Africanize America, for America has too much to teach the world and Africa. He would not bleach his Negro soul in a flood of white Americanism, for he knows that Negro blood has a message for the world. He simply wished to make it possible for a man to be both a Negro and an American, without being cursed and spit upon by his fellows, without having the doors of opportunity closed roughly in his face. This, then, is the end of his striving: to be a co-worker in the kingdom of culture, to escape both death and isolation, to husband and use his best powers and his latent genius.

It is exactly this form of complication that Gomez expresses and that also appears in the work of Nicolas Guillen. On black Cuban poet Nicolas Guillen, Glasco notes, "[he] explicitly rejected what he termed the North American 'road to Harlem', that of racial separation."[15] While Guillen is known for his admiration for *mestizaje* (or celebrating the harmonious mixed nature of the Cuban nation), his body of work and those of his contemporaries in the United States are far more complex. On one hand Glasco fails to understand that, for blacks in the United States during Du Bois's life, segregation was imposed and was not a choice. On the other hand, Glasco fails to note Guillen's stated admiration for Harlem Renaissance poet Langston Hughes and his profound respect for his ability to capture a certain image of blacks in the Diaspora. In fact, one of Guillen's poems, "The Ballad of My Two Grandfathers," shatters the myth of a single undifferentiated Cuba by emphasizing the specificity of the Afro-Cuban experience within the development of the Cuban nation. Guillen's words in the poem challenge notions of relativism and the harmonious incorporation of blacks into the Cuban national identity. Glasco fails to recognize the radical integrationist project within Du Bois' ideas. In short, for Glasco the predominant African American political stance is one of Black Nationalism and the Cuban stance is one of integrationism. This stands as an oversimplification of the politics of each group, both of which have pursued mixed strategies of integration and separation to achieve equality. This set of mixed strategies form a set of core concepts of black political organization and black ideology. Further, as previously noted, Glasco ignores Du Bois's call for African Americans to support U.S. intervention in World War I as a bridge to greater citizenship. Du Bois's call was for the exact same reasons that

Gomez supported black participation in the Cuban independence movement—it sought to advance citizenship claims through service to the nation in the military (see Christopher Parker, Helg, etc.). So what is gained by challenging the stark boundary posed by Glasco?

By emphasizing this day-to-day struggle over representation, we can draw critical similarities and differences between U.S. and Latin American racial politics. Du Bois offers a bridge to understanding the challenge of achieving recognition and overcoming oppression that faces subalterns. While the previous "elite" expressions of "double consciousness are clear, it is important that we also turn to everyday expressions. The struggle envisioned by Du Bois manifests itself in the contested terrain of popular culture and, more specifically, in popular music. It is in the genre of salsa music that, in the cases of Latin America and the United States, inclusion versus exclusion is not the central pivot of struggle but rather the struggle is over defining the terms of inclusion. Thus Afro-Latinos have strategically asserted black identity and the specificity of the black experience to critique racial and other forms of inequality in Latin American societies.

It is important to first establish the intellectual project of Du Bois and the similarity of his ideas with those of another important thinker from the African Diaspora, C. L. R. James. Each of their work seeks to disrupt accounts of national projects and the experience of racialized individuals by powerfully writing their experiences into the national project in a way that affirms both their humanity and their struggle. As Bogues (2003) notes, James and Du Bois sought to vindicate blackness by overcoming the erasure of black experiences in the writing of Western history and philosophy. He writes:

> What C.L.R. James and W.E.B. Du Bois did in their books *The Black Jacobins* (1938) and *Black Reconstruction* (1935) was to place squarely before us historical knowledge about two major events that reorder the narrative structures of Western radical history. Their projects created seismic shifts in twentieth-century radical historiography. One major development in twentieth-century historiography is the ways in which social theory became an integral part of historical understanding. . . . Both James and Du Bois were rewriting history from a stance in which experiences were not captures—or, if they were, they had been captured from a standpoint that the Black subject was a savage or, in the case of the Reconstruction, that Black male political equality led to the most corrupt and morally bankrupt state regimes in American history.[16]

Rewriting history also has its place in the contested terrain of popular culture in Latin America where these struggles can be witnessed. Turning specifically to salsa we can view the struggle over representations of blackness in history as it occurs in everyday cultural forms in Latin America. Thus, salsa music reflects the battle over representation described by Du Bois in his construction of double consciousness.

The salsa genre is one of the popular musical forms throughout Latin America. It is a form that was created out of an urban experience of Latinos (mostly Puerto Ricans) in New York but is based heavily on Afro-Cuban rhythms. It has become a Pan-Latin American musical genre connecting music with dance. The salsa movement, while connected to national projects, is also a part of a broader construction of Latin identity or *Latinidad*. That is, salsa music is simultaneously national and international in its formulation. There are differences of degree between Colombian salsa, Cuban salsa, and Puerto Rican salsa that sophisticated listeners can discern as well as different variations of salsa dancing connected to these subgenres. However, one thing that is ubiquitous within salsa music is the presence of *El Negro, La Negra*, or *La Mulata*. Salsa music, in its appropriation of African rhythms and fusion of more Western melodic structures, is frequently lauded as a living example of the tropical fusion of cultures. Salsa music is often posed as a manifestation of racial democracy because of its rhythms and themes of blackness within the lyrics. However, if we examine the role blacks have played in salsa music specifically and Latin music more generally, we find a series of problematic stereotypes.

The role these figures play in salsa music is most frequently as either sensual or playful icons. In this case, salsa music is not different from Latin musical genres that were its precursors (Moore 1997; Kutzinski 1993). In fact, there is a dominant focus on the black or mulatto female body (especially their hips and legs) that move to the music. Frances Aparicio argues that the focus on the black female body in the music obscures the existence of racial oppression. She writes:

> What is ironic in this synecdochal erasure is that it is precisely the Black woman's hips, her pelvis, and her genitalia, her vagina, that have been subjected historically to racism through rape and sexual violence. This displacement, then, is one of signifieds: by trivializing her hips *only* [*sic*] as a rhythmical and musical pleasurable entity, then Caribbean patriarchy can erase from the body of the mulata any traces of violence and racist practices for which it has been responsible throughout history.[17]

Salsa songs construct black bodies as objects of play and desire available for consumption. This consumption via the act of sexual intercourse and marked by the climax is frequently expressed using metaphors of food. The expression, "Ay que rico" in reference to black bodies and the moment of climax is common. Aparicio emphasizes that salsa music is a patriarchal form, but when it comes to this form of objectification the bodies of black men and women are cast in quite similar terms. In a related genre, merengue, the phenotypically white and blond Puerto Rican singer Giselle, in a song where she challenges her mother's scorn for her black lover, emphasizes that the source of her desire is, "Por que el tiene sabor el negro" (Because he has flavor, the black man). The reference to flavor again objectifies blackness and places the black male body, in this case, in a place of play, desire, and availability for consumption. These representations of blackness are frequently perceived to be positive and are used by black and white musicians alike. The presence of blackness is ubiquitous but can also be contested. Just as problematic representations of blackness in the American context are challenged by activists, academics, and popular culture figures, there are analogous challenges within the genre of salsa that cannot be read as impositions from African American forms of music, like soul or rap. Even in Du Bois's work on "Sorrow Songs" he highlights a similar politics of cultural production and its relationship to "authentic" representations of blacks in the United States.

Afro-Colombian salsa singer Joe Arroyo—like Du Bois and even more similar to James—attempts to rewrite the history of Colombian racial politics by recasting the image of La Negra in his song "Rebellion" (1999). The song parallels Du Bois's chapter in *Souls of Black Folks*, "Of the Coming of John," a tragic story of violence and revenge where a black man protects the virtue of a black woman. While Arroyo's story is different, there are striking similarities in the theme. The song, in contrast to other salsa music, places the black woman's body self-consciously as a site of violence, oppression, and contestation. Line by line, the song challenges notions of a unified Colombian history and foregrounds the history of oppression and struggle that mark the specific Afro-Colombian story. While the refrain, "No le pegue a la negra" (Don't hit the black woman), is patriarchal in that the black woman has no voice in the song and is protected by her husband, it is also a profound challenge to the presence of the black woman in salsa music. However, a close reading of Joe Arroyo's "Rebelión" is necessary to understand the strength of the challenge to representations of blackness in salsa music particularly and throughout Latin America more broadly.

The song begins with Arroyo speaking rather than singing. In his light tenor he explains that he is going to tell a story:

> Quiero contarle mi hermano un pedacito de la historia negra, de la historia nuestra, caballero.

[I want to tell you, my brother, a part of the black story, our story.]

It is important that he emphasizes that he is telling a *black* story. Through this the expression of Négritude is clear. Further, by emphasizing that it is *our* story Arroyo grasps a moment of empowerment for blacks to tell their own stories. Here he is expressing the vindicationism contained in the project. Arroyo goes on to set the scene in Cartegena, Colombia early in the development of the Colombian nation:

> En los anos mil seiscientos, cuando el tirano mando,
> las calles de Cartagena, aquella historia vivio.
>
> [In the 1600s, when tyranny ruled,
> in the streets of Cartegena, is where the story lived.]

In the next line Arroyo emphasizes the specificity of the black experience but disconnects it from the national experience by emphasizing blacks' Africanness. Further, by emphasizing that they were placed in chains, Arroyo highlights the specificity of the black experience. In the next few lines he expresses the contradiction of being glad to be on dry land, but experiencing the despair of their perpetual slavery:

Cuando alli llegaban esos negreros, africanos en cadenas,
besaban mi tierra, esclavitud perpetua
Esclavitud perpetua
Esclavitud perpetua

[When they arrived here, these blacks, Africans in chains
they kissed my land, perpetual slavery
Perpetual slavery
Perpetual slavery]

Later, Arroyo sets the stage for the powerful refrain by introducing a married couple and their Spanish master. Making the slave owner Spanish is a gesture that emphasizes his Europeanness in contrast to the Africanness of the enslaved couple. This binary is unlike the myth of racial democracy and constructions of *mestizaje*; Arroyo is emphasizing difference in his construction of the subjects of the song. The difference constructed by Arroyo culminates in violence not against the black man but against the black woman. Arroyo takes the more common representation of the black woman in salsa music and transforms her body into a site of overt political contestation and historical struggle. The slave owner, through his actions, has denied the couple any "normal" domestic relationship and in a sense has emasculated the black male by his actions:

Un matrimonio africano, esclavos de un español,
el les daba muy mal trato y a su negra le pego

An African marriage, slaves of a Spaniard,
he mistreated them, and hit the black woman.

In the next lines Arroyo's use of the term "guapo" to describe the black man carries several meanings. On one hand, *guapo* means handsome, emphasizing his sexuality. Other meanings for *guapo* include nice, tough, and brave. Arroyo casts the black man as a champion, a hero. From there, the hero then takes vengeance for his aggrieved wife while invoking the refrain, "Don't hit my black woman." From this point the chorus joins in, interrupting Arroyo's singular voice, and expressing a more general community voice, "Don't hit *THE* black woman." Here the men of the community are responding and the slave revolt has begun:

Y fue alli, se revelo el negro guapo, tomo
venganza por su amor y aun se escucha
en la verja, no le pegue a mi negra.
No le pegue a la negra.
No le pegue a la negra.
Oye man!!
No le pegue a la negra.

[And went there, and revealed himself to him, the handsome black man, and took vengeance for his love. And you can hear

at the gate. Don't hit my black woman.
Don't hit the black woman. Don't hit the black woman.
Listen, man!!
Don't hit the black woman.]

By examining this verse we see clearly that Arroyo seeks to disrupt versions of Colombian history specifically, and Latin American history more broadly, that marginalize the experience of blacks or obscure the specificity of black oppression. At the same time, he does it within a specific Latin American music genre that is known for its representation of the myth of racial democracy and mestizaje.

Challenging the usual representations of blackness within the genre defies critics who argue that expressions of black identity and resistance are imported from the African American experience. Arroyo articulates his point of view as an extremely popular and recognized Colombian artist with an international audience. His challenge expresses the existence of inclusionary discrimination in the music and in Latin American societies. Arroyo's authenticity as one of the leading figures in Colombian salsa suggests that expressions of black resistance are not contaminations from the U.S. black experience but fractal forms of response to oppression and culture as described by Paul Gilroy in his *Black Atlantic*. Arroyo invokes a second sight with regard to Colombian history and emphasizes difference, violence, and specificity of black oppression in the Americas. Salsa artists from Colombia of Arroyo's era self-consciously engaged the themes and traditions of soul music. Far from simply echoing their themes, they sought to address the politics that surround questions of race. Arroyo's expressions reflect challenges to representations of racial democracy within other forms of music like Spanish rap and the Puerto Rican rap-reggae fusion called "reggaeton," as well as the work of Brazilian groups like Olodum. These expressions are not entirely disconnected from emergent and continued social movements in places like Colombia, Brazil, Honduras, and Ecuador around issues of land rights, discrimination and other challenges that face populations of Afro-descendents. Critical to all of these struggles is a conception of black identity that invokes twoness or challenges a portrait of a harmonious national project. For example, Arroyo, in his work, not only delineates warring ideals but also gives voice to a violent struggle for liberation. Du Bois's work provides a useful lens to understand these challenges and offers the opportunity for creative comparisons across national borders.

Engaging Du Bois opens up critical possibilities to understand these cultural expressions and movements as they relate to national, subnational, and transnational racial projects and counternarratives. Those who obscure the relevance of Du Bois, in contrast, attempt to render expressions of blackness parochial and limited to a specific national milieu. Thus, we have no ability to discuss transnational black politics or cross national comparisons. Further, in emphasizing difference with the United States, these discourses mistakenly minimize the experiences of oppression of blacks in Latin America and deny the authenticity of their struggles for cultural, political, social, and economic rights and recognition. By not understanding the many components of the radial category of racial inequality, these authors fail to engage critical similarities between the struggles of blacks throughout the Americas. Ultimately, critical engagement with Du Bois illuminates voices in Latin America that are obscured by emphasizing differences with the United States. If we are to understand these movements and expressions beyond theories of "contamination," we must be guided by theorists like Du Bois and James as international intellectuals. In that sense, we are doing justice to their understanding of their own critical projects. We need theoretical guides for such comparison and we cannot a priori reject comparisons or concepts, because we can identify differences in our cases. Making concepts travel is a fundamental goal of social science. By identifying the components of racial oppression and resistance we can develop important modes of comparison.

NOTES

1. F. Aparacio, *Gender, Latin Popular Music, and Puerto Rican Cultures (Music/Culture)* (New York: Wesleyan University Press, 1998). Anthony Bogues, *Black Heretics, Black Prophets: Radical Political Intellectuals* (New York: Routledge); P. Bourdieu and L. Wacquant,"On the Cunning of Imperialist Reason," *Theory, Culture and Society* 16 (1999): 41–58.

2. D. Collier and James E. Mahon, Jr., "Conceptual Stretching Revisited: Adapting Categories in Comparative Analysis." *The American Political Science Review* 87, no. 4 (1993): 845–55.

3. M. Dawson, *Black Visions: The Roots of Contemporary African American Political Ideologies* (Chicago: University of Chicago Press, 2001); A. de la Fuente, "Race, National, Discourse and Politics in Cuba," *Latin American Perspectives* 25, no. 3 (1998), 43–70; de la Fuente"National Versus Racial Identity: Juan Gualberto Gomez of Cuba and W. E. B. Du Bois of the United States," unpublished manuscript, 2003.

4. W. E. B. Du Bois, *The Souls of Black Folks* (New York: Modern Library, 2003); Nadine Fernandez, "Race, Romance, and Revolution: The Cultural Politics of Interracial Encounters in Cuba." PhD Dissertation-University of California, Berkeley, 1996; P. Gilroy, *The Black Atlantic: Modernity and Double Consciousness* (Cambridge: Harvard University Press, 1994).

5. L. Glasco, "National Versus Racial Identity: Juan Gualberto Gomez of Cuba and W. E. B. Du Bois of the United States," in *Nord und Süd in Amerika, Vol. 1,* ed. Wolfgang Reinhard and Peter Waldmann, 410–84 (Berlin: Rombach, 1992).

6. M. G. Hanchard, *Orpheus and Power: The Movmiento Negro of Rio de Janeiro and Sao Paulo Brazil, 1945–1988* (Princeton, NJ: Princeton University Press, 1994).

7. M. G. Hanchard, "Black Cinderella?: Race and Public Sphere in Brazil," *The Black Public Sphere: A Public Culture Book* (Chicago: University of Chicago Press, 1995).

8. D. Hellwig, *African American Reflections on Brazil's Racial Paradise* (Philadelphia: Temple University Press, 1992).

9. A. Helg, *Our Rightful Share: The Afro-Cuban Struggle for Equality, 1886–1912* (Chapel Hill: University of North Carolina Press, 1995).

10. V. Kutzinski, "Sugar's Secrets: Race and the Erotics of Cuban Nationalism (New World Studies)", in *The Black Public Sphere Collective,* 169–78 (Charlottesville: University of Virginia Press, 1993).

11. Mara Loveman, "Comment: Is 'Race' Essential?" *American Sociological Review* 64, no. 6 (1999): 891–98.

12. Manning Marable, *Black Leadership.* (New York: Columbia University Press, 1998).

13. A. W. Marx, *Making Race and Nation: A Comparison of the United States, South Africa and Brazil*, 1st ed. (New York: Cambridge University Press, 1998).

14. R. D. Moore, *Nationalizing Blackness: Afrocubanismo and Artistic Revolution in Havana, 1920–1940* (Pitt Latin American Series) (Pittsburgh, PA: University of Pittsburgh Press, 1997); Melissa Nobles, *Shades of Citizenship: Race and the Census in Modern Politics* (Stanford: Stanford University Press, 2000).

15. L. Glasco, "National versus Racial Identity: Juan Gualberto Gomez of Cuba and W. E. B. Du Bois of the United States," in *Nord und Süd in Amerika, Vol. 1*, ed. Wolfgang Reinhard and Peter Waldmann, 410–84 (Berlin: Rombach, 1992).

16. Christopher Parker, *Fighting for Democracy: Race, Service to the State and Insurgency in the Jim Crow South* (unpublished manuscript, 2005).

17. J. Sidanius, Y. Peña, and M. Sawyer, "Inclusionary Discrimination: Pigmentocracy and Patriotism in the Dominican Republic," *Political Psychology* 22 (2001): 827–51.

CHAPTER 11

"LONG LIVE THIRD WORLD UNITY! LONG LIVE INTERNATIONALISM"*

HUEY P. NEWTON'S REVOLUTIONARY INTERCOMMUNALISM

BESENIA RODRIGUEZ

Whenever death may surprise us, it will be welcome, provided that . . . our battle cry reaches some receptive ear, that another hand stretch out to take up weapons. . . . Let the flag under which we fight represent the sacred cause of redeeming humanity, so that to die under the flag of Vietnam, of Venezuela, of Guatemala, of Laos, of Guinea, of Colombia, of Bolivia, of Brazil, to name only a few of the scenes of today's armed struggles, be equally glorious and desirable for an American, an Asian, an African, or even a European.

—Ernesto Guevara, cited by Huey P. Newton[1]

It is our goal to be in every single country there is. We look at a world without any boundary lines. We don't consider ourselves basically American. We are multi-national; and when we approach a government that doesn't like the United States, we always say, "Who do you like; Britain, Germany? We carry a lot of flags."

—Robert Stevenson, Executive President of Ford,
Business Week, cited by Huey Newton[2]

---

* Yuri Nakahara Kochiyama, *Passing It On—A Memoir*, ed. Marjorie Lee, Akemi Kochiyama-Sardinha, and Audee Kochiyama-Holman (Los Angeles: UCLA Asian American Studies Center Press, 2004), 173.

The first thing the American power structure doesn't want any Negroes to start is thinking internationally.

—Malcolm X[3]

IN 1992, filmmaker Spike Lee released *Malcolm X* to critical acclaim.[4] The film had an enormous cultural impact, with celebrities and youth donning hats and T-shirts with the letter "X."[5] Soon after, Gerald Horne's article, "Myth and the Making of Malcolm X," presented a critique of Lee's rendering, arguing that the film participated in constructing a mythology of figures like Malcolm X that "neglect[s] highly relevant and persuasive evidence because it does not necessarily comport with the contemporary lessons that one is to draw from these myths."[6] Lee's film, an intervention in a historiographical and popular record of the black freedom struggle that has "centered on Martin Luther King, Jr., with Rosa Parks and the Student Nonviolent Coordinating Committee (SNCC) playing pivotal supporting roles,"[7] creates an alternative mythology that, like the prevailing "King years" myth, fails to include the concurrent history of leftists like Claudia Jones, William Patterson, and Paul Robeson. Glaringly absent from Lee's film, Horne contends, is Malcolm's "embrace [of] a more progressive form of nationalism— when he followed in Patterson's footsteps seeking to take the Black Question to the United Nations." Also absent were his disputes with Louis Farrakhan, his relationships with "world leaders—especially on the African continent," and his early interpretations of the 1955 Bandung conference of African and Asian leaders "as spelling doom for the 'white devils.'"[8]

Readings of Malcolm, as Horne argues, have been "heavily influenced by a pervasive narrow nationalism" and, like the King myth, have largely failed to "encompass the international dimension."[9] "To do so," he asserts, "might necessitate encompassing a now taboo history involving the Council on African Affairs and the black Left."[10] Similarly, Farrakhan is completely absent from *Malcolm X*, Horne contends, because including him "would have been too difficult to explain and would have disturbed a major myth."[11]

Like his meeting with Fidel Castro in Harlem's Hotel Theresa in 1960, Malcolm's close relationship with Yuri Kochiyama was absent from *Malcolm X*.[12] This omission is particularly misleading because of what this friendship illustrates about Malcolm's allegiances and his ever-developing analysis of racism. The two met briefly in 1963 at the Brooklyn courthouse, where a timid Kochiyama told Malcolm that she admired his work but disagreed with his stance against integration. He invited her to his 125th Street office but was soon silenced by Elijah Muhammad for his infamous remarks following Kennedy's assassination. Three months later, Kochiyama invited Malcolm to a reception at her home for three atomic bomb survivors of the Hiroshima-Nagasaki World Peace Mission who were on a global tour speaking against nuclear arms proliferation.

More than anyone else, they wanted to meet Malcolm X. Although she never received a response, on June 6, 1964, Kochiyama organized a reception for Malcolm with the assistance of the Harlem Parents Committee.

Hopeful, "the Kochiyama family waited excitedly" in their Harlem apartment for Malcolm X to arrive.[13] Soon after, Malcolm arrived with three security guards; he apologized for not having responded to Kochiyama's letters and promised to write to her should he travel again. He thanked the guests for touring the "World's Worst Fair"—that is, Harlem, itself "scarred . . . by the bomb of racism," instead of the World's Fair in Queens.[14] He spoke of Europe's colonization of Asia and the similarities between African and Asian history, which he studied while in prison, and of his admiration for Mao Tse-tung for fighting feudalism and imperialism. An astute observer of international events, he spoke of the war in Vietnam two months before President Johnson ordered that additional military advisors be sent to Vietnam and nine months before the first U.S. combat troops landed. "If America sends troops you progressives should protest," he told them, insisting that "the struggle of Vietnam is the struggle of the whole Third World: the struggle against colonialism, neo-colonialism, and imperialism."[15] He kept his promise to Kochiyama, writing her eleven times from nine different countries, including Egypt, England, Saudi Arabia, Kuwait, Ethiopia, Kenya, and Nigeria before his assassination in early 1965.[16] In turn, Kochiyama joined his Organization of Afro-American Unity (OAAU). "Without question," her daughter and granddaughter wrote in the preface to her autobiography forty years later, "Malcolm X was and still is the single most influential person in Yuri's political development." Kochiyama herself has called Malcolm her "political awakening."[17]

Had Kochiyama been represented in the film, cradling Malcolm's head as his aides attempted to revive him, as in the photograph made famous by *Life* magazine, the image would have been largely unintelligible to an audience inundated by a mythology in which figures equated with "Black Power" like Malcolm and the Black Panther Party (BPP), are understood uncomplicatedly as "Black nationalists."[18] This mythology renders invisible aspects of their ideologies that fail to conform to existing frameworks. The BPP is often understood as the ideological heir to Malcolm X and explicitly claimed this tradition for itself. In revisiting Malcolm's deeply antiracist concept of world revolution, influenced by the Conference of Nonaligned Nations in Bandung, Indonesia, and the movements that emerged in its wake, this article examines the BPP's debt to Malcolm X through an alternative lens. It focuses on the construction of a tricontinental ideology and a set of solidarities beyond race and nation that belie the traditional interpretations of figures canonized as "Black Nationalists" or "Pan-Africanists," highlighting the extent to which these figures, at times explicitly, rejected these political and ideological categories.[19]

The Black Panthers espoused a critique that linked their own oppression with that of other racialized and oppressed groups in the United States and linked oppression at home with imperialism throughout the Third World. They promoted a variant of tricontinental socialism closely aligned with that of Castro and Mao Tse-tung and set up community programs throughout the nation to counteract the detrimental effects of racial capitalism. The BPP was influenced as much by the writings of Mao, Ho Chi Minh, Amilcar Cabral, and the events unfolding in Cuba, Vietnam, China, and Palestine as it was by the ideologies of Malcolm X and Robert F. Williams and events in Oakland, Detroit, and New York. Their shifting philosophies and internal ideological conflicts provide a valuable opportunity to examine the inherent multiplicity and instability of any discourse. For instance, they evolved from self-described "Black Nationalists" to "revolutionary nationalists," "internationalists," and finally, to "revolutionary intercommunalists" between 1966 and 1972.[20] Panthers have written extensively about the periods they spent traveling the globe and in exile; nonetheless this article limits its focus to the BPP's tricontinentalism, owing a considerable debt to Malcolm X as it did, through the lens of its chief theoretician, Huey P. Newton.[21]

The concept of the "Third World" has largely been associated with the period during which the Panthers emerged, borne of the 1955 conference in Bandung and the anticolonial movements in Asia, Africa, and Latin America of the 1960s and 1970s. Nonetheless a central feature of the erasure of this tricontinental tradition has been the oversight of the direct and indirect impact of two events one year apart on Malcolm X and the Black Panthers, the Bandung Conference and the Suez Crisis: "And once you study what happened at the Bandung conference . . . it actually serves as a model for the same procedure you and I can use to get our problems solved. At Bandung . . . there were dark nations from Africa and Asia. . . . Despite their economic and political differences, they came together. All of them were black, brown, red, or yellow."[22]

On November 10, 1964, at the Northern Grass Roots Leadership Conference in Detroit organized by Grace Lee Boggs, Malcolm X gave what would become one of his most celebrated speeches, "A Message to the Grassroots." In a fierce censure of the civil rights establishment, which he referred to as "house Negroes," Malcolm spoke of the March on Washington as a "picnic" and a "circus." In sharp contrast, he cited the Kenyan, Chinese, Algerian, and Cuban revolutions, which have been "bloody . . . hostile . . . [and which knew] no compromise."[23] The international focus of Malcolm's speech, which emphasized the need for political unity borne of a shared oppression, was all the more extraordinary given that much of the black press and leadership was focused on the events taking place in the U.S. South, including the passing of the Civil Rights Act and the murder of three civil-rights workers in Mississippi.[24]

In the wake of the proliferation of nuclear weapons and the cold war battle for the markets of the newly independent Asian and African nations, Indonesian Prime Minister Ali Sastroamidjojo delivered a speech in August of 1953 reflecting his view that "strong cooperation between [Asian and African] countries will strengthen the efforts in creating peace in the world." A few months later at a meeting later known as the Colombo Conference, leaders of four other Asian countries met in response to Sastroamidjojo's call and to discuss ways to address those problems created by racial colonialism and left unaddressed by the newly formed United Nations. This conference gave way to the establishment of the Asian-African Conference to be hosted by Indonesia and co-sponsored by India,

Ceylon (later Sri Lanka), Burma, and Pakistan. Among other aims, the Asian-African Conference sought to "discuss the matters, particularly related to Asian-African nations, for example matters related to national sovereignty, racialism, and colonialism" and "to observe the position of Asia and Africa, and their nations in the world, to observe what they can give to promote peace and cooperation in the world."[25]

The Indonesian government invited twenty-five countries, including the People's Republic of China, Egypt, Ethiopia, the then-Gold Coast, Syria, Iran, Lebanon, Libya, and Northern and Southern Vietnam, to the mountain city of Bandung April 16–24, 1955. Representatives of South Africa's African National Congress (ANC) also attended. In his opening speech, "Let a New Asia and New Africa Be Born," Indonesian President Sukarno highlighted themes that would become a central feature of tricontinentalism. He told the participants, who represented over half of the world's population from different nations, "social, cultural, religion, political background, and even different skin color," that they could be united by their experiences of colonialism and "by the same devotion to defend and strengthen the world peace." "The nations of Asia and Africa," he declared, "are no longer the tools and playthings of forces they cannot influence." He ended, "I hope [this conference] will give evidence of the fact that we, Asian and African leaders, understand that Asia and Africa can prosper only when they are united, and that even the safety of the world at large can not be safeguarded without a united Asia-Africa. . . . I hope that it will give evidence . . . that a New Asia and New Africa have been born!"[26]

The conference resulted in the "Ten Principles of Bandung," which affirmed the nations' commitment to promoting "respect for fundamental human rights, [for] . . . justice, [and] for the sovereignty and territorial integrity of all nations" as well as the "recognition of the equality of all races and of the equality of all nations large and small" and a call for all the signatories to "refrain from acts or threats of aggression" against any other territory.[27] Condemning "colonialism in all of its manifestations," including U.S. and Soviet neo-colonialism, Indian Prime Minister Jawaharlal Nehru declared, "If I join any of these big groups I lose my identity."[28] The conference's Final Communiqué emphasized the need for tricontinental nations to "loosen their economic dependence on the leading industrialized nations by providing technical assistance to one another through the exchange of experts and technical assistance for developmental projects, as well as the exchange of technological know-how and the establishment of regional training and research institutes."[29]

Fifty-years later, Indonesian and South African leaders issued the following statement: "following the historic Asian-African Conference in Bandung in 1955, the world has witnessed the emergence of new nations in the two continents and the births of a sense of kinship and solidarity between them. This 'Bandung Spirit,' brought about by the conference, subsequently became the underlying inspiration for these new nations to continue to strive towards the attainment of a just, peaceful, progressive and prosperous world order."[30] Bandung, and the

Nonaligned Movement it spawned six years later at its founding summit in Belgrade, constituted "the major political expression of the developing countries of the Third World." Nonaligned nations developed increasingly audacious condemnations of racism, imperialism, and neo-colonialism, emboldened by charter members like Yugoslavia, Cuba, and Egypt, each of whom hosted major conferences. For instance, in Belgrade, Cuban leader Osvaldo Dorticos charged, "If we dare to condemn the colonialist domination of Algiers and Angola, let us also condemn the colonialist domination of Puerto Rico," raising the ire of an already antagonistic United States.[31]

Harlem watched the events unfolding in Bandung with anticipation. Representative Adam Clayton Powell and novelist Richard Wright were in attendance, along with Margaret Cartwright (the first black reporter assigned to the UN) and journalist William Worthy; messages were read from Paul Robeson and W. E. B. Du Bois, whose passports were confiscated by the State department. Robeson's newspaper, *Freedom*, published articles on the conference, talking with Harlem residents on the significance of the declarations and solidarities forged in Bandung for their lives. In an article entitled, "War and Jim Crow Set Back at Bandung," journalist Kumar Goshal noted that, "since the conference laid great stress on political and economic imperialism and the evils of racialism, it was of supreme importance to all Americans, especially Negro Americans."[32] Goshal continued, "for as India's Nehru emphasized in his speech, it is the U.S-sponsored North Atlantic Trade Organization (NATO), the Southeast Asia Treat Organization (SEATO), and other military alliances that are today helping maintain what remains of imperialist rule in Asian and African colonies." He noted that "by and large, the Negro press in America was fully conscious of the importance and significance of Bandung. The Afro-American felt that the conference 'signaled the end of a centuries-old era of colonialism,' and asked 'Western statesmen . . . to heed this solemn warning that the old order of things no longer exists.'"[33]

In "Harlem Speaks . . . About Bandung," a *Freedom* journalist "took [a] camera and notebook to one of Harlem's busiest streets, to ask Mr. and Miss average Harlemite what they thought about the doings at . . . the historic Bandung [Conference]."[34] To varying degrees, the "average Harlemite," a homemaker, a photographer, a high school student, and a veteran, agreed that the conference would have a "good effect" on U.S. blacks, indicating that perhaps for many, the events in Bandung had a diffuse impact; yet the response from leading activist-intellectuals speaks to the centrality of the Bandung Conference for the development of a broader tricontinental imaginary.[35]

As cheers rang from the audience at Bandung, one participant recited Robeson's speech.[36] He expressed his "profound conviction that the very fact of the convening of the Conference . . . will be recorded as an historic turning point in all world affairs," stating that the gathering opened "a new vista of human advancement in all spheres of life."[37] In a resounding statement, Robeson declared, "the time has come when the colored peoples of the world will no longer allow the great natural wealth of their countries to be exploited and expropriated by the Western world while they are beset by hunger, disease and poverty."[38] He then discussed his "deep and abiding interest in the cultural relations of Asia and Africa" and lamented not being able to see his "brothers from Africa, India, China, Indonesia, and from all the nations represented at Bandung . . . old friends I knew in London years ago, where I first became part of the movement for colonial freedom—the many friends from India and Africa and the West Indies with whom I shared hopes and dreams of a new day for the oppressed colored peoples of the world."[39] "Fully endors[ing] the objectives of the Conference to prevent . . . [another H-bomb and] . . . the demand of Africa and Asia for independence from alien domination and exploitation," Robeson attempted to speak in behalf of U.S. and Caribbean blacks. He wrote, "Typical of my people's sentiments are these words from one of our leading Negro newspapers: 'Negro Americans should be interested in the proceedings at Bandung. We have found this kind of fight for more than 300 years and have a vested interest in the outcome.'"[40]

Two years later, Robeson, Du Bois, former Council on African Affairs secretary W. Alphaeus Hunton, and labor leader Jose' Santiago spoke at a Harlem event commemorating the anniversary of the Bandung Conference.[41] In his speech, Robeson acknowledged, "we American Negroes can no longer lead the colored peoples of the world because they far better than we understand what is happening in the world today. But we can try to catch up with them."[42] Once again challenging his audience to broaden its antiracism to include internationalism, anti-imperialism, and anti-capitalism, he urged them to "learn about China and India and the vast realm of Indonesia rescued from Holland . . . [and] of the new ferment in East, West, and South Africa. We can realize by reading . . . how socialism is expanding over the modern world and penetrating the colored world."[43]

From the Bandung Conference emerged the concept of the "Third World." As the antiracist labor leader James Boggs would note in 1974, it was not until after the 1939–45 war that the "people living in Latin America, Africa, Asia, and the Middle East . . . [began] to struggle as a social force to determine their own destiny and the future of the world. . . . There was no concept of Third World Peoples until approximately twenty years ago . . . created on the basis of the real historical struggles of the colonial peoples in Africa, Asia, Latin America and the Middle East."[44] He told his audience, "As independence was being fought for by many countries and won by a few, a conference was held in Bandung, Indonesia. . . . Out of the Bandung and subsequent conferences of Asia, Africa, Arab, and Latin American nations, came the concept of the Third World, referring to those countries struggling for independence, . . . which had been systematically damned into a state of underdevelopment by capitalism."[45] In the United States, Boggs observed, "the identification of black Americans with the Third World was really a question of solidarity, i.e., the feeling that progressive and revolutionary people all over the world should have for each other in their common struggle for the advancement of humanity."[46] Yet he criticized those who, "in order to express the fact that as non-whites they have all experienced discrimination and segregation in the United States, . . . adopted the phrase 'Third World' to describe themselves, even though this phrase and this concept came from a totally different set of historical circumstances."[47]

Four years earlier Yuri Kochiyama's article "Third World," printed in the *Asian Americans for Action*'s newsletter, provided an example of this tendency among U.S. radicals. "Whether in the Black, Brown, Red, or Yellow movement, the term Third World makes its appearance on flyers; is hurled about in speeches, pops up in raps; is unfurled on banners and carried on posters."[48] Kochiyama continued, "This world, Africa, Asia, and Latin America, is the world of the Black, Brown, and Yellow people seeking independence of the first two worlds." In a resounding statement illustrating the tricontinentalism that evolved from her Malcolm-inspired political awareness, she added,

> the cause of Palestine is that of Vietnam's; the liberation struggles of Mozambique and Angola are related to the guerrillas' involvement in Uruguay and Bolivia. The Philippines' sentiment against military colonialism is mutual with that of Puerto Rico's. The black man's [sic] degrading experience in Amerika corresponds with the treatment of the Indians, Eskimos, and Chicanos in both Amerikas, and the Aborigines in Australia. The uprisings in the Caribbean . . . and the demands of the Okinawans are all the voices of the Third World, the world's oppressed. . . . The Third World and its descendents in the western hemisphere are in motion. It is a life-and-death struggle to cast off the shackles of imperialism and colonialism, and obliterate racism. . . . The Third World must oppose, challenge, confront and halt transgression of imperialistically-inclined powers, including those in their own world, for imperialism knows no color or geographic lines. The Third World must offer an alternative—a more humane way of life, where diversity of peoples, cultures, religions, and ideologies will enhance civilization, rather than proscribe life.[49]

The second event, Gamal Abdel Nasser's nationalization of the Suez Canal Company in 1956, did not receive as much coverage in the U.S. black press as the Bandung Conference had, yet Nasser's "successful weathering of an invasion . . . made him a hero" not just for the "Bandung" or "Third World" nations but also "among many African Americans."[50] Over five thousand miles away, Du Bois wrote an acerbic poem entitled "Suez." In it Du Bois presented a caustic critique of the United States, the United Kingdom, and Israel—which he had supported in the past. He praised Nasser for rising in the face of "lions" with their "pockets full of gold" and declaring "what's mine is mine!"[51]

No U.S. black activist-intellectual had a more intimate connection with the Middle East or its postcolonial hero, Nasser, than Shirley Graham Du Bois.[52] Her book, written within the framework of Egyptian history, *Gamal Abdel Nasser: Son of the Nile*, was the first biography on Nasser to be published in the English language. Graham Du Bois discussed the Bandung Conference, criticizing the United States and Europe, including John Dulles's threats toward Egypt if it participated in the conference with "communist countries."[53] Nasser's comments at Bandung, pledging Egypt's commitment to "the welfare of peoples beyond our borders" and its "support . . . of self-determination for all peoples," led to his hailing as "Champion of Africa and Asia."[54] As Graham Du Bois wrote, "there was no doubt that [Nasser] emerged from the Bandung Conference a hero in the eyes of his own people and in the Middle East as a whole."[55] He emerged as "the most important leader of the Arab world and as one of the major figures of the non-aligned movement." Along with "Cuba's Castro and Ghana's Kwame Nkrumah," notes historian Melani McAlister, "Nasser represented an emotionally explosive convergence of anticolonial defiance and postcolonial global consciousness."[56]

In May of 1956 Nasser further defied Eisenhower by announcing that diplomatic relations had been established with the People's Republic of China.[57] The United States and United Kingdom reneged on their loans for the building of a High Dam at Aswan.

Two months later, Nasser announced, "We Egyptians will not permit any imperialists or oppression to rule us, militarily or economically. We will not submit to dollar dictatorship."[58] He vowed that the dam would be completed without Western assistance. On July 26, 1956, the anniversary of King Farouk's abdication, Nasser told a "wildly cheering crowd: 'Today we take over our Canal.'" The Suez Canal would be nationalized and managed by the Egyptian Canal Authority and Nasser would build the dam with its revenue.[59] Paraphrasing a Saidi proverb, Nasser shouted, "and if the imperialists don't like it, they can choke on their rage!"[60] According to Graham Du Bois, Egyptians "were hysterical with joy. They screamed! Men threw their arms about each other and wept. Not only Egyptians and Arabs, but subject peoples throughout the world were thrilled by Nasser's daring gesture."[61]

For many seeking tricontinental solidarity, Nasser's actions "represented a particular connection between black and Arab anticolonialism."[62] As McAlister convincingly argues, "Just as Egypt was geographically positioned at the intersection of the Middle East and Africa, in the years after Bandung, Nasser positioned himself as a leader in connecting African and Asian anticolonial movements."[63] For instance, Nasser, who had met Kwame Nkrumah at Bandung two years prior, sent a large delegation to Ghana's independence celebrations in 1957 and, upon Nkrumah's visit to Egypt shortly thereafter, Nasser awarded him Egypt's highest decoration. The two joined five other leaders in calling the First All-African Peoples Conference in Accra in 1958.[64]

Malcolm's relationships with U.S. activists of color like Kochiyama, his extensive knowledge of global events, and his travels throughout the Middle East and Africa contributed to his emergent "independent 'Third World' political perspective."[65] He showed signs of this perspective as early as the 1950s, when he gave a speech suggestive of "A Message to the Grassroots," comparing the developing crisis in Vietnam with the Mau Mau rebellion in Kenya, "framing both of these movements as uprisings of the 'Darker races' creating a 'Tidal Wave' against U.S. and European imperialism."[66] Yet, as historian Robin Kelley observes, "Africa remained his primary political interest outside black America."[67] In this way, Malcolm X's anticolonial Pan-Africanism differed from the tricontinental identity, inclusive of the nonblack world that many of his successors would develop: "Malcolm X was the first political person in this country that I really identified with. . . .We continue to believe that the Black Panther Party exists in the spirit of Malcolm."[68]

Black Panther Party co-founders Huey Newton and Bobby Seale did not seek to replicate Malcolm's OAAU. Nonetheless, his teachings were "fundamental in structuring the Black Panther Party for Self-Defense, as the group was originally named in October 1966."[69] Newton was introduced to Malcolm's speeches and writings while attending Oakland City College. The previous year, he and Seale witnessed eruptions in Watts as well as the Highway Patrol and Oakland Police's "determin[ation] to rule by force," carrying shotguns in full view.[70] "Out of this need [Newton and Seale] had no choice but to form an organization that would involve the lower-class brothers" or what Karl Marx referred to as the lumpen proletariat.[71]

Long before outlining the BPP's ten-point platform, Newton and Seale held informal political education sessions at Seale's house, studying "the literature of oppressed people and their struggles for liberation in other countries . . . to see how their experiences might help us to understand our plight."[72] These included the work of Frantz Fanon, Mao Tse-tung, and Ché Guevara.[73] Reflecting upon the BPP's early influences, Newton wrote, "We read these men's works because we saw them as kinsmen; the oppressor who had controlled them was controlling us, both directly and indirectly."[74] For Newton, learning how these freedom fighters confronted imperialism and racism was critical to creating a strategy for a U.S.-based struggle. Newton and Seale were careful not to "import ideas and strategies" and believed it necessary to "transform what we learned into principles and methods acceptable to the brothers on the block."[75] Thus, they studied Robert F. Williams's Negroes with Guns, which "great[ly] influence[d] . . . the kind of party we developed."[76] While they identified with Williams's calls for self-defense in Monroe, North Carolina, his appeal to working-class activists, and his analysis of the global effects of racial capitalism, Newton was dissatisfied with Williams's requests for assistance from the federal government, which Newton viewed as "an enemy, the agency of a ruling clique that the controls the country."[77] Malcolm, with his razor-sharp critiques of the U.S. government rooted in a sense moral superiority, provided a model ideological ancestor.[78]

While Newton considered the BPP "a living testament to [Malcolm's] life's work," which they studied carefully—Seale collected all of his speeches from papers like *The Militant* and *Muhammad Speaks*—the BPP remained largely the creation of its chief theoretician, Newton.[79] The two young men furthered Malcolm's ideology, "rejecting his black nationalism," as chief-of-staff David Hilliard notes, as well as his concentration on Africa, "while incorporating a class-based political analysis that owed much to the writings of" Fanon, Guevara, and Mao.[80]

Following Newton's July 1970 acquittal on charges in the death of Oakland police officer John Frey, the BPP entered a period of immense transition. As Hilliard notes, Newton's arrest in October 1967 galvanized the BPP's first wave of "political fervor"; "his release engendered a number of equally historic changes from 1971 to 1972."[81] Newton traveled to Africa, where he met Mozambique

President Samora Moises Machel, and to Asia, where he met Premier Chou En-lai, who, given events such as the United States' escalating encroachments upon Central American and southeast Asian sovereignty, helped shape Newton's most groundbreaking and prescient ideological formation, "intercommunalism."

When Newton and Seale formed the BPP in 1966, the organization was avowedly "Black Nationalist"—by this they meant that they "realized the contradictions in society, the pressure on black people in particular, and we saw that most people in the past had solved some of their problems by forming into nations."[82] They assumed that the sufferings of people of African descent would end "when we established a nation of our own." Newton and Seale sought a method of analysis that would fill the gaps left by existing groups. The two were involved in numerous organizations, some of which they joined, such as Oakland City College's Afro-American Association, many of which they did not, such as the Progressive Labor Party and the Nation of Islam. In addition to the "mystical or religious aspect," Newton paradoxically stated that he "found it difficult to accept some of the Black nationalist ways."[83] For Newton, Black Nationalism involved an "attitude of great hatred for . . . white people," an attitude that he could not fully internalize without feeling a "certain guilt about it," for which he was subsequently criticized.[84] Thus the BPP's initial self-identification as a Black Bationalist organization existed uneasily with Newton's own discomfort with the term and his 1971 statement that "from its very conception," the BPP "was meant as an antiracist party."[85]

Because of these contradictions, they soon redefined themselves as "revolutionary nationalists"—"that is, nationalists who want revolutionary changes in everything, including the economic system the oppressor inflicts upon us."[86] During this phase, Newton adopted the language of internal colonization, linking the oppression of "black communities throughout the country—San Francisco, Los Angeles, New Haven" with traditional colonies throughout the Third World.[87] The Panthers made explicit that they were "individuals deeply concerned with the other people of the world and their desires for revolution."[88] Allying itself with colonized peoples throughout the world, the BPP soon defined itself as an "internationalist" organization.[89]

At their first large-scale public appearance, the Panthers issued a speech that, for those who could look past the rifles and black uniforms, demonstrated the BPP's broad analysis of the interplay of global events and hinted at its budding commitment to developing ties that transcended racial and national borders. Executive Mandate Number One was written by Newton and read by Seale at the Sacramento State Capital on May 2, 1967 in protest of the Mulford Bill—or "Panther Bill" as it was alternately known—which criminalized open displays of loaded firearms.[90] While "racist police agencies throughout the country intensify the terror, brutality, murder, and repression of Black people," Newton argued, the California legislature was attempting to keep "Black people disarmed and powerless."[91] He wrote, "At the same time that the American Government is waging a racist war of genocide in Vietnam, the concentration camps in which Japanese-Americans were interned during World War II are being renovated and expanded." Newton continued, invoking a phrase made famous by Malcolm X: "Since America has historically reserved its most barbaric treatment for nonwhite people, we are forced to conclude that these concentration camps are being prepared for black people who are determined to gain their freedom by any means necessary." Reminding listeners of the racist treatment historically meted out to racialized people within and beyond U.S. borders at the hands of the U.S. government, Newton wrote,

> The enslavement of Black people at the very founding of this country, the genocide practiced on the American Indians and the confinement of the survivors on reservations, the savage lynching of thousands of Black men and women, the dropping of atomic bombs on Hiroshima and Nagasaki, and now the cowardly massacre in Vietnam all testify to the fact that toward people of color the racist power structure of America has but one policy: repression, genocide, terror, and the big stick. . . . As the aggression of the racist American Government escalates in Vietnam, the police agencies of America escalate the repression of Black people throughout the ghettos of America.[92]

Even as Newton struggled to classify his organization's burgeoning ideology, the speech delivered at the BPP's first attempt at gaining national attention illustrated a commitment to a "people of color"—or what I refer to here as a tricontinental—politics, one which holds the U.S. government accountable for its acts of aggression against racialized peoples wherever they are found. A central feature of this tricontinental politics, which Newton's statement evinces, is a sense of solidarity among peoples who have been subjected to this shared racist oppression. It is this shared exploitation, this understanding of race as primarily a tool of oppression rather than as a biological fact, that drives the BPP's alliances with the tricontinental region and with its anticolonial and antiracist politics.

The following year, the Panthers expanded their actions beyond patrolling police officers by taking their critiques and calls for justice to the UN "delegations of revolutionary countries" and calling a press conference to "alert the American people, the people of the world, and particularly the oppressed and colonized people of the world, to an already dangerous situation which is rapidly deteriorating."[93] By this, the BPP meant the "racist imperialist power structure's" attempted "silencing" Newton by falsely charging him with murder.[94] Citing Newton's opposition to both "the imperialist aggression of the United States in International affairs and the vicious oppression of black people in the domestic areas," the Panthers called upon "oppressed and colonized people to organize demonstrations before the embassies, consulates, and property of the imperialist exploiters of the United States whenever they have access to such installations and property, to show and manifest solidarity with Huey P. Newton and the black liberation struggle." They also called upon UN member nations to authorize the stationing of "UN Observer Teams" throughout U.S. cities "wherein black people are cooped up and concentrated in wretched ghettos."[95] In doing so, they drew on a tradition of antiracist activists who sought to combat the U.S. struggle before an international arena, including the National Negro Congress, the Civil Rights Congress, and the SNCC.

In addition, the BPP's newspaper published countless articles on the struggles of Chicanos, Puerto Ricans (particularly Young Lords), and Asian Americans, as well as anti-imperialist and anti-capitalist efforts in Africa, the Middle East, Latin America, and Asia, especially the Congo, Palestine, Bolivia, Cuba, and Vietnam.[96] In an October 1968 article reprinted from the Cuban daily newspaper, *Granma*, the BPP's Minister of Education, George Mason Murray, was quoted as having declared, "We have vowed not to put down our guns or stop making Molotov cocktails until colonized Africans, Asians and Latin Americans in the United States and throughout the world have become free."[97] Murray and New York Panther leader Joudon Ford were in Cuba as guests of the Organization of Solidarity of the Peoples of Africa, Asia, and Latin America (OSPAAAL) participating in the "Day of Solidarity with the Struggle of the Afro-American people." At a press conference, Murray, likely referencing the rebellions that reached Detroit the previous summer, stated, "We want to tell the people who are struggling throughout the world that our collective struggle can only be victorious, and the defeat of the murderers of mankind will come as soon as we create a few more Vietnams, Cubas and Detroits."[98] "In order to bring humanity to a higher level," U.S. blacks would "follow the example of Che Guevara, the Cuban people, the Vietnamese people and our leader and Minister of Defense, Huey P. Newton [and] . . . Malcolm X, Lumumba, Ho Chi Minh and Mao Tse-tung." Murray and Ford also referred to the armed struggles taking place in Bolivia, where Guevara had been assassinated, and told their audience of the coalitions between New York "blacks and Puerto Ricans," who are "aware" of the tactics used by "imperialists [to create] . . . divisions between minority racial groups" and were "uniting to fight oppression."[99]

In July 1970, in an interview from his prison cell, Newton characterized the BPP as an international and indeed, antinational, organization. Grounded in Marxist-Leninism, it was logical that the BPP would seek solidarity with anticapitalists throughout the world. "When the people start to move," he told interviewer Mark Lane, "I guarantee you it'll be international. The Black Panther Party is an international party. We have a coalition with all struggling people of the world. And we feel that we must be international because we're fighting an international enemy."[100] Newton reiterated the need to adapt "Marx-Leninism to our particular situation here in America." He cited North Korean premier Kim Il Sung's call for the "freedom for each nation to interpret Marx-Leninism according to its own needs, as he puts it, the chief thing is to fight against U.S. imperialism, World Enemy No. 1. . . . We struggle and support all of those countries and we want solidarity with them."[101] The embrace of internationalism also provided a way to circumvent the antiwhiteness that Newton associated with Black Nationalism and pointed to his increasingly sophisticated analysis of the very concept of nationhood. Noting the scarcity of inhabitable land, Newton believed that creating a new nation would require becoming a "dominant faction in this one," and yet the fact that oppressed people "did not have power was the contradiction that drove us to seek nationhood in the first place."[102]

Shortly after Newton's release from prison, the BPP held a two-part Revolutionary Peoples Constitutional Convention in Philadelphia on September 5, 1970, and at Howard University in Washington, DC, in November 27–29 of that year. According to Newton, this meeting of "Revolutionary Peoples from oppressed communities throughout the world" was convened "in recognition of the fact that the changing social conditions throughout the world require new analyses and approaches in order that our consciousness might be raised to the point where we can effectively end the oppression of people by people."[103]

The delegates gathered to create a document that championed women's self-determination and queer rights and proposed concrete strategies for battling sexism, heterosexism, and homophobia.[104] Organized as a series of mass assemblies, the convention held a symbolic purpose as well as a political one—those in attendance gathered together because they perceived, according to Newton, "a common enemy [and] a common goal." More importantly, it signified that "the geographical barriers which separated us from one another in the past are no longer obstacles to our revolutionary unity." Newton's welcome speech began, "Friends and comrades throughout the United States and throughout the world . . . we gather in the spirit of revolutionary love and friendship for all oppressed people of the world regardless of their race or the race and doctrine of their oppressors."[105] Newton then introduced his audience to the concept of intercommunalism, the notion that the rise of U.S. imperialism had "transformed all other nations into oppressed communities," making it impossible for revolutionaries like the Panthers to make their stand as nationalists or even as internationalists.[106]

Beyond recognizing "commonalities" among the peoples oppressed by racial capitalism, Newton contended that the very categories and concepts which defined peoples were becoming obsolete: "We once defined ourselves as nations because we had distinct geographical boundaries . . . We see, however, that the growth of bureaucratic capitalism in the United States transformed the nation."[107] Departing from Lenin's notion of imperialism as the highest stage of capitalism, Newton argued that capitalism, when traversing national boundaries to exploit the "wealth and labor of other territories," transformed both the capitalist nation and the subjugated territory.[108] The rapid development of technology led to a shift in the relationships within and between nations. The "swiftness with which their 'message' can be sent to these territories has transformed the previous situation," Newton argued. Beyond becoming a colony or a neo-colony, these territories, unable to "protect their boundaries . . . political structure and . . . cultural institutions," are no longer nations, just as the United States is no longer a nation but an empire whose power transcends geographical boundaries.[109]

In order to be classified as a neo-colony, he contended, a territory must have the capacity to return to its former state. "What happens when the raw materials are extracted and labor is exploited within a territory dispersed over the entire globe?," Newton asked rhetorically, "When the riches of the whole earth are depleted and used to feed a gigantic industrial machine in the imperialist's home?" He explained that "the people and the economy are so integrated into the imperialist empire that it's impossible to 'decolonize,' to return to the former conditions of existence." If colonies cannot "return to their original existence as nations," Newton contended, "then nations no longer exist. . . . And since there must be nations for revolutionary nationalism or internationalism to make sense . . . we say that the world today is a dispersed collection of communities." Newton defined a community as a small entity with a collection of institutions that exist in order to serve a small group of people. The global struggle, according to Newton's analysis, is between the small clique of individuals—and its ruling police force—that "administers and profits from the empire of the United States, and the peoples of the world who want to determine their own destinies."[110] He termed the current age in which a ruling circle uses technology to controls all other people as "reactionary intercommunalism."Grounded in dialectical materialism, Newton saw the contradiction inherent in the development of capitalism, which would eventually lead to its own demise.

The "communications revolution, combined with the expansive domination of the American empire, has created the 'global village.'"[111] As the U.S. empire "disperses its troops and controls more and more territory, it becomes weaker and weaker. . . . And as they become weaker . . . the people become stronger."[112] Technology has developed to the extent that it can provide material abundance—the "material conditions exist that would allow the people of the world to develop a culture that is essentially human. . . . The development of such a culture," Newton argued, "would be revolutionary intercommunalism."[113] Some communities, for instance in China, Cuba, Northern Korea, and Vietnam, have "liberated their territories and have established provisional governments."[114] Nonetheless, no territory can remain safe from the long arm of the U.S. empire indefinitely.

Newton's concept of intercommunalism emerged soon after he offered troops to the National Liberation Front (NLF) of Vietnam in 1970.[115] After making his offer, Newton remained dissatisfied with his statement of solidarity because of its central contradiction—even as he relinquished all claims to nationalism and expressed the BPP's ultimate goal as the "destruction of statehood itself," he continued to offer support for revolutionary nationalism among the NLF and all "developing countries."[116] Solidarity with the Vietnamese struggle seemed logical to Newton, for it "is also our struggle, for we recognize that our common enemy is the American imperialist who is the leader of international bourgeois domination. There is not one fascist or reactionary government in the world today," he argued, "that could stand without the support of United States imperialism."[117] The United States ceased being a nation as it became a "government of international capitalists and in as much as they have exploited the world to accumulate wealth, this country belongs to the world." On the contrary, the developing countries "have every right to claim nationhood because they have not exploited anyone. The nationalism of which they speak is simply their rightful claim to autonomy, self-determination and a liberated base from which to fight the international bourgeoisie." Because of their historic position as descendents of slaves without a sense of national belonging, U.S. blacks were in a vanguard position in the international struggle against racial capitalism and imperialism. As "the vanguard party—without chauvinism or a sense of nationhood," the Panthers offered troops to the NLF "and to the people of the world."[118]

Newton continued to hold U.S. blacks responsible for their government's exploitative nationalism, "because we are all guilty on one level or another of being the exploiter or accepting the bride of the exploiter if we are not at war with him."[119] The U.S. empire has taken the wealth of the communities of the world, centralizing it in a few hands; even those who did not participate in the exploitation have "reaped the benefits from this violation. Intercommunalism expresses our view that the people of the world own the wealth of the United States and our obligation to give them their just desserts."[120] Newton "disclaimed all of the black nationalists" in his statement, which caused conflict with other black organizations "because all of them, even the bourgeois ones, are somewhat nationalistic in tone and in goal." His rejection of nationalism for U.S. blacks and support of nationalism for Vietnamese rebels left him feeling as if he were "belittling them [or] being traitors to them."[121] Even for those who claimed simultaneous nationalism and internationalism, Newton noted a contradiction; a show of respect for national boundaries alongside a need for international solidarity arising from the U.S. continual violation of national boundaries. In spite of this tension, Newton sent the statement, remaining "very dissatisfied and unhappy for about a month."[122] He woke up in his Oakland apartment with the concept of intercommunalism emerging "like a vision. . . . I had solved the contradiction in my sleep."[123]

Newton's 1971 article, which appeared in *The Black Panther*'s Intercommunal News section, provided both a contemporary example of reactionary intercommunalism and evidence of his willingness to transcend race in his critiques of imperialism. He wrote, "In Addis Ababa, Ethiopia this week, Ethiopean [*sic*] overseer Haile Selassie sat down to lunch with Spiro Agnew, the reactionary intercommunal public relations man for the U.S. Empire."[124] The two leaders, which Newton charged, "expect god-like worship from the people," enjoyed a "European style luncheon . . . in the plush 'Jubilee Palace,' which was built with the blood and sweat of Ethiopia's oppressed citizens."[125] Agnew and "his overseer, Selassie," he wrote, shared similar goals "oppression of black and white."[126] He vowed that Panthers would "continue our struggle to free all people," including those led by a former icon of black independence, "from the bonds of oppression."[127]

Newton predicted that reactionary intercommunalism, having sown the seeds to its own destruction, would yield to revolutionary intercommunalism. Ever critical of cultural nationalists' romantic views of a precolonial past, he wrote, "Our hopes for freedom lie . . . in a future which may hold a positive elimination of national boundaries and ties; a future of the world, where a human world society may be so structured as to benefit all the earth's people (not peoples)."[128] Newton defined the transition to revolutionary intercommunalism as the phase in which the world's masses would control their own institutions and seize the means of production, allowing them the freedom to "re-create themselves and to establish communism, a stage of human development in which human values will shape the structures of society."[129] "Contradictions" such as "racism and all kinds of chauvinism" will not be resolved immediately, "but the fact that the people will be in control of all the productive and institutional units of society—not only factories, but the media too—will enable them to start solving these contradictions. It will produce new values, new identities; it will mold a new and essentially human culture as the people will resolve old conflicts based on cultural and economic conditions."[130] Subsequently, a qualitative change would occur and revolutionary intercommunalism would be transformed into communism—the point in history when people will "produce according to their abilities and all receive according to their needs."[131]

The most significant shift would be the creation of a universal identity. In February 1971 Newton participated in a public discussion with psychoanalyst Erik Erikson, during which the two spoke at length about the concept of collective identity. Ultimately, Newton argued that the creation of an identity that "extends beyond family, tribe, or nation—an identity that is essentially human" was critical for the survival of human beings. "If we do not have a universal identity," he declared to a somewhat hostile audience, "then we will have cultural, racial, and religious chauvinism, the kind of ethnocentrism we have now. Unless we cultivate an identity with everyone, we will not have peace in the world."[132] Much of the resistance to Newton's concept of intercommunalism—he was once booed while onstage—and the need for a "universal identity" was foreshadowed in the resistance many demonstrated to the BPP's alliances with white radicals, the international focus of its newspaper, and the offer of troops to the NLF.[133] He believed that most of his audiences were "not ready for many of the things we talked about" and considered the problem of simplifying his ideology for the masses to be the BPP's "big burden."[134] It was the BPP because its leadership was trying "to do everything possible to get [average people] to relate to us . . . [without being] too far ahead of the people['s] thinking. . . . We are being pragmatic . . . when that job is done, the Black Panther Party will no longer be the Black Panther Party," he told an audience member.[135]

In early 1969, the BPP's newspaper printed Claude McKay's famed 1919 poem, "If We Must Die." Like Guevara in this article's epigraph, McKay's poem speaks of facing a "common foe" and of a "brave," "noble," and decidedly masculine death. For Newton and the Black Panthers, the enemy, which "far outnumbered" oppressed peoples throughout the world, "press[ing them] to the wall" with "their thousand blows" was the "technology and military might of the United States, [which] made it possible . . . to violate the territorial integrity of the peoples of the world, creating an empire here and destroying all their qualities of nationhood and making them dispersed communities of oppressed people."[136] Newton defined "revolutionary intercommunalism" as a "higher level of consciousness than nationalism or internationalism," which "recognizes the need for unity and solidarity among the dispersed communities against" the United States.[137] Those who attended the Constitutional Convention where Newton first made public his concept of intercommunalism gathered to "organize our forces to move against the evils of capitalism, imperialism and racism" and "for the solemn purpose of formulating a new constitution for a new world."[138] Capitalists, Newton argued, had "used the philosophy of racism to support their wicked oppression. Through the philosophy of racism all those in this country have been taught that people are better than others because of differences in physical and social characteristics, and therefore they have a right to exploit the other."[139]

By the mid-1970s, the BPP reached the height of its influence. In spite of the numerous assassinations, raids, trials, and incarcerations, there were over forty U.S. chapters: coalitions with Asian Americans, Latinos, white antiwar activists, feminists, and lesbians and gay men; chapters in England, Israel, Australia, and India; and solidarity committees in Germany, China, Japan, and Peru.[140] Among the reasons for their international appeal was Newton's profoundly antiracist belief that "the physical and social characteristics of the people of our communities shall never be used as a basis for exclusion."[141]

NOTES

1. Huey P. Newton, "The Technology Question," in *The Huey P. Newton Reader*, ed. David Hilliard and Donald Weise (1972; repr. New York: Seven Stories Press, 2002), 261.
2. Ibid.
3. Malcolm X and Alex Haley, *The Autobiography of Malcolm X* (New York: Ballantine Books, 1964), 147.
4. The film was nominated for a Golden Globe and two Academy Awards, among several others.
5. Lee ends the film with photos of Tracy Chapman, Bill Cosby, and Janet Jackson wearing "X" caps.
6. Gerald Horne, "Myth and the Making of 'Malcolm X'," *American Historical Review* 98, no. 2 (1993): 441.
7. Ibid., 440.
8. Maulana Karenga discusses the absence of Malcolm's relationships with African leaders in New York, *Amsterdam News*, December 26, 1992, as cited in Horne, "Myth and the Making of 'Malcom X'," 445. Horne calls to task historians and scholars as well, stating that in its failure to give a "proper account of the play of international forces . . . Lee's flaw is not his alone."
9. Horne, "Myth and the Making of 'Malcom X,'" 446.
10. Ibid. The exceptional work produced in the ten years since Horne's article was published has begun to fill the void of which Horne speaks. See, for example, Brenda Gayle Plummer, *Rising Wind: Black Americans and US Foreign Affairs, 1935–1960* (Chapel Hill: University of North Carolina Press, 1996); Penny Von Eschen, *Race Against Empire: Black Americans and Anticolonialism, 1937–1957* (Ithaca: Cornell University Press, 1997); Cynthia Young, "Soul Power: Cultural Radicalism and the Formation of a U.S. Third World Left" (PhD diss., Yale University, 1999); Robin D. G. Kelley and Betsy Esch, "Black Like Mao: Red China and Black Revolution," *Souls* 1, no. 3 (1999); Robin D. G. Kelley, "'But a Local

Phase of a World Problem': Black History's Global Vision, 1883–1950," *Journal of American History* 86, no. 3 (1999); Nikhil Pal Singh and Andrew F. Jones, "Introduction," in *Positions, East Asia Cultures Critique; Special Issue: The Afro-Asian Century*, ed. Nikhil Pal Singh and Andrew F. Jones (Durham: Duke University Press, 2003); Brent Hayes Edwards, *The Practice of Diaspora: Literature, Translation, and the Rise of Black Internationalism* (Cambridge, MA: Harvard University Press, 2003); Gerald Horne, *Race Woman: The Lives of Shirley Graham Du Bois* (New York: New York University Press, 2000).

11. Horne, "Myth and the Making of 'Malcolm X'," 444.

12. For more on the meeting between Malcolm and Castro in Harlem, see Rosemari Mealy, *Fidel and Malcolm X: Memories of a Meeting* (Melbourne, Australia: Ocean Press, 1997).

13. Kochiyama, *Passing It On*, 68.

14. Ibid.; Yuri Kochiyama, "With Justice in Her Heart: A Revolutionary Worker Interview with Yuri Kochiyama," *Revolutionary Worker Online*, no. 986 (1998).

15. Kochiyama, *Passing It On*, 70.

16. Ibid., 69.

17. Ibid., 74, 71.

18. Ibid., xiv.

19. Anonymous, "The Violent End of the Man Called Malcolm X," *Life*, March 26, 1965.

20. I use the term "tricontinental" to describe both the region known as the "global South" or the "Third World" and the political formation that aligned itself with this region and its emergent anticolonial and antiracist politics. Borrowing from Robert Young, I also use the phrase to invoke identification with the 1966 Havana Tricontinental Conference, which initiated the first anti-imperialist alliance of the peoples of the three continents as well as the founding moment of postcolonial theory in its journal, the Tricontinental. While the term itself was not used by these activist-intellectuals, "tricontinentalism" offers a useful framework for understanding their global, antiracist, and anti-imperialist politics. Robert Young, *Postcolonialism: An Historical Introduction* (Oxford: Blackwell). See also Besenia Rodriguez, "'De la Esclavitud Yanqi a la Libertad Cubana': U.S. Black Radicals, the Cuban Revolution, and the Formation of a Tricontinental Ideology," *Radical History Review* 92 (Spring 2005): 62–87.

21. Huey Newton, ed., *To Die for the People: The Writings of Huey P. Newton* (New York: Random House, 1972), 32.

22. In early 1971, due to discord manipulated and intensified by Hoover's FBI, Newton expelled several members from the BPP, including the Intercommunal Section run from Algiers by Eldridge Cleaver. In hindsight, the early seeds of these differences can be appreciated, for instance, in the divergent ways in which Newton and Hilliard, on the one hand, and Cleaver, on the other, viewed Malcolm X's influence on their organization. Cleaver looked to Malcolm as one of his heroes, precisely because he saw him as "the father of revolutionary black nationalism." Eldridge Cleaver, *Revolution in the Congo* (London: Revolutionary Peoples' Communications Network, 1971), 7. Hilliard and Newton depart from Malcolm X precisely because of his nationalism and his focus on Africa to create a more expansive tricontinentalism, developing closer ties with antiracist groups of color in the United States and abroad. It is for these reasons that I focus on Huey P. Newton. For more on the internal divisions within the BPP and the FBI's role in fomenting them, see J. Edgar Hoover, "Untitled FBI Memo re. COINTELPRO," August 25, 1961. Dr. Huey P. Newton Foundation, Inc. and Black Panther Party Collections, Special Collections and University Archives, Stanford University, Stanford, California. Fred P. Graham, "F.B.I. Files Tell of Surveillance of Students, Blacks, War Foes," *New York Times*, March 25, 1971; Tim Butz, "COINTELPRO: Psychological Warfare and Magnum Justice," *Counter- Spy* (1976); Ernest Volkman, "Othello," *Penthouse*, April 1980. Ross K. Baker,

"Panther Rift Rocks Whole Radical Left," *Washington Post*, March 21, 1971; Earl Caldwell, "Internal Dispute Rends Panthers," *New York Times*, March 7, 1971; Black Panther Party, "Expelled," *The Black Panther*, February 13, 1971; Black Panther Party, "Enemies of the People," *The Black Panther*, February 13, 1971; Black Panther Party, "Intercommunal Section Defects," *The Black Panther*, March 20, 1971; Bobby Seale, "Bobby Seale: I am the Chairman of Only One Party," *The Black Panther*, April 3, 1971.

23. Malcolm X, "Message to the Grassroots" (paper presented at the Northern Grass Roots Leadership Conference, Detroit, MI, November 10, 1964).

24. Ibid.

25. Melani McAlister, "One Black Allah: The Middle East in the Cultural Politics of African American Liberation, 1955–1970," *American Quarterly* 51, no. 3 (1999): 633.

26. Republic of Indonesia and Republic of South Africa, Asian-African Summit 2005 and the Commemoration of the Golden Jubilee of the Asian-African Conference 1955 Asian African Summit, 2005), http://asianafricansummit2005.org/history.htm (accessed August 18, 2005).

27. Ibid.

28. Ibid.

29. Shirley Graham Du Bois, *Gamal Abdel Nasser: Son of the Nile, a Biography* (New York: Third Press, 1972), 148.

30. Indonesia and Africa, Asian-African Summit 2005 and the Commemoration of the Golden Jubilee of the Asian-African Conference 1955.

31. Ibid.

32. Wikipedia.org, Bandung Conference Wikipedia.org, August 18, 2005), http://en.wikipedia.org/wiki/Bandung_Conference (accessed August 18, 2005); Jawaharlal Nehru, Speech to Bandung Conference Political Committee Fordham University, 1955), http://www.fordham.edu/halsall/mod/1955nehru-bandung2.html (accessed August 18, 2005).

33. Wikipedia.org, Bandung Conference.

34. Indonesia and Africa, Asian-African Summit 2005 and the Commemoration of the Golden Jubilee of the Asian-African Conference 1955.

35. Ted Roberts, "Cuba and the Non-Aligned Movement," *Center for Cuban Studies Newsletter* 3, no. 4–5 (1976): 68–81.

36. Ibid.

37. Kumar Goshal, "War and Jim Crow Set Back at Bandung," *Freedom* (May–June 1955).

38. Ibid.

39. Unknown, "Harlem Speaks . . . About Bandung;" Ibid.

40. Anonymous, "Harlem Speaks."

41. Goshal, "War and Jim Crow Set Back at Bandung,"

42. Paul Robeson, "Greetings to the Asian-African Conference" (April, 1955). Paul Robeson Collection, Box 7:000150. Schomburg Center for Research in Black Culture, Bandung, Indonesia. Black History Matters 137.

43. Ibid.

44. Indeed, as *Reuters* and a *New York Times* reader noted, of the 2000 representatives in attendance at the Bandung Conference, no representatives and only two advisors were women. Laili Roesad, "Women Advisers at Bandung," *New York Times*, April 22, 1955; *Reuters*, "No Women Delegates Among 600 at Bandung," *New York Times*, April 18, 1955.

45. Robeson, "Greetings to the Asian-African Conference."

46. W. E. B. Du Bois, "The American Negro and the Darker World," (1957). Paul and Eslanda Robeson Collection. Moorland-Spingarn Research Center, Howard University, Washington, DC.

47. Ibid.
48. Ibid.
49. James Boggs, "Correcting Mistaken Ideas about the Third World" (March 14, 1974). James and Grace Lee Boggs Collection, Walter P. Reuther Library of Labor and Urban Affairs, Wayne State University Archives, Detroit, Michigan. Box 3: 16.4.
50. Ibid., 6.
51. Ibid., 7.
52. Ibid.
53. Yuri Nakahara Kochiyama, "Third World," *Asian Americans for Action Newsletter*, October 1970, 199.
54. Ibid., 199–200.
55. McAlister, "One Black Allah," 632–33.
56. S. G. Du Bois, *Gamal Abdel Nasser*, 163–64. The poem was published in several small progressive periodicals, including the December 1956 issue of *Mainstream Magazine*.
57. In a dissertation chapter, I discuss Graham Du Bois's tricontinental socialism, in particular, her interest in Egypt and the Middle East, which has often been understood as Pan-Africanist or dismissed entirely. I argue that her impassioned writings on the ongoing crisis in the Middle East provide significant insight into the global reach of her antiracist politics as well as her broader anti-imperialist project. See "Beyond Nation: The Formation of an Antiracist Tricontinental Discourse" (Yale University, in progress).
58. S. G. Du Bois, *Gamal Abdel Nasser*, 141.
59. Ibid., 152.
60. Ibid., 154.
61. McAlister, "One Black Allah," 632–33.
62. S. G. Du Bois, *Gamal Abdel Nasser*, 155; Erskine B. Childers, "The Road to Suez" (1962). Shirley Graham Du Bois Papers, Box 31:20. Arthur and Elizabeth Schlesinger Library on the History of Women in America, Cambridge, MA, 18.
63. S. G. Du Bois, *Gamal Abdel Nasser*, 159.
64. Ibid., 163.
65. Ibid.
66. Ibid., 163.
67. McAlister, "One Black Allah," 632.
68. Ibid. McAlister also discusses the Nation of Islam's endorsement of the Egyptian seizure of the Suez Canal.
69. Nasser also developed warm ties with Guinea's President Sékou Touré. S. G. Du Bois, *Gamal Abdel Nasser*, 176.
70. Robin D. G. Kelley, "House Negroes on the Loose: Malcolm X and the Black Bourgeoisie," *Callaloo* 21, no. 2 (1998): 431.
71. Ibid.
72. Ibid.
73. Huey P. Newton, David Hilliard, and Donald Weise, *The Huey P. Newton Reader* (New York; London: Seven Stories, 2002), 52.
74. David Hilliard, "Introduction," in *The Huey P. Newton Reader*, ed. David Hilliard and Donald Weise (New York; London: Seven Stories, 2002), 11.
75. Newton, Hilliard, and Weise, *The Huey P. Newton Reader*, 49.
76. Ibid.
77. Ibid., 50.
78. Ibid.
79. Ibid.

80. Ibid. Robert F. Williams, "Robert Williams Speaks at Panther Benefit," *The Black Panther*, December 27, 1969; Robert F. Williams, "Robert Williams Speaks at N.C.C.F. Panther Benefit; Detroit, Michigan," *The Black Panther*, January 3, 1970.

81. Newton, Hilliard, and Weise, *The Huey P. Newton Reader*, 50.

82. Each May, the BPP's newspaper, *The Black Panther*, published a commemorative issue celebrating Malcolm X's birthday. See, for example, Black Panther Party, "The Heirs of Malcolm have picked up the gun and now stand millions strong facing the racist pig oppressor," *The Black Panther*, May 19, 1970, cover.

83. Black Panther Party, "They Work Together to Oppress Us. We'll Work Together to Resist,' *The Black Panther Intercommunal News Service*, February 5, 1972.

84. Newton, Hilliard, and Weise, *The Huey P. Newton Reader*, 52.

85. Hilliard, "Introduction," 11. While Malcolm X identified largely with working-class black people and, often, with left-leaning causes, his increasing critiques of capitalism are rarely, if ever, discussed in Black Panther writings and references to his work. These critiques are highlighted to differing degrees by various scholars. See George Breitman, *Malcolm X: The Man and His Ideas* (New York: Merit Publishers, 1965); George Breitman, *The Last Year of Malcolm X: The Evolution of a Revolutionary* (New York: Merit Publishers, 1967); Eugene Wolfenstein, *The Victims of Democracy: Malcolm X and the Black Revolution* (1981; repr. London: Free Association Books, 1989, 1981); Kelley, "House Negroes on the Loose."

86. David Hilliard, "Part Three: The Second Wave," in *The Huey P. Newton Reader*, ed. David Hilliard and Donald Weise (New York; London: Seven Stories, 2002), 179.

87. Huey Newton, "Revolutionary Intercommunalism," in *Revolutionary Intercommunalism and the Right of Nations to Self-Determination*, ed. Amy Gdala (1971; repr. Newton, Wales: Cyhoeddwyr y Superscript, Ltd., 2004), 27.

88. Ibid., 49.

89. Ibid.

90. Ibid.

91. Ibid., 28.

92. Ibid., 49.

93. Ibid.

94. BPP Chief-of-Staff David Hilliard cites Oakland's history as pivotal in shaping Newton and the BPP's internationalism. Hilliard discusses the city's rich union tradition and its racially and ethnically integrated political environment: "Solidarity is the watchword, and we are surrounded by examples collectively asserting their power. The internationalism is emphasized by the fact that Oakland, like Mobile, is an integrated community. You don't simply find whites and blacks, but yellows, browns, Native Americans too. These groups coexist in a particular way. New York is famous for its many ethnic communities. But whenever I visit there, I'm surprised at how groups don't mix: the city is multiracial, not intraracial. But on July 4, when the young people of Oakland crowd the park by the bay to watch the fireworks, the array of skin shades is beautiful and impressive; couples claim five and six strains in their blood." David Hilliard and Lewis Cole, *This Side of Glory: the Autobiography of David Hilliard and the Story of the Black Panther Party* (Chicago: Lawrence Hill Books, 1993), 68.

95. Hilliard, "Introduction," 12.

96. Newton, Hilliard, and Weise, *The Huey P. Newton Reader*, 69.

97. Ibid., 70. In the years after these comments were made, Newton forged a relationship with allies representing, "oppressed people of Japan," who formed the "Committee to Support Black Panthers in Japan." These anti-imperialist activists, as critical of Japanese imperialism as they were of U.S. imperialism, corresponded with Newton, sending a monetary donation and expressing their support for the BPP's goals and most recently, Newton's

"ideas of intercommunalism," which they saw as "an enlightenment for many third-world people in Japan." Newton, in turn, expressed his desire to accept the Committee's invitation to Japan, regretting that he could not do so during his visit to the People's Republic of China. Newton wrote, "We were very glad to know that our Comrades in Japan have embraced the philosophy of revolutionary intercommunalism, for truly this will be a uniting factor between our people and yours. . . . Our struggles, as you yourselves clearly pointed out, are one struggle; our enemies are the same enemy; our victories shall be common" and reprinted the Committee's letter in the December 4, 1971, issue of *The Black Panther*. Matsuko Ishida and Japan Committee to Support the Black Panther Party to Huey Newton, September 25, 1971, Dr. Huey P. Newton Foundation, Inc. and Black Panther Party Collections, Box 7 (series 2): 3. Special Collections and University Archives, Stanford University; Matsuko Ishida and Japan Committee to Support the Black Panther Party to Huey Newton, September 25, 1971, Dr. Huey P. Newton Foundation, Inc. and Black Panther Party Collections, Box 7 (series 2): 3. Special Collections and University Archives, Stanford University; Huey Newton to Japan Committee to Support the Black Panther Party, November 29, 1971, Dr. Huey Black History Matters ^ 139 P. Newton Fd. Collection, Box 7 (series 2): 3; Huey Newton to Masao Omata, November 29, 1971, Dr. Huey P. Newton Fd. Collection, Box 7 (series 2): 3.

98. For an excellent treatment of the Panthers' policing of the police, which were aimed at capturing the imagination of local black communities by "subverting the state's official performance of itself . . . turning the police . . . into the 'symbols of uniformed and armed lawlessness,'" see Nikhil Pal Singh, *Black is a Country: Race and the Unfinished Struggle for Democracy* (Cambridge, MA: Harvard University Press, 2004), esp. 199–211.

99. Ibid.

100. Black Panther Party for Self Defense, "Panthers Move Internationally: Free Huey at the U.N.," *The Black Panther*, September 14, 1968; Black Panther Party, "Take Black Genocide Before U.N.," *The Black Panther*, March 21, 1970.

101. Black Panther Party, "Interview: With William Patterson and Charles Garry," *The Black Panther*, July 5, 1969; Party, "Take Black Genocide Before U.N."; Charles W. Cheng, "The Cold War: Its Impact on the Black Liberation Struggle Within the United States, Part 1 of 2," *Freedomways* 13, no. 3 (1973).

102. See, for example, Black Panther Party for Self Defense, "Chinese Government Statement," *The Black Panther*, July 20, 1967; Black Panther Party for Self Defense, "United States 'Democracy' in Latin America," *The Black Panther*, July 20, 1967; Black Panther Party for Self Defense, "Mexican- Americans Fight Racism," *The Black Panther*, May 4, 1968; Black Panther Party for Self Defense, "Eyes of the Third World on U.S. Racism," *The Black Panther*, May 4, 1968; Black Panther Party, "Chilean Workers Struggle Against Exploitation," *The Black Panther*, October 12, 1968; Black Panther Party, "Mexican Students Fight Against Repression," *The Black Panther*, October 12, 1968; Black Panther Party, "Palestine Guerrillas," *The Black Panther*, October 19, 1968; Black Panther Party, "Anti-U.S. Rallies," *The Black Panther*, October 19, 1968; Black Panther Party, "Che Guevara on Vietnam," *The Black Panther*, October 19, 1968; Black Panther Party, "Cubans Support Movement," *The Black Panther*, October 19, 1968; Huey Newton, "Los Siete de la Raza," *The Black Panther*, June 28, 1969; Black Panther Party, "Boycott Lettuce," *The Black Panther Intercommunal News Service*, September 23, 1972; Black Panther Party for Self Defense, "Bootlicker Tshombe Captured," *The Black Panther*, July 20, 1967; Black Panther Party, "Bolivians Fight," *The Black Panther*, February 2, 1969; Black Panther Party, "Bolivian 'Niggers' U.S.-Style Racism and Capitalism in Bolivia," *The Black Panther Intercommunal News Service*, April 22, 1972; Black Panther Party, "Cuban Revolution 10 Years Old," *The Black Panther*, February 2, 1969; Black Panther Party, "The Heroic

Palelestinian [sic] Women," *The Black Panther*, July 26, 1969; Black Panther Party, "The Week of the Heroic Guerrilla," *The Black Panther*, October 9, 1971. Black Panther Party, "Important Statements of a Brazilian Revolutionary Leader," *The Black Panther*, August 1, 1970; Black Panther Party, "International Communique No. 1," *The Black Panther*, October 12, 1968.

103. George Murray, "George Murray, Minister of Education, Black Panther Party, Relates Revolutionary History in the Making at Havana, Cuba Press Conference," *The Black Panther*, October 12, 1968.

104. Ibid.

105. Ibid., 27.

106. Mark Lane and Huey Newton, "Huey Newton Speaks," September 1, 1970, Dr. Huey P. Newton Fd. Collection, Box 57 (series 1): 7. Special Collections and University Archives, Stanford University, Stanford, 6.

107. Ibid.

108. Newton, "Revolutionary Intercommunalism," 27.

109. Huey Newton, "Revolutionary Peoples Constitutional Convention: Resolutions and Declarations, Washington, DC," November 29, 1970, Dr. Huey P. Newton Fd. Collection, Box 11 (series 2).

110. The idea for a Revolutionary Peoples Constitutional Convention emerged from the Conference for a United Front against Fascism, organized by the BPP and other community organizations and held in Oakland July 18–20, 1969. See Black Panther Party, "A United Front Against Fascism," *The Black Panther*, June 28, 1969; Eldridge Cleaver, "On the Constitution" (1970). Dr. Huey P. Newton Fd. Collection, Box 30 (series 2): 6.

111. Black Panther Party and Youth International Party, "Revolutionary Peoples Constitutional Convention" (1970). Publications Relating to the Black Panther Party. Tamiment Library, New York University, New York.

112. Huey Newton, "Huey's Message to the Revolutionary People's Constitutional Convention Plenary Session, Philadelphia" (September 5, 1970). Dr. Huey P. Newton Fd. Collection, Box 11 (series 2): 14.

113. Newton, "Revolutionary Peoples Constitutional Convention."

114. Ibid., 1.

115. Ibid.

116. Ibid., 2.

117. Newton, "Revolutionary Intercommunalism," 31.

118. Ibid., 36.

119. Ibid.

120. Ibid., 31.

121. Ibid., 32.

122. Huey Newton to Provisional Government of South Vietnam and National Liberation Front, 1970 Dr. Huey P. Newton Fd. Collection, Box 47 (series 1): 24. Nguyen Thi Dinh and South Vietnamese People's Liberation Armed Forces to Huey Newton, October 31, 1970, Dr. Huey P. Newton Fd. Collection, Box 7 (series 2): 3.

123. Ibid.

124. Ibid.

125. Ibid.

126. Newton, "Revolutionary Intercommunalism," 119.

127. Huey Newton, "Intercommunalism: A Higher Level of Consciousness" (n.d.). Dr. Huey P. Newton Fd. Collection, Box 48 (series 1): 4, 13.

128. Newton, "Revolutionary Intercommunalism," 119.

129. Ibid.

130. Ibid.

131. Within Black Atlantic political history, Selassie is perhaps most well known for being perceived as God incarnate among Rastafari and for leading an independent Ethiopia while it was under attack by Italy's Mussolini. By 1960, particularly after a failed revolutionary Marxist coup in December, Selassie became increasingly conservative, aligning with the United States, United Kingdom, and other Western nations. Black Panther Party, "Agnew Visits his Country Estate—Ethiopia," *The Black Panther*, July 19, 1971.

132. Ibid.

133. Ibid.

134. Ibid.

135. Huey P. Newton, "Uniting Against the Common Enemy"; *The Black Panther*, October 23, 1971.

136. Newton, "Revolutionary Intercommunalism," 33.

137. Ibid., 45.

138. Many Panthers, including Newton himself, have admitted that he was a far more effective thinker and writer than orator. He claimed that he received letters from "truly oppressed . . . welfare recipients . . . saying, "I thought the Party was for us; why do you want to give those dirty Vietnamese our life blood?" Ibid., 47.

139. 'If we must die—let it not be like hogs/ Hunted and penned in an inglorious spot,/ While round us bark the mad and hungry dogs,/ Making their mock at our accursed lot./ If we must die—oh, let us nobly die/ So that our precious blood may not be shed/ In vain; then even the monsters we defy/ Shall be constrained to honor us, though dead!/ Oh kinsmen! We must meet the common foe;/ Though far outnumbered let us show we're brave, And for their thousand blows deal one death blow!/ What though before us lies the open grave!/ Like men we'll face the murderous, cowardly pack,/ Pressed to the wall, dying, but fighting back!" Claude McKay, "If We Must Die," *The Black Panther*, January 4, 1969 [1919]. Newton, "Intercommunalism: A Higher Level of Consciousness."

140. Newton, "Huey's Message to the Revolutionary People's Constitutional Convention," 4, 6.

141. Newton, "Revolutionary Intercommunalism," 5. *Black History Matters*, 141

CHAPTER 12

"A FREE BLACK MIND IS A CONCEALED WEAPON"

INSTITUTIONS AND SOCIAL MOVEMENTS IN THE AFRICAN DIASPORA

ROBIN J. HAYES

MY EXAMINATION OF THE RELATIONSHIP BETWEEN AFRICAN LIBERATION AND BLACK Power engages discussions within the social movement literature about how sociall movements exchange ideas and tactics across national boundaries and the effects these exchanges have on the organizations and activists involved. Sociologists have confirmed that transnational exchanges between social movements are usually based on some form of mutual identification between the organizations or activists involved. According to social movement theorists such as Hanspeter Kriesi, Doug McAdam, and Dieter Rucht, this process, defined as "cross-national diffusion," involves the unidirectional transmission of concepts from one social movement organization or activist to another.[1] The manner in which social movement leaders choose to promote their work or strategically borrow from the repertoires of organizations in other countries, in addition to "identity transforming mechanisms" that can change the political perceptions of social movement activists or constituents and their relationships with other groups, can facilitate the process of cross-national diffusion.[2] However, interviews with Black Power activists and archival data reveal that institutions indigenous to historically marginalized communities also play a critical role in facilitating transnational interactions between social movements. Institutions indigenous to the African Diaspora are identity transforming mechanisms that serve as communication networks, draw upon their communities' cultural frameworks to delineate (and sometimes expand) the boundaries of group membership, and validate the work, "performances," and "claims" of social movement actors.[3] My research provides empirical evidence of how institutions in the African Diaspora encourage black social movements to identify with each other in spite of national boundaries and reciprocally exchange their ideas and tactics.

Institutions indigenous to historically marginalized communities are perceived by their constituencies to be intended for community consumption and controlled by esteemed community members. Strong institutions can provide the resources necessary for social movements to interact across national boundaries, take advantage of political opportunities, and collect and disseminate information.[4] By providing sites for formal and informal discussions about issues that are relevant to their communities, indigenous institutions also help frame cultural and political discourses in historically marginalized communities and facilitate the construction of interpersonal networks that are important to the survival and success of community members.[5] The environments created and sustained by these institutions help foster political protest because, "institutional life . . . draws people into the settings within which collective action can erupt."[6] In her groundbreaking work *The Boundaries of Blackness*, political scientist Cathy Cohen demonstrates how indigenous institutions can also exclude the most vulnerable members of their communities by defining certain behaviors and beliefs as guidelines for membership in historically marginalized groups.[7] Anthropologist Kamari Clarke agrees that institutions in the black Atlantic have "substantive rules and regulations that legitimate norms" and "protocol and procedures by which norms can be derived."[8] Therefore the identity shaping the work of institutions that are indigenous to the African Diaspora is grounded in dynamic conceptions of racial kinship and group consciousness. My research reveals how these conceptions affect the way indigenous institutions provide information about the ideas and tactics of social movements in other countries.

Scholars of the African Diaspora have established that black institutions express solidarity with communities of African descent in other countries, because they recognize how the ideology of race and mechanisms of white supremacy operate globally in a variety of contexts.[9] The transnational interests of these institutions are grounded in a tradition of engagement with people, institutions, and ideas in the black Diaspora and a desire to address the wounds caused by the traumatic forced dispersal of their people from their African homeland, which is both geographically "real" and culturally, politically, and historically "imagined."[10] Historians Sidney Lemelle and Robin D. G. Kelley explain that black institutions construct and maintain the systems of knowledge that reproduce this "imagined community" and thus have a significant effect on the construction of African diasporic identity.[11] These transnational imaginings, and the meanings that are invested in these visions by institutions, are the consequences of social practices that involve forms of work, negotiations between individuals and "globally defined fields of possibility."[12] The research I have conducted illustrates that indigenous institutions mediate transnational relationships between black social movements

by articulating to their constituencies how the ideas and tactics of these movements embody the strategies of resistance best suited for what they define as the
common diasporic good.

METHOD

This study draws from semi-structured interviews that I conducted with participants in the American Committee on Africa (ACOA), Black Panther Party
(BPP), Organization of Afro-American Unity (OAAU), and Student Nonviolent Coordinating Committee (SNCC). I chose each of these interview subjects
because my prior examination of these organizations' records, memoirs of Black
Power activists, and secondary source data confirmed that they would be able to
provide information about the issues most relevant to this research. I met with
each subject in person for sessions that lasted between ninety minutes and three
hours between June 2004 and June 2006. The interviews took place at locations
of the subjects' choosing (usually their workplaces or homes) in the New York
City metropolitan area, Philadelphia, Jackson, New Haven, and Lusaka. Each of
these subjects spoke with me extensively about their work as antiracist activists,
their involvement with African American institutions, and their observations of
interactions between Black Power organizations and African independence movements between 1957 and 1971. During interviews, former participants in the
ACOA, BPP, OAAU, and SNCC explained how they received information about
African independence movements through historically black colleges and universities and *emancipated spaces* and how this information influenced the ideas and
tactics of the organizations with which they were involved.

This study also draws from archival sources. I reviewed the *Dr. Huey P. Newton
Foundation Collection* at Stanford University Library's Division of Special Collections and University Archives, which includes Newton's correspondence, BPP
records, and a variety of audiovisual materials. In addition, I reviewed the *Student
Nonviolent Coordinating Committee Papers, 1959-1972* and *The Student Voice*.
My research for this chapter also included writings and eyewitness accounts by
Eldridge and Kathleen Cleaver, two of the leaders of the International Section of
the BPP in Algiers, speeches by Malcolm X and Stokely Carmichael, and memoirs
by activists who were involved in the leadership of the BPP, OAAU, and SNCC,
including David Hilliard, Malcolm X, Stokely Carmichael (Kwame Ture), James
Forman, and John Lewis.

This chapter explores how historically black colleges and universities (HBCUs)
and *emancipated spaces*—helped facilitate exchanges between Black Power and
African independence organizations. Among the most influential institutions
in the African American counterpublic, historically black colleges and universities and *emancipated spaces* help organize and develop resources that are significant to social movement emergence, including leaders, solidarity incentives, and
communication networks.[13] My research revealed that HBCUs create physical
and discursive environments in which dominant narratives of black inferiority

are discredited through curriculum and practices that emphasize the heritage of communities of African descent and their successful struggles against racism. This work also examines how emancipated spaces fostered cross-national exchanges between African independence and Black Power, because interpersonal exchanges between activists are critical to communication throughout the Diaspora.[14] I use the term "emancipated spaces" to describe physically and intellectually safe areas within the African Diaspora that are organized by antiracist activists, such as conferences, salons, or more intimate gatherings, where frank dialogues and earnest interpersonal exchanges between activists of African descent can take place. These spaces are emancipated because they offer activists opportunities to devise and exemplify alternatives to white supremacist practices. HBCUs and emancipated spaces are important to the African American tradition of resistance and are able to facilitate mutual identification between black social movements across national boundaries by providing information to their constituencies about organizations and activists in other countries and offering resources to support transnational political activism.

"A FREE BLACK MIND IS A CONCEALED WEAPON"

Interviews with activists indicate that historically black colleges and universities create environments that facilitate the political involvement of students by offering information about antiracist strategies, fostering group consciousness, and exposing them to social movement activity. I learned about Peter Bailey from my research about Malcolm X and the OAAU. He served as the editor of the OAAU's newsletter, the *Blacklash*, and authored *Seventh Child: A Family Memoir of Malcolm X*. He recently retired from his position as editor of *Vital Issues: a Journal of African American Speeches*. From the office of *Vital Issues* and the National Council of Black Women in Silver Spring, Maryland, which is decorated with a stately portrait of Sojourner Truth, Peter Bailey informed me that his early political involvement occurred when students organized solidarity demonstrations at Woolworth's in Washington, DC, to support the original sit-in demonstrators from Greensboro, North Carolina: "My first political activism was at Howard . . . they had some serious, in the '60s . . . you know, students there. You know, when we used to go . . . [to] Woolworth's it was not a whole bunch of us. Maybe about twenty-five, thirty of us would go over there to do the picketing. . . . I thought that was a major move and my first political activism. And I think that was because of the fact that you had . . . some very, very sharp and well aware students at Howard."[15]

Mr. Bailey's recollections of the discursive environment at Howard are corroborated by Stokely Carmichael, former chairperson of the SNCC, who explained in his autobiography that he decided to attend Howard University after meeting members of the Nonviolent Action Group (NAG) who impressed him "as smart, serious, political, sassy—and they were black."[16] Both Carmichael and Marian Wright Edelman, an alumnus of Spelman College and former SNCC activist, recall building close relationships with HBCU professors who supported their political activism.[17] During the late 1950s and 1960s, HBCUs served as part of an antiracist communication network that linked young, educated African Americans from different economic and national backgrounds through geographic and discursive space.

Although HBCUs were critical to the development of the SNCC, students, faculty, and administrators at historically black colleges and universities had a variety of opportunities to interact with other Black Power organizations and activists. The Nashville consortium of black institutions of higher learning, which included Fisk, Tennessee State College, American Baptist Theological Seminary, and Meharry Medical School, developed into a movement center that devised tactics and strategies and trained a variety of future SNCC leaders to implement them, including Diane Nash, James Bevel, Marion Barry, Cordell Reagon, Bernard Lafayette, and Mathew Jones.[18] Tuskegee students also invited Malcolm X, principal of the OAAU, to speak on the campus on February 3, 1965, just three weeks before he was assassinated.[19] While he was still a minister with the Nation of Islam, Malcolm X also participated in a legendary debate at Howard University with March on Washington coordinator Bayard Rustin, which was entitled "Integration or Separation?" and orchestrated by Rustin to present X's ideas to the Howard community.[20] In this period, the manner in which historically black colleges and universities provided resources to support group consciousness-raising activities appears closely related to the emergence and sustenance of radical African American activism.

HBCU student activists eventually adapted the slogan "a free black mind is a concealed weapon," and even wary administrators were unable to deny the support these students enjoyed from the grassroots of the African American community.[21] The theoretical grounding and political interests of HBCU students affected the Black Power movement in a variety of ways. In August 2005 I interviewed Bob Moses, whom civil rights movement historians and activists identify as one of the most committed and inspirational SNCC organizers. Bob Moses is the founder and head of the Algebra Project, which aims to advance the struggle for equality by assisting students in inner city and rural areas to achieve mathematics literacy. From his classroom in Jackson, Mississippi at the historically black Lanier Public High School, a few minutes drive from where fellow antiracist organizer Medgar Evers was assassinated, Bob Moses observed that HBCU students who were involved in SNCC had received a political education, which informed their political praxis:

> many of them . . . helped lead SNCC into the whole country . . . there's not just style but there's an issue of argumentation . . . who can stand up in public meetings and actually argue and persuade . . . a lot of the students . . . are learning how to do this, but they're doing it in a style, which evolves from the work itself, and not from the university . . . and it's not just academic, it's in the sense of youth who are oriented to real political discussion, not something they've done in the classroom. It's something that they've gathered together and are doing as radicalized young students. . . . They're radicalized on the one hand by movement and

they're radicalized on the other hand by intellectuals who are interested in these kind of movements.[22]

The impact of HBCU students' political education on their activism with the SNCC exemplifies how institutions indigenous to the African Diaspora build community and arm their constituencies with the intellectual tools to challenge white supremacist discourses and practices.

During the 1960s, HBCUs also facilitated connections between this new generation of activists and African anticolonialism in a manner that further enhanced the image of Africa as a redemptive and inspirational icon. Former SNCC chairperson and HBCU alumnus John Lewis recalled how the success of African independence movements raised the stakes for African American insurgency:

> When I came back to Nashville in the fall of 1958 . . . [t]here was a sense of urgency and awareness spreading . . . among black students throughout the city. There was a growing feeling that this movement for civil rights . . . *demanded* our involvement. This wasn't even just an American movement anymore. Amazing changes were happening in Africa . . . and we couldn't help being thrilled . . . but also a little ashamed. Here were black people thousands of miles away achieving liberation and independence from nations that had ruled them for centuries . . . and we couldn't even get a hamburger and a Coke at a soda fountain.[23]

HBCU students had a variety of opportunities to interact with representatives from African independence movements on campus. For example, Stokely Carmichael remembered that Bayard Rustin brought Kenneth Kuanda, leader of the Zambian independence movement, to a meeting of NAG shortly after he had been elected president and that Howard's proximity to the new African diplomatic corps enabled interactions between the student activist group and African independence representatives.[24] In August 1968, Howard University hosted the Eighth Annual Conference of the Pan-African Students Organization in the Americas (PASOA) entitled "African Solidarity at Home and Abroad," which featured presentations by African "diplomats and celebrated artists."[25] The 1970 Congress of African People, which "aimed to make a sharp and penetrating analysis of the conditions that oppress [people of African descent] and develop strategies and programs for liberation," was held at several Atlanta University Center (AUC) campuses, including Morehouse, Spelman, and Morris Brown.[26] Conference organizers featured a screening of the film "Angola Liberation Struggle" and invited independence leaders Julius Nyerere and Sékou Touré as keynote speakers.[27] The Tuskegee Institute bestowed honorary degrees at special convocations upon Nnamdi Azikiwe, Kwame Nkrumah, Tom Mboya, and Leopold-Sédar Senghor, who respectively helped lead nationalist struggles in Nigeria, Ghana, Kenya, and Senegal.[28] HBCUs also hosted several South African independence activists, including African National Congress Youth League co-founder Oliver Tambo and Pan-Africanist Congress co-founder Vusumzi Linda Make, who were traveling throughout the United States on speaking tours organized by the ACOA.[29] These events illustrate how historically black colleges and universities, as institutions indigenous to the African Diaspora, used their resources to provide information to their constituencies in a manner that could facilitate transnational exchanges between social movements. Black Power activists who were involved with HBCUs were politicized by these educational institutions and exposed to the ideas and tactics of social movements in other parts of the Diaspora.

"A Whole Window Has Been Opened"

The manner in which Black Power activists encountered African independence movements within emancipated spaces in the black Diaspora also affected their political ideas and tactics. Emancipated spaces enable the physical and discursive articulation of Diaspora through the relationships and correspondence that they engender by actualizing African descent as a frame of cultural identity and antiracist praxis.[30] Historian Penny Von Eschen pointed to the Pan-African Congresses of the early twentieth century (1900, 1919, 1921, and 1927) and the Bandung Conference in 1955 as examples of diasporic gatherings in which transnational strategies for combating white supremacist practices were discussed and relationships between activists were developed and strengthened.[31] Through the boundaries of group membership that emancipated spaces delineate with decisions about invited participants and programmatic content, these institutions also valorize and criticize certain kinds of antiracist politics. Similar to other institutions indigenous to the African Diaspora, emancipated spaces help build group consciousness and enable antiracist dialogues and practices to take place.

Following his engagement with African independence activists in discursively and geographically emancipated spaces during his 1964 tour of decolonized African nations, Malcolm X's political and tactical strategies demonstrated that he believed African anti-colonialism were directly relevant to the struggles of African Americans. In May 2005, I interviewed former OAAU participant and *Muhammad Speaks* editor Sylvester Leaks in the Crown Heights brownstone where he has lived for forty years. Since his involvement with the OAAU, he has continued his work as a journalist and community activist. Mr. Leaks informed me that the founding charter of the OAAU was directly influenced by the Organization of African Unity (OAU): "we went to the United Nations and got a copy of [the constitution of the OAU] [*laughs*] And [copied it] almost verbatim. . . . Unite. Unite. Unite. Unite black people whatever his or her persuasion. . . . Unity was the first, foremost and last thing that was needed."[32]

Peter Bailey observed that the audiences of OAAU weekly rallies at the Audubon Ballroom in Harlem, which were as large as six or seven hundred people, were attended by "students . . . working class brothers and sisters, some black professionals . . . real regular black folks who had become disaffected from the traditional civil rights organizations and were looking for another way of doing things."[33] Mr. Leaks also observed that the OAAU's consciousness-raising work during its weekly rallies helped address the African American community's desire to learn about Africa.[34] For example, Malcolm X introduced Tanzanian Minister of Economic Development Muhammad Babu to his constituents: "[He was] the one who came forth and suggested that the African summit conference pass a resolution thoroughly condemning the mistreatment of Afro-Americans in America and thoroughly supporting the freedom struggle for human rights of our people in this country was President Julius Nyerere. I was honored to spend three hours with him, when I was in Dar es Salaam and Tanganyika."[35]

The considerable impact of Malcolm X's experiences in emancipated spaces in postcolonial Africa on his political ideas and strategies illustrates how this indigenous institution offers valuable resources and information that encourage activists to identify with and emulate organizations in other countries of the African Diaspora.

In June 2005, I interviewed George Houser, who served as the executive director of ACOA between 1954 and 1981, at the woodland home that he shares with his wife, Jean, in a cooperative community in Pomona, New York. ACOA became one of the most prominent organizations that advocated for U.S. foreign policies that supported autonomy and economic justice in Africa. ACOA developed strong ties with African American institutions and activists during its decades of activism against colonialism and neo-colonialism.[36] Although he is semi-retired, Mr. Houser's home office is filled with manuscripts, newspaper clippings and photographs of him with African dignitaries, including Nelson Mandela. During our meeting, he informed me that SNCC activists accepted an invitation that was sent through ACOA by Harry Belafonte to visit Guinea in 1964, because they were interested in Guinean independence leader Sékou Touré, "[who] was looked upon as one of the radical leaders in Africa, along with Nkrumah [and] Lumumba."[37]

SNCC activists recognized newly independent Guinea as an emancipated space partly because of the widespread presence of black Africans in leadership positions. Bob Moses, who participated in the SNCC delegation to Guinea, along with James Forman, John Lewis, Bob and Dona Moses, Prathia Hall, Julian Bond, Ruby Doris Robinson, Bill Hansen, Donald Harris, Matthew Jones, and Fannie Lou Hamer, remembered how members of the delegation eagerly engaged and learned from people during their trip: "I think certainly, for Fannie Lou Hamer, it was an astonishing experience . . . she was wonderful because language was no barrier. She didn't speak a word of French, but she never stopped talking . . . she would just grab people and talk to them and talk to them. And, of course, when you do that and you can do that from some place very central and deep within you, then you're communicating . . . it's like a whole window has been opened."[38]

While meeting with SNCC activists, Sékou Touré encouraged them to take a broad view of the goals of their struggle, which appreciates the close relationship between what SNCC did in the United States and what happened in Africa.[39] Touré also told the delegation, "it is fundamental that you see the problem as exploitation. While you should speak to black people first of all, it is the entire community that must be liberated."[40] The impact of these experiences on the delegates is exemplified by Mrs. Hamer's observations: "All my life I've lived in America and no president ever came to see me. I go to Guinea, and the first day, the president himself come to where we staying to say hello. . . . It was *ordinary* people like me, who were running *everything*. The ones who were left back and never had no chance. Now they have the chance to run they own country."[41]

The SNCC delegation's experiences in postcolonial Guinea illustrate how Black Power activists in emancipated spaces engaged African independence movements. After their return to the United States, the SNCC incorporated an increasingly transnational perspective and interest in building coalitions with African liberation movements.[42] This evidence suggests that transnational encounters between activists in emancipated spaces can have a significant influence on the antiracist praxis of black social movements.

The BPP engaged African independence movements in the emancipated space of postcolonial Algeria. Former Minister of Information Kathleen Cleaver observed that Algerian interest and support of the BPP was enhanced by their belief that "the liberation of Blacks from racist oppression and capitalist exploitation required a social revolution to transform the economic and political institutions of the United States."[43] Following the *Front de Libération Nationale*'s bloody victory over the French, the organization used its newly acquired state authority to transform Algiers into a subversive metropole. The 1969 Pan-African Cultural Festival in Algiers—which was attended by hundreds of people from all over Africa, the Middle East, and Europe, including representatives from the SNCC, other political figures, African American jazz musicians, playwrights, actors and actresses, poets, scholars, writers, and political activists (among them Nina Simone, Archie Shepp, Ed Bullins, Dr. Nathan Hare, and Haki Madhubuti)—provided the BPP with an opportunity to both learn from and engage African anti-colonialism.[44] Muhammed Ben Yaya, Algeria's Minister of Information, allowed the BPP to set up an exhibition in connection with the festival that featured an Afro-American Information Center, the art work of *Black Panther* illustrator Emory Douglass as well as large framed posters of Black Panther martyrs and brightly colored drawings showing Afro-Americans holding guns or fighting the police.[45] In his autobiography, former BPP Chief of Staff David Hilliard recalled his emotional response to spending time at the festival: "I've gone through the cobblestone streets of the Casbah with a guide showing us the houses where Ali La Pointe—the hero of the great film *The Battle of Algiers*—actually hid out, met some of the revolutionary leaders who knew Fanon. My stomach clenches with its familiar ache. We've come a long way . . . we've become a part of history."[46]

The BPP's successful participation in the Pan-African Cultural Festival led to the establishment of an officially recognized International Section that was headed by Kathleen and Eldridge Cleaver in Algiers and would eventually include Don and Barbara Easley Cox, Michael Cetewayo Tabor, Connie Matthews, Pete and Charlotte O'Neal, Sékou Odinga, Denise Oliver, and Larry Mack.

Political scientists Charles Jones and Michael Clemons explained that Algeria became a "valuable political sanctuary for Panther expatriates" in which they had interactions with representatives from the African Party for the Independence of Guinea-Bissau and Cape Verde Islands (PAIGC), African National Congress (ANC), Pan-Africanist Congress (PAC), and the Zimbabwe African People's Union (ZAPU).[47] The official recognition that the *Front de Libération Nationale* (FLN) granted to the International Section came with the ability to receive entrance and exist visas for guests and members of the organization, a monthly stipend, and political credibility within the activist community in this emancipated space.[48] Eldridge Cleaver explained to a *New York Times* reporter: "In Algiers, the Panthers are respected as one of approximately a dozen liberation movements accredited by the Algerian Government and provided with assistance and support in their task of overthrowing the governments in power in their respective countries. This . . . is the first time in the struggle of the black people in America that they have established representation abroad . . . Here in Algiers . . . we internationalize our struggle, we show that oppression is an international problem."[49]

By validating the BPP as a revolutionary organization on par with other African nationalist liberation movements, the FLN encouraged these activists to further radicalize their political ideology and organizing strategies.

I contacted Barbara Easley Cox through her husband, Don Cox, who is currently in exile in the French Pyrenees. Historical accounts confirm that the Coxes were part of the leadership of the Oakland chapter of the BPP before they joined the International Section in Algeria. In July 2005, I interviewed Ms. Cox at home in north Philadelphia, which is decorated with a quote from SNCC co-founder Ella Baker that states, "We who believe in freedom cannot rest." Her home is near the church in which she eulogized fellow Panther Fred Hampton, who was assassinated by members of the Chicago police department in 1969. Barbara Easley Cox told me that while she lived in Algeria, her friendships with African and Arab women as well as other representatives of African liberation movements profoundly influenced her political worldview:

> I met women from other liberation movements, and that's where my knowledge of the world became more focused, because when people say, "I'm from so and so and so," you start looking at the world map, whether it's Asia, Africa, South America, and then you start listening to other people's historical battles, and then you realize that you're not the only group of little black people, this select group of African-Americans in America that are fighting for that freedom. . . . Once you travel abroad and you deal with other peoples, you start realizing it is not a white-skinned person that is the enemy. It is a class distinction, and I think the class distinction is, you know, based on a few people having a whole lot and everybody else getting nothing.[50]

Ms. Cox's experiences further illustrate how, by attracting activists from throughout the African Diaspora and providing resources to social movements, emancipated spaces facilitate transformative exchanges of antiracist ideas and tactics.

Conclusions

This article illustrates how institutions indigenous to the African Diaspora can help facilitate cross-national mutual identification between black social movements by devoting their resources to providing information about strategies for resisting white supremacy, facilitating interpersonal exchanges between activists and offering valuable resources for transnational political activism. HBCUs and emancipated spaces educated their African American constituencies about African independence movements, enabled African independence and Black Power activists to dialogue with one another, and offered a variety of resources to support the work of both movements. The work of these institutions led to transnational exchanges between the OAAU, SNCC, BPP and African anti-colonial organizations that directly affected the political ideas and tactics of Black Power.

This chapter points toward a theory of transnational engagement between social movements in the African Diaspora that I am currently exploring through further research. My work suggests that institutions indigenous to the African Diaspora help facilitate transnational exchanges between black social movements in three ways. First, indigenous institutions devote their resources to gathering and sharing information about social movement organizations and activists throughout the Diaspora. Second, these institutions filter this information through the *master injustice frames* that their communities use to continuously oppose their marginalization. *Master injustice frames* are the philosophies, narratives, symbols, and actions that historically marginalized communities most consistently utilize over time to articulate their subjugation as unjust and build collective identity. Third, indigenous institutions authenticate the claims and work of black social movements in other countries that are consistent with the parameters of their communities' *master injustice frames*. In performing these functions, institutions indigenous to the African Diaspora enhance and perpetuate the mutual identification across national boundaries that is necessary for transnational exchanges between social movements.

Notes

1. Hanspeter, Kriesi, "Cross-National Diffusion of Protest," in *New Social Movements in Western Europe*, ed. Hanspeter Kriesi, et al. (Minneapolis: University of Minnesota., 1993), 181–206; Doug McAdam and Dieter Rucht, "The Cross-National Diffusion of Movement Ideas," *Annals of the American Academy of Political and Social Science* 528 (July 1993): 56–74.

2. Doug McAdam, Sidney Tarrow, and Charles Tilly, *Dynamics of Contention*, ed. Doug McAdam, Sidney Tarrow, and Charles Tilly, *Cambridge Studies in Contentious Politics*, paperback ed. (Cambridge: Cambridge University Press, 2001), David A. Snow and Robert D. Benford, "Alternative Types of Cross-National Diffusion in the Social Movement Arena," in *Social Movements in a Globalizing World*, ed. Donatella della Porta, Hanspeter. Kriesi, and Dieter Rucht (New York: St. Martin's, 1999).

3. McAdam, Tarrow, and Tilly, *Dynamics of Contention*, 157–58.

4. Margaret E. Keck and Kathryn Sikkink, *Activists Beyond Borders: Advocacy Networks in International Politics* (Ithaca, NY: Cornell University Press, 1998).

5. Melissa Harris-Lacewell, *Barbershops, Bibles and Bet: Everyday Talk and Black Political Thought* (Princeton, NJ: Princeton University Press, 2004).

6. Frances Fox Piven and Richard A. Cloward, *Poor People's Movements: Why They Succeed, How They Fail* (1979; repr. New York: Vintage, 1979), 11.

7. Cathy Cohen, *The Boundaries of Blackness: Aids and the Breakdown of Black Politics* (Chicago: University of Chicago Press, 1999).

8. Kamari Maxine Clarke, *Mapping Yoruba Networks: Power and Agency in the Making of Transnational Communities* (Durham, NC: Duke University Press, 2004), 40.

9. Brent Hayes Edwards, *The Practice of Diaspora: Literature, Translation and the Rise of Black Internationalism* (Cambridge, MA: Harvard University Press, 2003); Penny M. Von Eschen, *Race against Empire: Black Americans and Anti-Colonialism, 1937-1957* (Ithaca, NY: Cornell University Press, 1997).

10. Robin D. G. Kelley, "'But a Local Phase of a World Problem': Black History's Global Vision, 1883–1950," *The Nation and Beyond: Transnational Perspectives on United States History; A Special Issue* 86, no. 3 (December 1999): 1045–77.

11. Benedict Anderson, *Imagined Communities: Reflections on the Origin and Spread of Nationalism* (London: Verso, 1983); Sidney Lemelle and Robin D. G. Kelley, "Introduction Imagining Home: Pan-Africanism Revisited," in *Imagining Home: Class, Culture and Nationalism in the African Diaspora*, ed. Sidney. Lemelle and Robin D. G. Kelley (London: Verso, 1994), 7.

12. Arjun Appadurai, *Modernity at Large* (Minneapolis: University of Minnesota Press, 1996), 31.

13. Harris-Lacewell, *Barbershops, Bibles and Bet*; Aldon Morris, *Origins of the Civil Rights Movement: Black Communities Organizing for Change* (New York: Free Press, 1984).

14. Edwards, *The Practice of Diaspora*; Kelley, "'But a Local Phase of a World Problem'"; Ronald W. Walters, "The Pan African Movement in the United States," in *Pan Africanism in the African Diaspora: An Analysis of Modern Afrocentric Political Movements* (Detroit: Wayne State University Press, 1997).

15. Robin Hayes, interview with A. Peter Bailey, Silver Spring, Maryland, 2005.

16. Stokely Carmichael and Michael Ekwueme Thewell, *Ready for Revolution: The Life and Struggles of Stokely Carmichael (Kwame Touré)* (New York: Scribner, 2003), 112.

17. Ibid., 137; Harry G. Lefever, *Undaunted by the Fight: Spelman College and the Civil Rights Movement, 1957-1967* (Macon, GA: Mercer University Press, 2005), 167.

18. Morris, *Origins of the Civil Rights Movement*, 176.

19. New York Federal Bureau of Investigation, "Organization of Afro-American Unity Internal Security—Miscellaneous," in *FBI File on Organization of Afro-American Unity* (New York: 1965).

20. Jervis Anderson, *Bayard Rustin: Troubles I've Seen: A Biography* (Berkeley: University of California Press, 1998).

21. Carmichael and Thewell, *Ready for Revolution*, 254.

22. Hayes, interview with Robert (Bob) Moses, Jackson, Mississippi, 2005.

23. John Lewis and Michael D'Orso, *Walking with the Wind: A Memoir of the Movement* (New York: Simon and Schuster, 1998), 81.

24. Carmichael and Thewell, *Ready for Revolution*.

25. Pan-African Students Organization in the Americas, "Press Release: 8th Annual Conference at Howard University," in *Student Nonviolent Coordinating Committee Papers, 1959-1972* (Sanford, NC: Microfilming Corporation of America, 1968).

26. "Congress of African People," in *Dr. Huey P. Newton Foundation Inc. Collection, 1968-1994* (Palo Alto, CA: 1970), http://content.cdlib.org/view?docId=tf3k40032t&chunk.id=c02-1.9.6.3.10&query=audio%20&brand=oac.

27. Ibid.

28. Federal Bureau of Investigation, "Organization of Afro-American Unity Internal Security."

29. Hayes, interview with George Houser, Pomona, New York, 2005.

30. Stuart Hall, "Cultural Identity and Diaspora," in *Identity: Community, Culture, Difference*, ed. Jonathan Rutherford (London: Lawrence & Wishart, 1990).

31. Von Eschen, *Race against Empire*.

32. Hayes, interview with Sylvester Leaks, Brooklyn, New York, 2005.

33. Hayes, interview with A. Peter Bailey.

34. Hayes, interview with Sylvester Leaks.

35. George Breitman, ed., *Malcolm X Speaks: Selected Speeches and Statements* (New York: Grove, 1965), 101.

36. In 2001, the American Committee on Africa merged with the Africa Fund and the Africa Policy Information Center (APIC) to become Africa Action (http://www.africaaction.org).

37. Hayes, interview with George Houser.

38. Hayes, interview with Robert (Bob) Moses.

39. Clayborne Carson, *In Struggle: SNCC and the Black Awakening of the 1960's* (Cambridge, MA: Harvard University Press, 1995), 104.

40. James Forman, *The Making of Black Revolutionaries: A Personal Account* (New York: Macmillan, 1972).

41. Carmichael and Thewell, *Ready for Revolution*, 613; emphasis original.

42. Donna Richards, "Memo Re: A Sync African Project," in *Student Nonviolent Coordinating Committee Papers, 1959-1972* (Sanford, NC: Microfilming Corporation of America, 1961–1967 [n.d.]).

43. Kathleen Neal Cleaver, "Back to Africa: The Evolution of the International Section of the Black Panther Party (1969–1972)," in *The Black Panther Party (Reconsidered)*, ed. Charles E. Jones (Baltimore: Black Classic, 1998), 212.

44. Conrad Clark, "Pan-African Cultural Festival Draws Blacks," *New York Amsterdam News*, July 26, 1969; "Ed Bullins Attends First Pan-African Festival," *New York Amsterdam News*, August 30, 1969.

45. Cleaver, "Back to Africa," 213.

46. David Hilliard and Lewis Cole, *This Side of Glory: The Autobiography of David Hilliard and the Story of the Black Panther Party* (Boston: Little, Brown and Company, 1993), 267.

47. Michael L. Clemons and Charles E. Jones, "Global Solidarity: The Black Panther Party in the International Arena," in *Liberation, Imagination and the Black Panther Party: A New Look at the Panthers and Their Legacy*, ed. Kathleen Cleaver and George Katsiaficas (New York: Routledge, 2001).

48. Ibid.

49. Sanche de Gramont, "Our Other Man in Algiers," *New York Times Magazine*, November 1 1970.

50. Hayes, interview with Barbara Easley Cox, Philadelphia, 2005.

PART IV

EUROPE AND ASIA ON THE COLOR LINE

CHAPTER 13

TOKYO BOUND

AFRICAN AMERICANS AND JAPAN CONFRONT WHITE SUPREMACY

GERALD HORNE

IN THE PERIOD BEFORE WORLD WAR II, Japan was probably the nation most admired among African Americans. W. E. B. Du Bois, Marcus Garvey, Booker T. Washington, and others may have had conflicts among themselves, but they all looked to Tokyo as evidence for the proposition that modernity was not solely the province of those of European descent and that the very predicates of white supremacy made no sense. This is an important point to consider for many reasons. Those who have focused on the appeal of the former Soviet Union to Americans need to consider that the choice was not necessarily between the *herrenvolk* democracy of the United States and the imperfect socialism of Moscow, but Imperial Japan was also considered as an alternative. Furthermore, historians have increasingly begun to point to external factors as a major reason for why Jim Crow began to crumble in the United States. This is usually put in the context of the cold war, Soviet aid to African liberation movements, and the indisputable point that Washington had difficulty winning hearts and minds in Africa and elsewhere among the world's majority as long as peoples of African descent in the United States were faced with Jim Crow. This focus on external factors as a cause for the erosion of Jim Crow is important, because it sheds light on why progress toward racial equality tends to flag when external pressure seems to lessen. But in assessing this external factor, we must take into account the specter of Japan, particularly in the first four decades of the twentieth century and not just the USSR from 1917 to 1991.

In addition, scholars on the left have been criticized for not treating race as an independent variable or an unmediated factor. Bringing Japan into the equation suggests the difficulty of seeking to treat race as an independent variable, just as the fact that scholars doing historical research on race—even those examining the first four decades of the twentieth century—commit scholarly malpractice when they fail to take Tokyo into account.

To be fair, part of the difficulty in unraveling Japan's influence is the reticence of European and Euro-American elites when it comes to confronting the race question beyond the black–white dyad. For example, in fighting the inaugural war of U.S. imperialism—that is, the war against the Philippines at the turn of the twentieth century—one general order of the U.S. Army declared that "such delicate subjects as . . . the race question, etc. will not be discussed at all except among ourselves and officially."[1] This trend continued during the Pacific War. Theodore White, one of the most highly regarded U.S. journalists of the twentieth century, acknowledged during his tenure in China during the war that "the ethic of the time forbade one from reporting in terms of race."[2] Frank Furedi, who has authored one of the more salient books on race in recent years, writes that not only was there reticence but also that "[It] is striking how little racist thinking was questioned before the Second World War. Even radical critics of imperialism were reluctant to criticize the racist justification for national expansion." Referring mostly to Europe, he adds. "It is striking to note how much more willing writers were to discuss class rather than race." The fear of racial revenge, which, unlike class revenge, conceivably did not have limits at least as far as Europeans were concerned, "was a major reason for this relative silence."[3]

James Belich, the leading scholar of the titanic wars that led to a stalemate between the British invaders and the indigenous people of New Zealand, argued that as a result of this humbling episode, Great Britain resorted to its "final safety net," which was "to forget."[4] John Dower writes,

> If one asks Americans today in what ways World War II was racist and atrocious, they will point overwhelmingly to the Nazi genocide of the Jews. When the war was being fought, however, the enemy perceived to be most atrocious by Americans was not the Germans but the Japanese and the racial issues that provoked greatest emotion among Americans were associated with the war in Asia, . . . Japan's aggression stirred the deepest recesses of white [supremacy] and provoked a response bordering on the apocalyptic.[5]

The war with Japan awakened the idea of racial revenge, that is, that Japanese in league with African Americans and other Asians would seek retribution for a racialized colonialism and imperialism. So provoked, European and Euro-American elites moved, even as the war was unfolding, to begin the reluctant and agonized retreat from apartheid, though like a child awakening from a nightmare, they largely chose to forget a major reason why they were taking this monumental step.

Such a retreat was far from the mind of Commodore Matthew C. Perry of the U.S. military when he stepped onto the shores of Japan in 1853, prying that nation out of more than two centuries of self-imposed isolation. Interestingly, at this turning point in world history, Perry decided to wade ashore, "marching between two orderlies, both tall and stalwart Negroes."[6] Other than the fact that in the nineteenth century a disproportionate number of U.S. sailors were black, it is unclear why Perry chose to be accompanied by blacks at this fraught moment. Perhaps he thought that the fact that Euro-Americans could subordinate and subjugate "tall and stalwart Negroes" would convince the Japanese of the invaders' power while warning them of what fate awaited them if they did not acquiesce.

Japan took the hint and over the next few decades engineered an amazing turnaround that led to the construction of the first major non-European power by the end of the century. Rather quickly, Japan became a beacon of attraction for African Americans, who thought they could learn lessons from Tokyo in how to subdue white supremacy. Du Bois was among the many Africans and Asians who saw the beginning of the end for white supremacy in Japan's defeat of Russia in 1905, since, as he wrote, "the Negro problem in America is but a local phase of a world problem."[7] In 1912, Washington told an inquiring Japanese correspondent, "Speaking for the masses of my own race in this country I think I am safe in saying that there is no other race outside of America whose fortunes the Negro peoples of this country have followed with greater interest or admiration . . . in no other part of the world have the Japanese people a larger number of admirers and well-wishers than among the black people of the United States."[8] A few years later, the FBI reported nervously that Garvey "preached that the next war will be between the Negroes and the whites unless their demands for justice are recognized, and that with the aid of Japan on the side of the Negroes, they will be able to win such a war."[9]

Du Bois, Washington, and Garvey were simply expressing the widespread admiration for Japan that permeated Afro-America. To cite one example, members of the African Methodist Episcopal (AME) Church, according to one scholar, "believed that the ability of Japanese to compete with Europeans and Americans on their own terms dispelled the myth of white superiority. Thus, AME leaders wholeheartedly supported the Japanese in the war against Russia. . . . Japan fascinated church members, who demanded information on every aspect of Japanese life and culture."[10]

Of course, the organization that was to become the Nation of Islam was probably the most zealous of the pro-Tokyo elements in black America, with its leader Elijah Muhammad even going so far as to claim that Negroes were "Asiatic," not African. But even before he arrived at this conclusion, others had beaten him to the punch. Harry Dean was a grandson of the legendary Paul Cuffee. In the early nineteenth century, Cuffee may have been the most prominent African American in the nation and certainly one of the most affluent in that he controlled a number of ships. In the late 1890s, Dean, who stressed, "I am an African and proud of it," sailed to southeast Africa where he encountered a chief whose "name was Teo Saga" and who was "more Japanese than African." Dean was told that, "before the cataclysm South Africa, Madagascar, Sumatra, Java and even Korea and Japan were all connected by land, and formed a great, illustrious, and powerful empire. The people were highly cultured, the rulers rich and wise. When the great flood came over the land it left only the remote provinces. However that may be, one may still find such Japanese names as Teo Saga on the coast of Africa to this very day."[11] This admiration for Japan was also reflected in literature. In 1913, The Crisis, journal of the NAACP, published a story that imagined a military alliance of Japan and Mexico against the United States, further supported by black deserters from the U.S. Army and the secession of Hawaii, led by angry Japanese Americans. The U.S. president was forced to appeal to Jed Blackburn, a Jack Johnson-type character who led a force of ten thousand black soldiers on a suicidal counterattack of Japan's invasion of Southern California.[12]

This literary provocation was matched across the Pacific. General Sato Kojiro's 1921 potboiler, *Japanese-American War*, imagined the surprise destruction of the U.S. Pacific fleet, occupation of Hawaii and an invasion by Japanese forces of the U.S. mainland supported by ten million blacks led by Marcus Garvey. There was more about Garvey and black unrest in the 1924 nonfiction book *The Negro Problem* by Mitsukawa Kametaro.[13]

Interestingly, these stories mirrored real life events. In 1916, the Plan of San Diego was revealed. This plan allegedly involved an abortive attempt by Chicanos and Mexico in league with Japan and other foreign powers, to dismember the United States, kill all the white males in the west and southwest and establish independent black and Indian republics, while reclaiming territory for Mexico that had been lost to the United States when Mexico itself was dismembered seventy years earlier.

As this episode suggests, like other foreign nations seeking leverage over the United States, the Japanese catered to disaffected minorities. This was not a new tactic. France had a well-merited reputation for brutality in colonizing West Africa, yet African Americans as ideologically diverse as Josephine Baker, Richard Wright, James Baldwin, and others too numerous to mention viewed Paris as a welcoming second home. Of course, there are those in Tel Aviv today who view the Arab minority in that nation as something of a security threat, subject to being wooed by less-than-friendly neighbors.

The Diplomatic Archives in Tokyo reveal that the Japanese paid close attention to African Americans. They maintained details on blacks in the U.S. military, the racial breakdown of various states, and material on black illiteracy, death rates, occupational status, as well as lists of "influential Negro Leaders" and "important Negro publications."[14]

This attention from Japanese elites was mirrored among the Japanese masses. Walter White's novel *The Fire in the Flint* was translated into Japanese with the title changed to *Lynching*; this new edition was a best seller, due in no small part to a publicity campaign by the Japanese government pointing out that the novel pictured the kind of barbarian acts that were tolerated and even encouraged in a nation (the United States) that was criticizing Japan's policies in China.[15] Even Japanese opposed to the policies of their government, for example, Katayama Sen—a founder of the Communist Parties of Japan, the United States, and Mexico, a man who had matriculated at Fisk University and was a friend of the Jamaican-American poet Claude McKay—likewise found U.S. racial policies abhorrent.

Increasingly, on both sides of the Pacific, a perception was growing among peoples of color that they had a common enemy in white supremacy; certainly this was the viewpoint of Du Bois. He argued that the exclusion of Japanese from the United States had resulted from a deal between the South and the West in which the former endorsed the Oriental Exclusion Act of 1924 in exchange for the sacrifice of the Dyer federal antilynching bill. A similar analysis could be made concerning U.S. opposition to Japan's attempt to insert a clause concerning racial equality in the post–World War I Treaty of Versailles. White Southerners feared what this might mean in terms of blacked united with those in the Far West who were concerned about what this might mean about Asian Americans and Native Americans. And of course, Japan's bitter experiences at Versailles and with the 1924 anti-immigrant bill worsened relations between Tokyo and Washington.[16]

Of course, Japanese and Negroes faced similar racist rationales. Tom Ireland, a EuroAmerican who was regarded widely as one of Cleveland's finest men, with a B.A. from Princeton and an L.L.B. from Harvard, wrote in 1935, "the Mongolian race is too divergent from a biological standpoint to intermarry or to assimilate with the Caucasian for the good of either. . . . Such miscegenation," he added, "[is] invariably bad." Hence, unlike Europeans, Asians should be barred from the United States.[17]

Bruised by the indignities of white supremacy, Tokyo adroitly played on these sentiments and made a concerted and not unsuccessful attempt to win over the black community to pro-Japan positions. The popular historian of the black experience J. A. Rogers traveled to Ethiopia to cover the Italian invasion and brought the eager readers of his newspaper columns stories about a possible merger through marriage of the Japanese and Ethiopian royal families; of course, it would have been ludicrous—perhaps even an offense worthy of a lynching—to even suggest a comparable merger through marriage of, say, the British and Ethiopian royal families. Supposedly Rogers was entertained royally in Japan and allegedly promised Tokyo "favorable publicity" when he returned to the United States.[18]

On the other hand, Rogers may have had an incentive to provide favorable publicity, even setting aside the courting he supposedly received. Rogers wrote in the *Amsterdam News* in 1934, even before the Italian invasion of Ethiopia, that Japan was aiding Africa by selling cheap clothes there. Before the arrival of their cotton goods, he argued that Africans wore clothes until they were filthy, thus breeding "lice, typhus and other diseases." He quoted an "overworked doctor" in Tanganyika who said the "purchase of cheap Japanese rubber soled shoes has done more to check hookworm here than all the efforts of the health department." The flood of Japanese imports into British-controlled territory was one of the many factors exacerbating tensions between the two island nations and accelerating the drive toward war.[19]

Because of their sympathy for Japan, many African Americans were less than sympathetic to China after the Japanese invasion in 1931. The *Pittsburgh Courier* writer Ira Lewis summed up the sentiments of many when he argued in a front-page article that "between the Japanese and the Chinese, the Negroes much prefer the Japanese. The Chinese are the worst 'Uncle Toms' and stooges that the white man has ever had." With barely concealed rancor, he added, "as soon as he gets a chop suey place which is anything like decent, the first thing he does is put up a color bar."[20] Du Bois tended to agree, as he contrasted invidiously what he saw as China's tepid response to racist U.S. immigration laws with Japan's robust response. He too referred the Chinese as "Asian Uncle Toms" of "the same spirit that animates the white folks in the United States."[21] Many African Americans in the Far West held firmly to the perception that Japanese Americans were much more willing to flout the dictates of Jim Crow and serve black customers in their restaurants and hotels than Chinese Americans, who were seen as much more willing to observe the dictates of antiblack racism.

Moreover, many blacks were overreacting to the denunciation in the mainstream U.S. press of the Japanese invasion in China, which they saw as hypocritical in light of these same papers' failure to condemn the white supremacy that Europe and the United States had imposed on Shanghai, for example. Du Bois summed up the thoughts of many African Americans when he posed this query on arriving in China in the 1930s: "Why is it," he inquired, "that you hate Japan more than Europe when you have suffered more from England, France and Germany than from Japan." Du Bois announced in late 1941 that "the British Empire has caused more human misery than Hitler will cause if he lives a hundred years . . . it is idiotic to talk about a people who brought the slave trade to its greatest development, who are the chief exploiters of Africa and who hold four hundred million Indians in subjection, as the great defenders of democracy."[22] Months after Pearl Harbor, Adam Clayton Powell, Jr., concurred with Du Bois when he noted that the "difference between nazism and crackerocracy is very small," since "crackerocracy [too] is a pattern of racial hatred."[23] Parenthetically, as I examine Chinese foreign policy in the 1970s in southern Africa, which featured alliances with apartheid South Africa and U.S. imperialism, it is difficult to escape the idea that African American intellectuals may have been on to something when they complained about China's willingness to collaborate with European and Euro-American imperialism. Thus, as tensions rose between the United States and Japan in the 1930s, Tokyo came to realize that it might have an ally in African Americans.

This sentiment was cultivated assiduously by a number of Japanese nationals who resided in black communities like Harlem. A keen example was Yasuichi Hikida, an animated, graying man who always showed up at Negro social functions accompanied by a Negro woman. Like many Japanese nationals in New York who were collaborating with Tokyo, he worked for an affluent white family in Forest Hills while maintaining a residence in Harlem. He wrote an unpublished biography of Toussaint L'Ouverture and had one of the finest collections of books on blacks in New York short of the Schomburg Library (he was also a close friend of Arthur Schomburg). As late as 1941, he appeared at a debate in

Harlem where, as he put it idiomatically, "Chinese were Jim Crowed by whites." He also referred to blacks as "our darker brothers." Hikida was not unusual. W. C. Handy, composer of the "St. Louis Blues," recalled a Japanese cook who traveled about the country for five years as a member of his vaudeville troupe and who later turned out to be an eavesdropping Japanese army officer. The Japanese valet of the actor Charlie Chaplin also turned out to be an agent for Tokyo.[24]

The U.S. government was not totally oblivious to these maneuvers. Before Pearl Harbor, U.S. intelligence agencies intercepted a Japanese message that spoke of their use of a "Negro literary critic" whose purpose was to "open a news service for Negro newspapers. The Negro press is so poor that it has no news service of its own and as I have told you in various messages, [we] had been getting relatively good results . . . because of the advantage we have in using men like this . . . as an experiment," the message went on ominously, "I am now instructing Mr. [name deleted] of the National Youth Administration, and a graduate of Amherst and Columbia to be a spy." The message continued, "in organizing our schemes among the Negroes . . . Washington . . . should be our hub"; though it was added, "in the arsenals of Philadelphia and Brooklyn there are also a few unskilled Negro laborers, so I would say that in the future there will be considerable profit in our getting Negroes to gather military intelligence for us . . . we have already established connections with very influential Negroes to keep us informed with regard to the Negro movement."[25] Scholar Tony Matthews has argued that Tokyo turned for spying toward the "American Negroes, a massive force of largely disgruntled citizens, many of whom had a special racial axe to grind."[26]

Of course the antifascist tendencies among African Americans should not be underestimated, though I should add that I find it striking that the man considered the "brains" of U.S. fascism—a man who met with Mussolini, attended the Nazi Party gathering at Nuremberg in 1937, and wrote voluminously—was a black graduate of Exeter and Harvard: Lawrence Dennis. On the other hand, it is arguable that pro-Japan organizations attracted many more adherents among blacks than their pro-Soviet counterparts. Thus Robert Hill, the leading scholar of Garveyism, writes that the Pacific Movement of the Eastern World "gained a substantial black nationalist following during the 1930s. One longtime leader, David D. Erwin, claimed that it had forty thousand members in 1936, while other estimates went as high as one million."[27] Nor should we forget that one of the more influential organizations among African Americans today, the Nation of Islam, is a direct descendant of this pro-Tokyo movement.

Robert Leonard Jordan was one of the key leaders of this movement. Born in Jamaica in 1900, he moved to England at the age of fourteen and at eighteen left on a Japanese ship; by 1920, like so many other Jamaicans, he was in Harlem. By 1936, he was President General of the Ethiopia Pacific Movement with an office at 204 Lenox Avenue in Harlem. That year he addressed a lengthy letter to the Japanese foreign minister, Hachiro Arita; the stationary listed one "T. Kikuchi," a Japanese national, as the group's "chief business advisor." This letter, which can be found in the Diplomatic Archives in Tokyo, noted that "we the dark race of the Western Hemisphere through the Ethiopia Pacific Movement . . . are putting our entire confidence in the Japanese people with the hopes that in the very near future, we will desire a very close relationship with the Japanese government."[28]

Along with Elijah Muhammad, Jordan was among the most prominent of the pro-Tokyo spokesmen in the United States; certainly he was the most prominent of this group in Harlem. The FBI reported that, while toiling for the aforementioned Japanese shipping company and residing in Japan, he "found the Japanese to be very friendly to the Negroes and that he had the privilege of studying the customs of the Japanese and becoming a member of . . . society in Japan."[29]

By early 1942, Japan's largely successful invasion of Asia revealed that many Asians saw no reason why they should fight for colonial masters who openly professed white supremacy in London and The Hague. African Americans like Jordan were arriving at the same conclusion. In January of that year, a meeting of black leaders voted thirty-six to five (with fifteen abstaining) that African Americans were not 100 percent behind the war. Before that, in 1939, the FBI reported that "'enlightened Negro leaders' had told them 'that between eighty and ninety per cent of the American colored population who have any views on the subject at all, are proJapanese as a result of the intensive Japanese propaganda among this racial group.'" Right after Pearl Harbor, one-half of the blacks interviewed in New York City told black interviewers that they would be better off, or at least no worse off, under Japanese rule.[30]

Inferentially on December 13, 1941, days after the attack on Pearl Harbor, the *Amsterdam News* revealed some of the linkages that tied blacks to Tokyo:

> Immediately following [Pearl Harbor, the NYPD] invaded Harlem and began rounding up all Japanese suspects. . . . In view of the fact that with the exception of marked facial distinction there is somewhat of a striking similarity in hue between the [Japanese] and many Harlemites . . . colored policemen played an invaluable role in the mass arrests . . . the area on Lenox Avenue between 110th and 116th street [is] noted for [some] Japanese restaurants. . . . Many of the sons of Nippon . . . declared "me colored man too" when they were tabbed.

Yet it was men like Jordan and also Carlos Cooks, a man venerated by Elombe Brath, one of Harlem's leading black nationalists of today and a host of a popular radio program on Pacifica Radio, who were in the vanguard of this trend. But between Jordan and Cooks, it was the former who was probably better known in Harlem at the time. The *People's Voice*, a popular front newspaper that despised these nationalists, conceded that Jordan, whom it called the "Harlem Mikado," had an "eloquence [that] is said to have driven a number of competing street speakers to introduce [Japanese] propaganda in their talks to hold audiences."

This evident ideological hegemony of Jordan apparently drove the popular front in Harlem to distraction. The *People's Voice* told its readers that Japan had a "BB Plan," that is,

> Black Followers of Buddhism [which] preached Buddhism as the religion of people of color the world over [and] the key to racial success. . . . Under the BB

Plan, American Negroes who become Buddhists automatically won Japanese citizenship, would get chances to visit Japan, study sciences and professions, receive military and naval training. . . . Success of the plan would mean establishment of a black empire in Africa . . . PV's investigations have uncovered the fact that there may be some connection between the world B plan and the activities of Duse Mohammed Ali.

In trumpeting this alliance between Buddhists and Muslims and followers of Moorish Science, this left-wing paper noted that "the scope of the world B Plan of the Japanese is almost unbelievable"; it scornfully denounced the "cunning of an Oriental group" that has "gone back to the wars of the Crusaders in the interest of Christianity."[31]

What had driven the popular front to the point of hysteria was the fear that pro-Japan sympathies among African Americans, carefully cultivated over the years by Tokyo and propelled by a vile white supremacy, could complicate the war effort and lead to the victory of the anti-Comintern Axis. In the fall of 1942, many of these pro-Tokyo blacks were arrested. The indictment of James Thornhill, one of these leaders in New York, charged that he said that "colored United States soldiers should not fight the Japanese" and that, like the man who became Malcolm X, he might "shoot the wrong man" if drafted and given a rifle. Thornhill was also born in the Caribbean (U.S. Virgin Islands); like Jordan he derisively referred to the United States as the "United Snakes of America." Repeatedly he told Harlemites, "you should learn Japanese"; "when they tell you to remember Pearl Harbor, you [should] reply 'Remember Africa.'" With fervor he added, "the white man brought you to this country in 1619, not to christianize you but to enslave you. This thing called Christianity is not worth a damn. I am not a Christian, we should be . . . Moslems."[32]

With evident anxiety, the FBI reported in 1943 that "numerous complaints have been received that the American Negroes favor a Japanese victory in the present war."[33] This sentiment was not unique to black Americans. The Colonial Secretary in Kingston, Jamaica was told in 1941 that at one "Cold Supper Shop" there was frequent "anti-British talk" heard via the "wireless."[34] It should not be forgotten, even when analyzing communists' approach to the war, that it was hard for many, particularly those of African descent, to accept the argument that Britain was the fountainhead of democracy, particularly when Churchill already had announced that the Atlantic Charter's promise of democracy did not apply to those subjected to a racialized colonialism. Even Hugh Mulzac, a member in good standing of the popular front, wrote that there was a "strong feeling among colored Americans in 1941 that the colonial powers be allowed to destroy each other. As a former British subject I felt this keenly."[35]

Thus, just as black nationalists expressed outright sympathy for Tokyo, some blacks on the left found it difficult to provide unalloyed support for the Allies in light of the latter's white supremacy. As Du Bois put it, "If Hitler wins, down with the blacks! If the democracies win, the blacks are already down."[36] The Allies were well aware of this black hostility to white supremacy that made Africans worldwide susceptible to Tokyo's siren call. Hence, during the war, the British Colonial Office was reluctant to initiate an anti-German campaign among West Africans because officials calculated that such propaganda might encourage a revolt against rule as such. "Having been encouraged to hate one branch of the white race, they may extend the feeling to others," warned one memorandum.[37]

Strikingly, the ideological ancestors of today's black conservatives also were both sympathetic to Tokyo and highly critical of the Allies. George Schuyler is the main example in that regard. A prolific journalist and novelist, he too had been courted by Tokyo and wrote a series of articles on Japan in the 1930s that were so pro-Nippon that his publisher refused to print them.[38]

In September 1940, as many black communists were scoring Tokyo because of its policies in China, Schuyler took an opposite tack; he saw this invasion as an exemplar of the:

> progressive deflation of white supremacy and arrogance in the Orient. Where white men once strutted and kicked coolies into the street, they now tread softly and talk in whispers. . . . The Japanese have done a fine job in making the white man in Asia lose "face" and shattering the sedulously nurtured idea of white supremacy. Of course the white people hate them because they fear them.[39]

In his withering denunciations of white supremacy, the conservative Schuyler often used rhetoric that would have made black nationalists proud. By 1944 the fervor in the black community for the Double V campaign against fascism at home and abroad had dissipated as many shifted to a pro-Allies stance, but not George Schuyler. Days after D-Day in Europe, he wrote acerbically that:

> the Europeans have been a menace to the rest of the world for the past four hundred years, carrying destruction and death wherever they went. . . . True, their system of world fleecing directly benefited only a handful of Europeans, but indirectly it ben-efited millions of supernumeraries, labor officials and skilled workers. . . . Europe has been a failure and a menace. The European age is passing. One can derive a certain pleasure from observing its funeral.[40]

Schuyler was likely the staunchest critic of the internment of Japanese Americans in the United States, returning to this subject again and again in his work. That this internment, he asserted, was "a scheme to grab [Japanese-American] holdings and hand them over to white people is shown by the efforts to prevent Negroes from taking them over. . . . This may be a prelude to our own fate. Who knows? . . . Once the precedent is established with 70,000 Japanese-Americans," he added ominously, "it will be easy to denationalize millions of Afro-American citizens."[41]

Ironically, after the war concluded, U.S. elites cracked down on the black communists, who were harshest in their condemnation of Tokyo, whereas those like Schuyler who took an opposing position were promoted. Interestingly black communists, who were the most consistently anti-Tokyo force among blacks during the war, suffered most after Tokyo was defeated. The assault on black Reds created favorable conditions for the rise of black nationalists, who had been pro-Tokyo; thus the organization that was to become the Nation of Islam rose, just as the popular front and the left in the black community diminished. The end of the war also marked the decline of pro-Tokyo sentiments among African Americans as Japan became a reliable ally of Washington. Indeed, the kind of sympathy for Asia that these pro-Tokyo movements symbolized did not arise again until the era of Maoist China.

On the other hand, this veritable race war, in which those who defined themselves as "white" seemed to be losing in the early stages of the conflict, helped to convince sober-minded elites in Washington to retreat from the more egregious aspects of white supremacy. Even as the war was unfolding, the United States sought to do away with the "white primary," which limited black voting rights, and struck from the books most of the Chinese exclusion laws. There is nothing like the prospect of losing a race war to convince even the most obtuse of the necessity of eroding racism.

Still, in 1944 during the height of the war, Du Bois concluded, "the greatest and most dangerous race problem today is the problem of relations between Asia and Europe."[42] Yet in the United States today, race discourse not only focuses on the black–white dyad but also refuses to stray beyond the shores of this nation. Indeed, though black intellectuals of an earlier era wrote voluminously about Asia, few do so today. This may be because blacks of an earlier day were effectively denied citizenship and therefore were compelled to be internationalist; ironically, part of the downside of full citizenship rights has been the erosion of black internationalism. Moreover, the decision made decades ago to shroud the race question in Asia has borne fruit by helping some to think that the vaunted "color-blind" approach characterizes relations between those of Asian and European descent. This misconception is heightened by the use of the vague term "westerner," which is used to describe Europeans who reside to the west of Asia as well as Australians and New Zealanders who live east of there. African Americans are no longer "Tokyo bound," but race remains a global concern.

NOTES

1. Brian McAllister Lin, *Guardians of Empire: The U.S. Army and the Pacific, 1902–1940* (Chapel Hill: University of North Carolina Press, 1997), 60.
2. Frank Furedi, *The Silent War: Imperialism and the Changing Perception of Race* (London: Pluto, 1998), 164.
3. Ibid., 6, 101.
4. James Belich, *The New Zealand Wars and the Victorian Interpretation of Racial Conflict* (Auckland, New Zealand: Auckland University Press, 1986), 235, 321.
5. John Dower, *Japan in War and Peace: Essays on History, Culture and Race* (London: Fontana, 1996), 258–59.
6. Samuel Etiot Morison, *"Old Bruin": Commodore Matthew C. Perry, 1794–1858* (Boston: Little, Brown, 1967), 332.

7. W. E. B. Du Bois, "The Color Line Belts the World," in *Writings by W.E.B. Du Bois in Periodicals Edited by Others*, ed. Herbert Aptheker (Millwood, NY: KrausThomson, 1982), 330 (from *Collier's Weekly*. October 20, 1906).

8. Booker T. Washington to Naoichi Masaoka, December 5, 1912, in ed. Louis Harlan and Raymond W. Smock, *Booker T. Washington Papers, Volume 12: 1912–1914* (Urbana: University of Illinois Press, 1982), 84.

9. "Bureau of Investigation Reports," New York, December 5, 1918, in ed. Robert A. Hill, *The Marcus Garvey and Universal Negro Improvement Association Papers, Volume 1, 1826–August 1919* (Berkeley: University of California Press, 1983), 306.

10. Lawrence C. Little, "A Quest for Self-Determination: The African Methodist Episcopal Church During the Age of Imperialism. 1884–1916" (PhD diss., Ohio State University, 1993), 227, 252.

11. Harry Dean, *Umbala: The Adventures of a Negro Sea-Captain in Africa and on the Seven Seas in His Attempt to Found an Ethiopian Empire* (London: George Harrap, 1929), 93.

12. Kevin K. Gaines. *Uplifting the Race: Black Leadership, Politics and Culture in the Twentieth Century* (Chapel Hill: University of North Carolina Press, 1996), 206.

13. David Levering Lewis, *W. E. B. Du Bois: The Fight for Equality and the American Century, 1919–1963* (New York: Henry Holt, 2000), 392.

14. Reports, November 28, 1933, circa 1930s and 1940s, 1460-1-3, Diplomatic Archives, Tokyo.

15. Walter White, *A Man Called White: The Autobiography of Walter White* (Athens: University of Georgia Press, 1995), 69.

16. Lewis, *W. E. B. Du Bois*, 416.

17. Tom Ireland, *War Clouds in the Skies of the Far East* (New York: G. P. Putnam, 1935), 288, 292.

18. Patrick Washburn, *A Question of Sedition: The Federal Government's Investigation of the Black Press during World War II* (New York: Oxford University Press, 1986), 261.

19. *New York Amsterdam News*, June 9, 1934.

20. Washburn, *A Question of Sedition*, 110; *Pittsburgh Courier*, March 28, 1942.

21. *New York Amsterdam News*, November 18, 1931; Lewis, *W. E. B. Du Bois*, 413–15.

22. *People's Voice*, May 2, 1942.

23. *People's Voice*, September 12, 1942.

24. Roi Ottley, *Inside Black America* (London: Eyre & Spottiswoode, 1948), 256.

25. Intercepted messages, July 2, 1941, May 9, 1941, Box 2, Franklin D. Roosevelt Papers, Franklin D. Roosevelt Library, Hyde Park, New York.

26. Tony Matthews, *Shadows Dancing: Japanese Espionage against the West, 1939–1945* (London: Robert Hale, 1993), 27.

27. Robert A. Hill, *The Marcus Garvey and Universal Negro Improvement Association Papers, Volume 7: November 1927–August 1940* (Berkeley: University of California Press, 1991), 506.

28. Robert Jordan to Hchiro Arita, November 18, 1936, A461, ET/II, Diplomatic Archives, Tokyo.

29. J. Edgar Hoover to Jonathan Daniels, August 11, 1943, Box 6, 4245g, Official File, Franklin D. Roosevelt Papers, Franklin D. Roosevelt Library, Hyde Park, New York.

30. Marc Gallicchio, *The African American Encounter with Japan and China: Black Internationalism in Asia, 1895–1945* (Chapel Hill: University of North Carolina Press, 2000), 107, 121, 142.

31. *People's Voice*, February 28, 1942; March 7, 1942; March 21, 1942.

32. *USA v. James Thornhill*, Southern District Court of New York, Box 1049, R33.18.2.5, C-113–264, 1942, National Archives and Records Administration, New York.

33. Hill, ed., *The FBI's RACON: Racial Conditions in the United States during World War II* (Boston: Northeastern University Press, 1995), 81.

34. To Colonial Secretary, September 4, 1941, "Fifth Column Activity," 1B/5/77/49, CSO 750, 1941, National Archives of Jamaica.

35. Hugh Mulzac, *A Star to Steer By* (New York: International Publishers, 1963), 129.

36. Ibid.; Lewis, *W. E. B. Du Bois*, 467.

37. Mulzac, *A Star to Steer By*, 129; Furedi, *The Silent War*, 184.

38. George Schuyler, *Black Empire* (Boston: Northeastern University Press, 1991), 281.

39. *Pittsburgh Courier*, September 14, 1940.

40. *Pittsburgh Courier*, June 17, 1944.

41. *Pittsburgh Courier*, April 25, 1942; May 29, 1943.

42. Du Bois, "Prospect of a World without Race Conflict," *American Journal of Sociology* 49, no. 5 (March 1944): 450–56, 451.

CHAPTER 14

FEMME NÉGRITUDE

JANE NARDAL, *LA DÉPÊCHE AFRICAINE*, AND THE FRANCOPHONE NEW NEGRO

T. DENEAN SHARPLEY-WHITING

From henceforth there would be some interest, some originality, some pride in being Negro, to turn oneself towards Africa, the cradle of the Negro, to remember a common origin. From these new ideas, new words, have come the revealing terms: Afro-American, Afro-Latin.

—Jane Nardal, "Black Internationalism," *La Dépêche africaine*

It is a peculiar sensation, this double consciousness, this sense of always looking at one's self through the eyes of others, of measuring one's soul by the tape of a world that looks on in amused contempt and pity. One ever feels his two-ness-an American, a Negro; two souls, two thoughts, two unreconciled strivings; two warring ideals in one dark boy, whose dogged strength alone keeps it from being torn asunder.

—W. E. B. Du Bois, *The Souls of Black Folks*

COINED BETWEEN 1936 AND 1937 BY THE MARTINIQUAN POET AIMÉ CÉSAIRE during the writing of his now-celebrated *Cahier d'un Retour au Pays Natal* [*Notebook of a Return to My Native Land*], Négritude as a poetic, literary, cultural, and intellectual movement, signaled the birth of a Pan-Africanist cultural nationalism among black Francophone writers, a "New Negro" from the Francophone world. Although the neologism is readily traceable to Césaire, mapping the concept of Négritude as the inauguration of black humanism, as a "theory of black cultural importance and autonomy,"[1] remains the stuff of a panoply of critical works.

Before the 1935 publication of *L'Etudiant Noir*, a one-issue journal sponsored by the Association des Etudiants Martiniquais en France that featured, according to Georges Ngal, at least, "les deux textes fondateurs du mouvement de la

Négritude" [the two founding texts of the Négritude movement],[2] by Aimé Césaire and the Senegalese poet Léopold Sédar Senghor, there were a number of
black Francophone novelistic and journalistic precursors that treated the themes
of assimilation, colonialism, race consciousness, and identity. The most notable
among those texts were Rene Maran's 1921 Prix Goncourt-winning *Batouala*;
veritable *roman negre*; Suzanne Lacascade's 1924 *Claire-Solange, une Africaine*; the
journals *La Dépêche Africaine*, *Le Cri des Negres*, and *La Revue du Monde Noir*;
and the 1932 Marxist-Surrealist pamphlet Légitime Défense. Other race-related
journals appeared and disappeared in the early to late 1920s. But it was not until
September 1931 that Senghor made the acquaintance of Césaire and the Guyanese Leon-Gontron Damas, who was the third voice of this poetic trilogy, thus
setting the stage for their collective exploration of their conflicting identities, the
"tormenting question," in the words of Senghor, of "Who am I?" their experiences of being black, African, African-diasporic, and French. For Césaire and
Damas, "in meeting Senghor, [they] met Africa." Through Damas and Césaire,
Senghor's horizon was opened to the dynamism of the literary and cultural worlds
of West Indians and African Americans living in Paris in the 1930s. For their
part, Césaire, Senghor, and Damas, the designated founders of this poetics in
the French-speaking world, provide a conspicuously masculine genealogy of their
critical consciousness. They credit the writers of the Harlem Renaissance, specifically Claude McKay, Langston Hughes, James Weldon Johnson, and Sterling
Brown; Du Bois' *Souls of Black Folk*; the philosopher Alain Locke's 1925 anthology
The New Negro: An Interpretation; Carter G. Woodson's *Opportunity*, the popular
organ for the Urban League; and the National Association for the Advancement
of Colored People's (NAACP) political vehicle *The Crisis* as primary influences on
their consciousness about matters of race and identity. As Senghor revealed in an
article in *Présence Africaine* entitled "*Problematique de la Négritude*," "the general
meaning of the word [Négritude], the discovery of black values and recognition
for the Negro of his situation, was born in the United States of America."[3]

   The masculinist genealogy constructed by the poets and shored up by literary
historians, critics, and Africanist philosophers continues to elide and minimize
the presence and contributions of black women, namely their Francophone counterparts, to the movement's evolution. In effect, if African American writers of
the 1920s radicalized the consciousness of these young and aspiring Francophone
black writers, if the race-conscious New Negro of the United States planted the
seeds of Négritude in their collective imagination, then the three future Négritude
poets also received inspiration from Mesdemoiselles Jane and Paulette Nardal.[4] In
a letter written in February 1960, Senghor revealed: "We were in contact with
these black Americans during the years 1929–34, through Mademoiselle Paulette
Nardal, who, with Dr. Sajous, a Haitian, had founded *La Revue du Monde Noir*
[*Review of the Black World*]. Mademoiselle Nardal kept a literary salon, where
African Negroes, West Indians, and American Negroes used to get together."[5]
In correspondence also dated in the year 1960 and sent to Senghor's biographer
Louis Jacques Hymans, Paulette Nardal "complained bitterly" of the erasure of

her and Jane Nardal's roles in the promulgation of the ideas that would later become the hallmarks of Césaire, Damas, and Senghor. They "took up the ideas tossed out by us and expressed them with flash and brio." In effect, Nardal wrote, "we were but women, real pioneers—let's say that we blazed the trail for them."[6]

The *soeurs* Nardal, with their Sunday literary salon and review, did more than provide a cultured place (their apartment) and literary space (*La Revue du Monde Noir*) for the intellectual coming-of-age of the trio. In the words of Guadeloupean writer Maryse Condé, "It is an accepted fact that French Caribbean literature was born with Négritude."[7] But what is not such a widely accepted or acknowledged fact is that women writers and thinkers of the Négritude era, such as Jane Nardal, were at the movement's literary and philosophical centers and, often too, at the vanguard.

On February 1928 the journal of the Comité de Defense des Interets de la Race Noire (CDIRN), *La Dépêche Africaine*, was published in Paris under the editorial direction of the Guadeloupean Maurice Satineau, secretary of the comité, with a prestigious multiracial board of collaborators.[8] The newspaper's motto, "Defendre nos colonies, c'est fortifier la France" [To defend our colonies, is to fortify France] sums up the interesting patchwork of militant colonial reformism, assimilationism, and cultural Pan-Africanism found in its monthly columns. Serving as a means of correspondence "between Negroes of Africa, Madagascar, the Antilles and America," *La Dépêche Africaine* maintained that, in the assimilationist and colonial reformist fashion of the era, "the methods of colonization by civilized nations are far from perfect; but colonization itself is a humane and necessary project."[9] As the spiritual inheritor of writer René Maran's defunct bimonthly *Les Continents, La Dépêche Africaine* consistently evoked the ideas of 1789 France and Schoelcherism, that is, the liberal principles of Victor Schoelcher, a French administrator responsible for abolishing slavery in the French colonies. In this *après-guerre* France, where unemployment emerged unchecked alongside xenophobia, racism, and paternalism and where primitivism (the realm of "Negro") became the rage in Paris as people attempted to forget the war-ravaged and morally bankrupt Europe and exotic literature amply filled in the spaces in between, black French-speaking intellectuals wanted to revive France as a paragon of liberty, fraternity, and equality—as a civilizing and civil nation.

With its global readership, *La Dépêche Africaine* was quite popular during its four year publication run. In the November 1928 issue, under the rubric "*Une Bonne Nouvelle*" ("A Piece of Good News"), the organ announced that by January 1929 the monthly would become a bimonthly periodical, appearing the first and the fifteenth of each month: "The increasingly favorable reception that the colonial and metropolitan public reserves for our organ obliges us to augment our frequency."[10] According to a report filed by the police prefecture with the Ministry of Colonies in November 1928, *La Dépêche Africaine* had nearly ten thousand copies in circulation.[11] But by January 1929, the journal apologized to its readership, explaining that it would "momentarily be unable to appear two times per month as previously announced due to administrative reorganization."[12] And by

May 1929, the journal's management settled on publication on the thirtieth of every month.[13]

But *La Dépêche Africaine* continued to inconsistently publish on the first, fifteenth, or thirtieth of any given month. Such inconsistency would later be revealed in reconnaissance reports as financially tied. According to a report on the activities of the journal filed by government officials on May 30, 1930, the editors were in need of funds: "Les dirigeants de *La Dépêche Africaine* sont parvenu a rassembler les fonds necessaire pour faire paraitre leur feuille" [The managers of *La Dépêche Africaine* came up with the necessary funds to publish the newspaper], and "Le numero date Fev-Mars est sorti des presses de l'imprimeur au debut d' Avril" [The issue scheduled for February-March left the printing presses at the beginning of April].[14]

The journal's mission was to address social, political, and economic issues: "*La Dépêche Africaine* [is] . . . an independent journal of correspondence between blacks, for the moral and material interests of the indigenous populations, through the objective study of the larger colonial questions considered from political, economic, and social points of view."[15] In this diasporic vein, the periodical infrequently published a section in English subtitled "United We Stand, Divided We Fall," under the editorship of Fritz Moutia.

Delving into anglophone black politics, the organ took on the Scottsboro case in the United States. The Scottsboro case involved nine black men who allegedly raped two white women. Coerced by authorities into giving false statements, the women later retracted the allegations. Seven of the men were nonetheless facing execution. In the April 1, 1932, issue, *La Dépêche Africaine* ran a front-page article and letter to President Herbert Hoover. Under the headline "Un supreme appel au President Hoover: L' execution des sept Negres serait un crime contre l'Humanité" [A Supreme Appeal to President Hoover: The Execution of the Seven Negroes Would Be a Crime Against Humanity], Maurice Satineau and Georges Forgues, president of the CDIRN, appealed thus: "Profoundly moved by the sentence that the Supreme Court of Alabama is about to hand down against the seven Negroes accused of rape, we are making an appeal to your greater conscience, to your sense of justice, humanity and equity in order to prevent such an execution which would be considered, by all races and notably, the black race . . . as a crime of which one cannot predict the repercussions."[16]

The official French government response to *La Dépêche Africaine* was dogged monitoring of its editor, Satineau, and the newspaper's contents. With its Pan-Africanist politics, *La Dépêche Africaine* was linked, according to officials, in spirit, politics, and content to René Maran's reformist journal *Les Continents* and Garveyism—thus a potential threat to the colonial powers that be.

According to one agent's report, "The address and telephone number are presently those of the management of the journal *La Dépêche Africaine*. It is thus no longer possible to deny the relations that unite this paper and the pan-black organization [United Negro Improvement Association, UNIA]. This explains as well why we remarked in the note of last February, page 10: '*La Dépêche Africaine* reminds one of the defunct pan-black organ *Les Continents*.'"[17]

The fact that after his expulsion from the United States, Marcus Garvey set up offices in the seventh *arrondissement* of Paris at 5 rue Paul Louis Courieur, the same address as *La Dépêche Africaine* and with the same telephone number, certainly gave the administration canon fodder. It was also more than pure coincidence for the government that *La Dépêche Africaine* announced that it would augment publication following Garvey's October 1928 visit: "le journal announce qu'il va devenir bientôt bi-mensuel: serait-ce une consequence du passage de Marcus Garvey à Paris?"[18] Internal divisions also plagued *La Dépêche Africaine*. Although some of the collaborators "energetically conformed to Garveyist ideas," others, like Satineau, wanted to strike a balance between Garvey's pan-black agenda—to maintain UNIA financial support—and curry favor from the colonial administration.[19] In response to being targeted as Garveyists, the management shifted the organ's focus to Guadeloupean politics and began to feature sympathetic articles on colonial administrators, which government agents regarded as mere "camouflage."[20] After 1930, the political content was viewed less cynically and as nonsubversive to French colonial interests. Garveyism, with its glorification of the black race and the recognition of ancient African civilizations, Du Bois's racially imbued Pan-Africanism, and the cultural emphasis of Locke's New Negro Movement would nonetheless continue to be filtered through a Francophone lens and appropriated in *La Dépêche Africaine*'s cultural and literary criticism.

The sections "La Dépêche Politique," "La Dépêche Economique et Sociale," and "La Dépêche Litteraire" owed their cultural and literary pan-noirisme to the global literacy of Mademoiselles Jane and Paulette Nardal. *La Dépêche Africaine*'s first issue in February 1928 presented its prestigious collaborators. A photograph of Jane Nardal listed her qualifications as "Licencée-des-lettres." Paulette Nardal joined the magazine's roster in June 1928. She was duly titled "Professeur d'Anglais," and her contribution to the journal would be to "write a series of articles on the economic and literary evolution of Black Americans."[21] Paulette Nardal's specialties were, however, more literary and artistic than socioeconomic. Both sisters were Sorbonne-educated, bilingual Martiniquans. Their race-conscious transnational finishing school was at the salon of René Maran where they met various African American artists and writers, such as Alain Locke, Claude McKay, Augusta Savage, and Langston Hughes. Paulette, the elder of the two, wrote seven pieces for the journal, including two short stories, a comparative essay on Antillean and black American music entitled "Musique Negre: Antilles et AfraAmerique," and an extensive article (with photographs) of Harlem Renaissance sculptor Augusta Savage's work.[22] Jane, on the other hand, wrote two provocative essays on literary exoticism and black cultural internationalism and coined the neologism "Afro-Latin." Although their work has been often referred to as "proto-Négritude" and, hence, as setting the stage for the veritable movement, such an assessment is primarily tied to the fact that the word itself would not come into being for at least another eight years. But Jane Nardal's "Internationalisme Noir," an essay that discusses race consciousness among the African Diaspora and cultural *métissage*, would provide an essential kernel of the philosophical foundation for the literary and cultural movement later celebrated the world-over as Négritude.

Jane Nardal's article appeared in the "Dépêche Economique et Sociale" section of the first issue of *La Dépêche Africaine*. Her name and an appraisal of her essay as politically "modest" in tone also appeared in a March 1928 agent's résumé of activities of "Blacks from the colonies in the Metropole."[23] Her article outlines in broad brush strokes several concepts that would become pivotal in early Négritude parlance: *après-guerre nègre*, global community, Black Internationalism, Afro-Latin, *conscience de race*, New French-speaking Negro, *esprit de race* (spirit of race). In an *après-guerre* commentary, Nardal begins by suggesting that one of the aftereffects of World War I was an attempt to break down barriers between countries. The peace conference at Versailles, as well as the meeting of "four unobtrusive gentlemen" in Paris in 1919 to settle the "destinies of mankind," as Du Bois remarked, underscore Nardal's observation.[24] Like Du Bois, Nardal links the broader implications of the formation of a Euro-American community after the war to that of the global black community in formation. If the war led European and American world powers to envision themselves as a human community with common interests, the war also gave rise to "a vague sentiment" among blacks that they too were part of that human community who "in spite of everything belong to one and the same race."[25] The renewed spirit of humanism, which weighed heavily in the post–1918 air that by turns attempted to exclude and marginalize blacks, had actualized among black elites un esprit de race. Although the concept of a unified black race was not a necessarily novel idea among those who helped to construct and shore up the idea from Buffon to Jefferson to Gobineau, the subjects—"blacks"—of their philosophies of race did not, according to Nardal, accept in principle such a notion: "Previously the more assimilated blacks looked arrogantly upon their brothers of color, believing themselves to be clearly from another species; on the other hand, those blacks who had never left African soil, who had never experienced slavery, looked upon those who had at the whim of whites been enslaved, then freed, then modeled into the imago of the white as vile cattle."[26] This Nardalian vista of ethnic schisms between the Francophone African Diaspora is a shocking one that underscores conflict rather than identification.

World War I and its attendant hardships, the unfulfilled promises of citizenship and dashed hopes of equal rights, the presence of black soldiers of various ethnic backgrounds in Europe, the discovery of *l'art negre* by European votaries of the "primitive" and African literature, civilizations, religions, and sculpture, Negro spirituals, and the publication of philosopher Alain Locke's "New Afro-American Poetry" in *Les Continents* combined to evoke sentiments of diasporic connectedness and sentimental glances toward the ancestral homeland of Africa.

Africa was no longer the European stereotyped and promulgated land of savages but rather a cultural mecca, henceforth to be proudly claimed by its Diaspora.

From the rubble of enslavement, bestowed freedom in the French West Indies, and the tangle of French assimilation, a new consciousness about Africa and race inspired the invention of a new identity: the Afro-Latin. Nardal imagines a symbiosis between her Frenchness, Latinness. and Africanness into a new identity, a new self-consciousness. She would not renounce her *latinité* or her *africanité*. This symbiosis was to bring into existence an authentic self-consciousness. The Afro-Latin shares the experience of the Afro-American, of Du Boisian double consciousness, of double-appartenence ("double belonging") in, "which one merge[s] his double self into a better and truer self. In this merging he wishes neither of the older selves to be lost. He would not Africanize America, for America has too much to teach the world and Africa. He would not bleach his Negro soul in a flood of white Americanism, for he knows that Negro blood has a message for the world. He simply wishes to make it possible for a man to be both a Negro and an American."[27] Similarly, in its use of the masculinist language of the time as well as in its philosophic content, "Black Internationalism" is not merely about the formation of a global black community or emergent race consciousness but also the synergy of the "Afro," or African and the Latin world, an embracing of cultural *métissage*, of double-appartenance, in order to return "en soi" [to the self]:

> From these new ideas, new words, have come the revealing terms: Afro-American, Afro-Latin. They confirm our thesis all in casting a new meaning on the nature of this Black Internationalism. If the Negro wants to be himself, to affirm his personality, not be the copy of such and such type of another race (which often earns him contempt and mockery) it does not follow however that he becomes resolutely hostile to all contributions of another race. He must to the contrary profit from the acquired experience, from the intellectual richness, through others, but in order to better know himself and affirm his personality. To be Afro-American, Afro-Latin, that means to be an encouragement, a comfort, an example, for the blacks of Africa in showing them that certain benefits of white civilization do not necessarily lead to a denial of their race. . . . Africans on the other hand can profit from this example by reconciling these teachings with their ancient traditions of which they are rightly proud.[28]

Nardal's cultural mixing does not dismiss African traditions or reserve for Africa the gift of emotion rather than reason—a racialist distortion of her theorizing that would later crop up in Léopold Senghor's "Humanisme et Nous" and *Liberté: Négritude et Humanisme*. Her fatalism, if you will, emerges around the French colonial project. In turn, however, she calls for symbiosis as a path of cultural and racial resistance to passive assimilation. In her analysis, the Afro-Latin will assimilate the Latin rather than be wholly assimilated by it. And yet hostility to "une autre race," an allusion to the more strident interpretations of Garveyist thinking, clearly undermines the new humanist spirit that she attempts to outline. The return to the self, the excavation of the glories of the black race with European critical tools of engagement, allows for a better understanding of who one is as "black" and "French," "Afro," and "Latin." In keeping with the "'benefits of colonialism" ideological paradigm of *La Dépêche Africaine*, she shifts the concept of Afro-Latinité from the West Indian context to the Franco-African terrain, where mass indigenous resistance to the colonial project had proven most formidable. Versed in the ethnological and sociological literature of the day on Africa, Maurice Delafosse, Leo Frobenius, and Jean PriceMars's *Ainsi Parla l'Oncle*, Nardal writes that "the cultivated man" would not treat Africans *en masse* as savages since "sociological works have made known to the white world the centers of African civilizations, their religious system, their forms of government, their artistic riches."[29]

Although sympathetic to the "bitterness" that the Africans express in seeing the effects of colonization on these "millenaire" traditions, she argues conversely that colonial policy in Africa could be a source of "racial solidarity" and race consciousness among the "different African tribes."[30] She further offers up "Afro-Americans" as pioneers of cultural resistance, innovation, preservation, and race consciousness blended with their fierce Americanism, suggesting to her Franco-phone readers that they immerse themselves in Locke's *New Negro*. The combined socioeconomic, political, and historical realities of black life under the yoke of American racism contributed to the formation of a unique Afro-American culture that speaks to such experiences that chronicle the coming-into-being of the "New Negro." For Nardal, without the hardships of those experiences as marginalized Americans there would not have been the creation of the other: a race-conscious new American Negro. The French-speaking black, the "Afro-Latin," reasons Nardal, "in contact with a race less hostile to the man of color than the Anglo-Saxon race, has been retarded in this path" of race consciousness, "authentic" cultural development, and race solidarity.[31] The Afro-American will, for Nardal, serve as a model for the Afro-Latin, for Afro-Latin literature, art, and culture as such is nonexistent. Nardal encourages the Afro-Latin to recognize that Africa and "*le pays latin*" are not "incompatibles."[32] She then prophesies the emergence of New Francophone Negroes, who "encouraged by the example of black American intellectuals, will distinguish themselves from the preceding generation . . . [and], schooled in European methods, will make use of them to study the spirit of their race, the past of their race, with the necessary critical verve."[33]

As Promethean around issues of identity and race in the black Francophone context as Nardal's concepts of Afro-Latinité and cultural *métissage* were in 1928, they were also importantly subversive and indeed went philosophically counter to the univeralism supposedly inherent in French humanism, culture, and colonial policies on "nos colonies et nos indigenes" [our colonies and our natives]. Nardal suggests a decentering of Frenchness. Yet, for France, the essence of the words "culture" and "civilization" are French. With sword or gun in hand, France believes it is imparting the "souverainete de ses lois et la marque de son genie" [the sovereignty of its laws and the mark of its genius] to the conquered.[34] They will "do them [the natives] good in spite of themselves" through the gift of Frenchness, as Frantz Fanon noted in his critique of French colonialism in Algeria.[35] Indeed, Jules Michelet wrote in his *L'Introduction a l' Histoire Universelle*:

The Frenchmen wants above all to imprint his personality on the vanquished, not because it is his, but because it is the quintessence of the good and the beautiful; this is his naive belief. He believes that he could do nothing that would benefit the world more than to give it his ideas, customs, ways of doing things. He will convert other peoples to these ways, sword in hand, and after the battle, in part smugly and in part sympathetically, he will reveal to them all that they gain by becoming French.[36]

Ideas of difference, integration, symbiosis, synergy, syncretization, and hyphenations were threatening to *l'esprit français national*, the body politic, and the belief in the "plus grande France."[37] And although Nardal advocated racial identity politics for specifically cultural and literary purposes, politically such race-conscious politics would prove to be the bane of French colonialism in Africa in the coming decades, for it would set in motion the historical processes of decolonization. Was Nardal aware at the time of the insurgent subtext of her black humanism and cultural pan-blackness? Years later, she and her sister Paulette would insist on their unique intentions to create a cultural, not political, movement similar to that of Locke's *New Negro* in America. Jane Nardal's "Black Internationalism," however, with its somewhat embellished résumé of a nascent race consciousness among Francophone blacks, which would lead to a cultural explosion *à la mode* American, that is, to Négritude, interestingly mapped out the first stage in the dialectics of liberation for the colonized. Such a politically charged stand would arouse government suspicions and shadow the two sisters in their future race consciousness-raising cultural collaboration, the bilingual *La Revue du Monde Noir*.

NOTES

1. Aimé Césaire, *Notebook of a Return to My Native Land* [Cahier d'un retour au pays natal], trans. Abiola Irele (Ibadan, Nigeria: New Horn Press, 2005).
2. Georges Ngal, "Lire," *Le Discours sur le Colonialisme* (Paris: Presence Africaine, 1994), 12.
3. Janet Vaillant, *Black, French, African: A Life of Leopold Sédar Senghor* (Cambridge, MA: Harvard University Press, 1990), 1.
4. Georges Ngal, "Lire," *Le Discours sur le Colonialisme* (Paris: Presence Africaine, 1994), 13.
5. Leopold Sédar Senghor, "Problematique de la négritude," *Présence Africaine* 78 (1971): 12–14.
6. The three future Négritude poets also received inspiration from Cuban writer Nicolas Guillen and Haitian writers Jacques Roumain and Jean Price-Mars and employed as tools of critical engagement the ethnology of Frobenius and Delafosse.
7. A 1960 letter written by Senghor cited in Lilyan Kesteloot, *Black Writers in French: A Literary History of Négritude*, trans. Ellen Conroy Kennedy (Washington, DC: Howard University Press, 1991), 56.
8. Louis Jacques Hymans, *Leopold Sédar Senghor: An Intellectual Biography* (Edinburgh: University Press of Edinburgh, 1971), 36.

9. Maryse Condé, "Language and Power," *College Language Association Journal* 3, no. 1 (1995): 19.

10. *La Dépêche Africaine* stopped publication in 1932, then resumed in 1938 for one year before it eventually folded for good.

11. "Notre But-Notre Programme," *La Dépêche Africaine*, February 15, 1928, p. 1. All translations provided by the author.

12. "Une Bonne Nouvelle," *La Dépêche Africaine*, November 15, 1928, p. 2.

13. Archives d'Outre Mer in Aix-en-Provence (ADO), Slotfom V, 2 *La Dépêche Africaine*, November 29,1928.

14. *La Dépêche Africaine*, January 15, 1929, publicity page.

15. *La Dépêche Africaine*, May 30, 1929, p. 3.

16. See ADO, Slotfom V, 2 *"La Dépêche Africaine* et reports policiers."

17. "Notre But-Notre Programme," *La Dépêche Africaine*, February 15, 1928, p. 1.

18. "Un supreme appel," La Dépêche africaine, April 1,1932, p. 1.

19. ADO, Slotfom III, 81, no. 595. The original French report notes: "C'est l'addresse et ce numero de telephone sont presentement ceux de la direction du journal *'La Dépêche Africaine.'* Il n'est plus possible de nier les relations qui unissent cette feuitle et l'organization pan-noire. Ceci explique pourquoi, ainsi qui nous le signalions dans la note du Fevrier demier page 10, *'La Dépêche Africaine'* a evoqué le souvenir du defunt organ pan-noir *'Les Continents."*

20. ADO, Slotfom III, 81, no. 678, Decembre 18, 1928: "Le numero 9 de la "Dépêche africaine" apparu le 15 Novembre. Le camouflage persiste: le portrait du nouveau ministre des colonies est publie en premier page avec de commentaire sympathique." ("Issue number 9 of 'La Dépêche africaine' appeared on November 15th. The camouflage persists: the portrait of the new Minister of Colonies is published on the front page with sympathetic commentary").

21. "La Dépêche Iitteraire," *La Dépêche Africaine*, June 15, 1928, p. 4.

22. Paulette Nardal, "Musique negre: Antilles et AfraAmerique," *La Dépêche Africaine*, June 15, 1930, p. 5; "Vne femme sculpteur noire," *La Dépêche Africaine*, September 15, 1930, p. 5.

23. ADO, Slotfom III, 81, no. 132, March 13, 1928.

24. W. E. B. Du Bois, "Negro at Paris," in *Writings by W. E. B. Du Bois in Periodicals Edited by Others*, vol. 2, 1910–1934, ed. Herbert Aptheker (Millwood, NJ: Kraus-Thomson Ltd., 1982), 127.

25. Jane Nardal, "Internationalisme noir," *La Dépêche Africaine*, February 15, 1928, p. 5.

26. Ibid.

27. W. E. B. Du Bois, *The Souls of Black Folk: Essays and Sketches* (New York: Fawcett, 1961), 17.

28. Nardal, "Intemationalisme noir," 5.

29. Ibid.

30. Ibid.

31. Ibid.

32. Ibid.

33. Ibid. The original French reads: "jeunes AfroLatins" ("young Afro-Latins") who, "aides, encourages par les intellectuels noir americains, se separant de la generation precedente . . . les formes aux methodes europeenes, ils s' en serviront pour etudier l' esprit de leur race, le passe de leur race, avec tout l' esprit critique necessaire."

34. Raoul Girardet, *Le Nationalisme franrais: Anthologie, 1871–1914* (Paris: Seuil, 1983), 86.

35. Frantz Fanon, *A Dying Colonialism* (New York: Grove, 1965), 63.

36. Jules Michelet, *L'Introduction a l'histoire universelle* (Paris, 1834), 78–79.

37. Girardet, *Le Nationalisme*, 86.

CHAPTER 15

REGIONALISM AGAINST RACISM

THE TRANSEUROPE STRUGGLE FOR RACIAL EQUALITY

CLARENCE LUSANE

EUROPE IS POISED TO MAKE THE MOST SWEEPING CHANGES IN ANTI-DISCRIMINATION legislation since the beginning of unification. While a few states have specific legislation against discrimination, most current members of the European Union (EU) and those that will be joining over the next few years do not. The obligation to implement the EU's Article 13 Race Directive and Employment Directive will radically transform discourse and policy regarding racism and equality across much of Europe.[1] Whether this effort will be met with a serious commitment by states to enforce the new legislation will determine to what degree Europe will genuinely become multicultural and multiracial in content as much as in form.

Despite decades of aggressive activism and legislation, premature declarations that racism has been addressed and official denial that racism even exists, grassroots organizing and global conferences, in nation after nation racial and ethnic inequalities, marginalization, and psychological and physical violence doggedly persist. At the same time there is a growing number of regional laws and policies against racism. The increased number and range of these policies, which are mandatory under regional agreements, provide an important resource for activists and policy makers who are seeking remedies for issues related to racialized social exclusion and discrimination. No other regional political machinery has generated the breadth of legislation, policies, and enforcement mechanisms to address discrimination. Recent political developments in Europe, specifically the defeat and withering of Social Democratic governments, might erode or even reverse this progress. The emerging debate in the EU on developing a cooperative cross-national immigration policy, deep differences on Iraq, and growing economic competition have generated fissures in the alliance that are likely to last for some time to come. For the moment, however, the forces of antiracism, antiracist legislators, and non-governmental organizations (NGOs) are having significant success as new legislation continues to be created and passed.

Attempts at supra-national governance are rarely effective in addressing issues of racism and discrimination. Although UN declarations and condemnations have been plentiful, the capacity to enforce an antiracist policy framework has had only limited success at the international level.[2] European regionalism, with a stronger coercive ability due to the willingness of states to concede significant sovereignty in many areas, perhaps offers the best way of addressing racism beyond the limits of the nation-state.

THE WORLD AGAINST RACISM

Since at least the end of World War II, many policymakers, activists, and scholars have seen the struggle against racism in supra-national terms.[3] In the immediate postwar period, following the racist rampage that the Nazis unleashed upon the world, the international community concluded that racism and extreme intolerance should no longer be considered solely a matter of national concern. As the founding charter of the United Nations states in chapter 1, article 1, one of the main purposes of the United Nations is "To achieve international cooperation in solving international problems of an economic, social, cultural, or humanitarian character, and in promoting and encouraging respect for human rights and for fundamental freedoms for all without distinction as to race, sex, language, or religion." The struggle against racism is also highlighted in articles 2 and 16 of the 1948 United Nations Universal Declaration of Human Rights (UDHR). More ideologically muscular than the charter or the UDHR were the declarations and resolutions passed in the 1960s and 1970s. The most important of these was the International Convention on the Elimination of All Forms of Racial Discrimination (ICERD) passed in 1965 and went into effect in 1969. The ICERD has one of the largest number of signatures and ratifications of any of the UN conventions, more than 165 states at last count. Article 1, section 1 of the ICERD proposes a broad definition of racial discrimination as "any distinction, exclusion, restriction or preference based on race, color, descent, or national or ethnic origin which has the purpose or effect of nullifying or impairing the recognition, enjoyment or exercise, on an equal footing, of human rights and fundamental freedoms in the political, economic, social, cultural or any other field of public life." Although its definition of racial discrimination is problematic and difficult to use in legislation, countless activists and scholars have employed and cited it.[4]

The convention also created the Committee for the Elimination of All forms of Racial Discrimination (CERD), which serves as the monitoring body for the ICERD. Signatories are required to present bi-annual reports to the CERD on the status of racial equality in their states and the actions taken to address inequality and discrimination. This committee of nonpartisan experts reviews the reports and then makes recommendations to states on improvements. The value of this reporting scheme is that it forces states to confront, at least officially, the issue of racism and to offer some solutions.[5]

At the same time, a fundamental contradiction has always existed in this effort. The politics of antiracism clashes with the equally recognized and sanctioned politics of nation-state sovereignty, which declares that extranational, regional, and international authorities and actors cannot cross political borders and interfere in a state's "internal affairs." Broadly speaking, sovereignty has won the battle between the two and the UN has lacked the legal, political, and economic authority to enforce its policies, achieve its aims, or have states live up to their commitments regarding racial discrimination.[6]

The UN World Conference against Racism (WCAR), held in Durban in the fall of 2001, reflects both the strengths and limits of UN efforts to tackle this question. Bringing together thousands of activists, NGOs, policymakers, world leaders, and others, the WCAR placed the issue of equality on a global stage and drafted a wide range of progressive recommendations through its Final Declaration and Program of Action. Few believe, however, that most states will follow through on their commitments to the recommendations or that the UN will have the power or resources to ensure that effective follow-up occurs.

A TRANS-EUROPE APPROACH

This dilemma was recognized fairly early by many outside observers and thus has shaped how regional and non-UN transnational bodies approached the issue. In Europe, the EU, the Council of Europe (COE), the European Parliament, and even the Organization for Security and Cooperation in Europe (OSCE) have addressed these concerns.[7] Learning from the UN experience, governing regional institutions in Europe have struggled to create binding agreements that, like international treaties and domestic laws, would be enforceable either through legal measures or through political or economic sanctions or incentives.

It must be emphasized that a robust legal regime is not in itself a cure for racism. In the United States, where relatively strong civil rights laws have existed for half a century and constitutional protections have existed for twice as long, racism in the form of social and economic disparity, racial profiling, political disfranchisement, media stereotypes, Islamophobia, and institutionalized discrimination continues. The same can be said for other racially tense states that have laws against racism, such as Canada, South Africa, Brazil, and Colombia.

Legal remedies nevertheless play an important and often determinant role in the context of a broader antiracism policy. Particularly in Europe, where anti-discrimination law has had little currency in weak states with vacuous legal systems, regional policies are a potential bridge to democratic rights for the socially marginalized. Legal remedies, as opposed to political will, may be the most expeditious and realistic means for redress and justice for racial and ethnic minorities in many European states. As James Goldston points out, "For members of minority groups . . . strategic litigation aimed at politically independent judges may be the only means of vindicating fundamental rights in societies where numerical majorities retain the political power to deny those rights at will."[8] The creation of these laws in Europe with nearly universal support from member states and a relentless effort by NGOs indicate a will to address the issue of racism even if only in limited terms and in the name of national interest. In turn, the laws become potential mobilizing vehicles for regional and domestic antiracist forces. The laws, in significant ways, legitimate the antiracist movement, even as conservative media and politicians decry them as "race baiting" and "political correctness."

In post-9/11 Europe, those who are or are believed to be Arab or Muslim have come under media slander (fear of the "bogus" asylum seeker rhetoric) and verbal and physical attacks.[9] Politicians of all parties, the media, and the extreme right have demonized (and racialized) immigrants and refugees, especially asylum seekers. The electoral advances of the Far Right in a number of states reflect a growing tension over the development of growing racial, ethnic, and religious diversity in the region. Disturbing long-term trends of racial and ethnic disparities in housing, health care access, policing and criminal justice, educational attainment and exclusions, negative media images, and lack of political representation underscore these relatively recent events.

The struggle for a unified Europe takes place on many terrains, including that of human rights. Human rights activism in Europe, seen in its broad context, includes addressing the tension between the majority white populations of the region with the permanent and growing communities of different colors, voices, and religions. The face of Europe has changed dramatically in the last two generations. It is a fact that many people embrace these changes as positive and welcome the diversity and amalgamated cultural changes that they bring. However, a significant number of political leaders and individual citizens, the mass base for "Fortress Europe," are fiercely resistant to these changes and are mobilizing across the region. It is these forces of obstruction that antiracism must confront.

CRISES OF TRANSITION:
RACE AND RACISM IN CONTEMPORARY EUROPE

Europe finds itself in a very different political and social world than where it was a decade and a half ago. The earthshaking political and economic transitions—the dissolution of the Soviet Union, the war in the Balkans, unprecedented immigration from the global south, and the growing policymaking role of the EU—have generated xenophobia and political opportunism. Furthermore, this crisis of transition is long-term and growing more complex and multifaceted as intersecting forms of oppression, identity, and interest search for an elusive common ground.

A major factor driving the antiracist legislative agenda in the EU and the COE is the rapid rise and electoral success of far-right political parties in a number of states. More clever and politically sophisticated than in the past, these parties use the coded language of anti-immigration, crime, and "cultural cohesion" to build broad mass support. While explicit references to particular racial groups as undesirables are mostly avoided and some groups even include people of color, it is clear that the anti-immigrant rhetoric is not concerned with white refugees from Canada, the United States, or New Zealand. However, when it comes to Muslims, explicit racist language is often employed without hesitation. Such verbiage was the political capital of the late Dutch right-wing political leader Pim Fortuyn among others. Most worrisome is the rise of these parties in the so-called liberal welfare states (Norway, Denmark, and Sweden), as well as the Netherlands and Belgium. More than fifteen million Europeans have voted for far-right parties in the last ten years.

Three racial episodes in 2002 gave antiracist activists in the region reason to pay attention to Europe's swing rightward: the second-place finish of the far-right candidate Jean-Marie Le Pen in the May first-round presidential election in France; the rise, success, and subsequent assassination of Pim Fortuyn in the Netherlands; and the election of three British National Party candidates to local office in the United Kingdom, a number that has expanded by five times in 2003. In each of these campaigns, the right used racialized anti-immigration and xenophobia to their advantage.[10] Unlike their peers of a generation or two ago, the extremists no longer stand outside of the mainstream political system and often tone down their rhetoric or disguise their aims. Strong antiracism laws in the region effectively prevent overtly racist parties from being accepted into mainstream political life. The willingness of the EU to censure Austria after the Freedom Party gained enough votes to become part of a coalition government demonstrates the success of anti-racist activists and policymakers in outlawing explicit racism. Virtually all far-right political leaders with aspirations to national office are forced to declare themselves not racist and even to bring a token number of racial or ethnic minorities into their public circles.[11]

The danger is not that the far-right is going to dominate electoral politics, since in no European state has any far-right party won more than 30 percent of the popular vote nationally. The ideological and political danger of the moment is that far-right parties are determining political and policy agendas across the region. It is their spin on immigration, jobs, crime, education, and culture that is shaping events and social intercourse. In all of these areas, new initiatives are being appropriated from the playbook of the Right, (sometimes) massaged by spinmeisters and made more palatable, and then transformed into mainstream policy by conservative, centrist, and even Social Democratic governments. In the areas of crime and immigration, for example, conservative and even ostensibly liberal parties have used rhetoric and proposed harsh policies very similar to those advocated by the Right.

Even as these parties assert themselves, a rash of racist and deadly violence from neo- Nazi and ultra-fascist groups continues in the region. In the last two years alone, there have been brutal beatings and murders in the United Kingdom, Norway, Spain, Germany, Russia, and elsewhere.[12] These attacks are tied to the discourses from the Right on crime, immigration, and cultural purity as they helped create an atmosphere of intolerance where such assaults are acceptable. The September 11th attacks in the United States have given new energy to the Far Right and to racist forces in Europe and elsewhere. Exploiting the fears of millions of citizens, they not only offer easy and bigoted solutions to complex and deeply rooted problems, but they also pose a challenge to democracy that drives current European efforts at antiracism.

EUROPEAN ANTIRACISM LAWS AND POLICIES

Europe has responded to these challenges in a manner that is more regional than domestic. While only a few states have actually changed their domestic laws to specifically outlaw particular racist behavior (such as hate language, use of certain symbols, or distribution of hate materials), regional institutions have taken giant steps in that direction. In addition, they have made it mandatory for membership in the EU and the COE that states have, at a minimum, signed the key international and regional human rights, anti-discrimination, and antiracism conventions and covenants. Though the efforts are under-resourced and politically tenuous, the institutions have also backed up this legislation with judicial power and monitoring groups. The European Court of Human Rights, which was founded by the COE and handles cases that are in violation of the European Convention on Human Rights (ECHR), has binding authority making its rulings enforceable.[13] The Council and the EU have the power to sanction, fine, or even expel recalcitrant members who refuse to abide by the court's rulings or decisions.

The regional anti-discrimination laws, as a whole, are more progressive than domestic laws in the various nations, which in some instances have attempted to replicate inappropriately U.S. policies. In the United Kingdom, for example, the Race Relations Act (1976) was modeled, to a significant degree, on U.S. civil rights legislation. While a "one size fits all" approach is not tenable, a genuine cross-national effort by the regional institutions makes sense in helping states with little experience in constructing anti-discrimination legislation or who are reluctant to do so. The strength of regional antiracist laws lie in their collective and cumulative impact. The body of conventions, laws, protocols, policies, and monitoring bodies together constitute a broad and comprehensive, though still incomplete, legislative regime against racism. No one instrument alone does the job, but together they complement each other.

While the Race Directive (described later in this chapter) does not specifically protect asylum seekers (article 3.2), the proposed Protocol 12's prohibition against "discrimination on any ground such as sex, race, colour, language, religion, political or other opinion, national or social origin, association with a national minority, property, birth or other such status" does (article 1.1). The ECHR does not technically allow for antiracist complaints to stand alone (article 14), but Protocol 12 (article 1.1 and article 3) corrects that problem. Both Protocol 12 and the ECHR do not adequately address indirect discrimination, however, the Race

Directive strongly outlaws all forms of racism, including more covert, oblique versions (article 2.2b).

EUROPEAN UNION

What began as primarily an economic effort, the European Economic Community, has evolved into a much wider body that also addresses issues of social policy, such as human rights. Article 13 of the 1997 Treaty of Amsterdam, long fought for by European human rights groups and antiracist organizations, provided the EU with a legal basis to authorize direct and formal action against discrimination based on sex, disability, religion or belief, sexual orientation, age, and racial or ethnic origin. As a consequence, in 2000 the European Commission adopted two directives that required the members and candidate members of the EU to make substantial changes in their domestic laws and policies regarding this discrimination. The Race Directive (RD) and the Employment Directive (ED)—which cover age, religious, sexual orientation, and disability—afford protectionin the areas of employment, self-employment, working conditions, membership in workers' organizations, and whistle-blowing. The RD also provides protection in the areas of education, social security, access to goods and services, and cultural life.

According to James Goldston, Senior Counsel of the European Roma Rights Center, the RD "is the most important and far-reaching development in the field of European discrimination law in recent times."[14] All EU member states (and those seeking accession) must create laws, policies, regulations, and administrative bodies that reflect the principles and guidelines outlined in the directive. Both direct and indirect discrimination are addressed and defined in article 2 of the RD. The RD defines indirect discrimination more forcefully than the existing definition in the UK Race Relations Act, one of the few states with such legislation. Article 3 defines the scope of the legislation as covering both the public and private sectors with no small firm exemption. This provision covers many job occupations as well as social areas and the provision of goods and services. In section 3.2, it is noted that the RD "does not cover difference of treatment based on nationality" or apply to "any treatment which arises from the legal status of the third-country nationals and stateless persons concerned." A broad interpretation of this statement would mean that little or no protection exists for refugees and asylum seekers. However a more narrow reading, given the principles of the RD and its explicit declaration in the preamble that "prohibitions against discrimination should also apply to nationals of third countries," implies that this provision should only concern the area of immigration law where states are usually given broad leeway in international law. As states transpose the RD into domestic law, it will be critical to ensure that this provision does not become a cover for draconian and racist policy attacks on refugees and asylum seekers.

In article 5, the notion of "positive action," as opposed to "positive discrimination," is promoted. This means that states can take steps to voluntarily promote

and institute means of generating racial equality (positive action) but are pro-
hibited from imposing quotas or penalizing members of the majority population (positive discrimination). In the past, European law was not too keen on
the stronger positive discrimination when it came to race. In carefully worded
language, article 5 notes that states can adopt "specific measures to prevent or
compensate for disadvantages linked to racial or ethnic origin."

This has been one of the most controversial approaches in the field of antiracist policy-making. An extensive and sometimes successful campaign against
"affirmative action" has existed in the United States for many years. In many
nations in Europe, where minority communities are vastly underrepresented in
countless occupations and education, positive action will probably not be enough
to address these concerns. Activists will find that they must push for stronger
positive discrimination or affirmative action to achieve their equality goals. The
importance of protecting those who expose racist behavior (a protection written
into article 9) is made clear by the numerous cases of harassment launched against
employees who speak out against racist language and behavior.

One continuing problem with the European antiracist laws is that relatively
few people know about them. Addressing this concern, article 10 calls for the
promotion of a public media campaign as well as the training of public bodies
around the policies and principles of the RD. It will be important that antiara-
cist activists are involved in both governmental and non-governmental education campaigns around the RD and other laws. Article 12 requires governments
to consult and communicate with relevant NGOs and other organizations
around the implementation and interpretation of the RD. As with many other
regional and international legal instruments, there is a reporting obligation,
which in this case is every five years. This process is open to comments and views
of relevant NGOs.

Activists have also called for aggressive implementation of the RD, funding
access to legal aid, extension of the scope for positive action, meaningful and
effective consultation, effective publicity, and communication. The Starting Line
Group, a coalition of Europe-wide antiracist and human rights NGOs, which
worked for a number of years to develop and lobby for the passage of the RD,
has offered a number of additional recommendations to strengthen the directive.
These include a call for extending the RD to cover religious bias; inclusion of a
ban on incitement or pressure to discriminate; express protection for legal persons
and third-country nationals against discrimination; coverage in national implementing legislation of all public bodies, particularly the police and immigration
authorities; provisions in national implementing legislation that allow relevant
organizations to bring cases in their own name and on behalf of other complain-
ants; and provisions in national implementing legislation that allow relevant organizations to initiate investigations of alleged discrimination among others.[15]

A large number of human rights, antiracist, and minority rights NGOs have
struggled for a progressive transposition of the Race Directive into domestic law.
The European Network Against Racism (ENAR) is one network that has been

active throughout the region in mobilizing grassroots involvement around the new laws. It is comprised of about six hundred NGOs from around the region. In January 2003 and February 2004, it held conferences to assess the progress in transposing the directive into national law and to share organizing experiences.

The target date for implementation by member states of the RD was to be no later than July 2003. However only about half of the states of the fifteen-member EU—Belgium, Denmark, France, Italy, the Netherlands, Portugal, Sweden, and the United Kingdom—had full or partial legislation either in place or drafted by that time.[16] Activists will have to continue to monitor and lobby states vigorously to achieve full implementation of the directive. They will also have to monitor and work with the anti-discrimination efforts of the COE.

COUNCIL OF EUROPE

The COE has historically expressed great concern regarding the issue of racism. The clearest example is in article 14 of the 1950 ECHR that reads, "The enjoyment of the rights and freedoms set forth in this Convention shall be secured without discrimination on any grounds such as sex, race, colour, language, religion, political or other opinion, national or social origin, association with a national minority, property, birth or other status." In addition, other provisions of the ECHR can also be interpreted as protections against racism, including article 3 (right to liberty and security), article 6 (right to a fair trial), article 7 (the legal principle of "no punishment without law"), article 9 (freedom of thought, conscience, and religion), and article 10 (freedom of expression). Other relevant legal instruments and policies from the COE include the European Social Charter (1961), European Convention on the Legal Status of Migrant Workers (1983), Framework Convention for the Protection of National Minorities (1995), European Convention on Nationality (1997), Convention on the Participation of Foreigners in Public Life at the Local Level (1997), and EuropeanCharter for Regional and Minority Languages (1998).

The biggest crack in the political armor of the ECHR rests in article 14, previously cited. Article 14 is not "self-executing" and can only be invoked in relation to another violation of the Convention. In other words, a charge of race or sex discrimination, for instance, cannot be made separate from another rights violation in the Convention, such as the provisions against torture (article 3) or enslavement (article 4). While this linkage is used on occasion, article 14 has been vastly underused for the most part. Furthermore, Article 14 does not define what discrimination entails or what grounds are covered; that is, it is an open-ended provision that has also made it difficult to bring cases under its ambit.[17]

The definition of a violation of article 14 based on the cases that the European Court of Human Rights has accepted, argues Livingstone, consists of three elements: different treatment, lack of legitimate justification, and disproportionate implementation.[18] In the Court's rulings, different treatment—discrimination—is allowed if there is legitimate justification and there is proportional application of that different treatment. It is important to note that different treatment need not mean that a particular injury has occurred. This has led to rather complex interpretations of the balance between justification and proportionality and an unstable and ill-defined threshold.

Another area of considerable debate is the definition and constitution of racial groups. This is not covered by the ECHR and leads to some legalistic contradictions. In the United Kingdom, for example, where the ECHR has been translated into domestic law—the Human Rights Act—this vagueness has led to legal bias. The current definition, as laid down by Lord Fraser in Mandela v. Dowell, means that Sikhs and Jews are defined as racial groups and, therefore, receive full protection under the Race Relations Act while Rastafarians, who are not seen as having a long history, and Muslims, who are not seen as homogeneous, are not defined as legal racial groups and not covered under the Act.

While it is arguable that other UK laws, particularly regarding religious rights, offer protection, the glaring distinction between what many would consider groups that fall into the same category underscores the need for clarity in definition of terms. Activists have long sought to have Article 14 as a "stand alone" provision. In 2000, to address this concern, the COE proposed Protocol 12. Unlike the RD and ED, which became law upon passage, Protocol 12 must be signed and ratified by ten member states before it becomes law. More than twenty-five member states of the COE have signed but not ratified Protocol 12. Georgia was the first state to ratify it in June 2000. While the United Kingdom has thus far refused to sign, let alone ratify the Protocol, a campaign has started to push for its passage.

The European Commission against Racism and Intolerance (ECRI), the Council's monitoring body on racism, operates in a way similar to the CERD. It investigates and publishes reports on the status of race relations in all forty-three member states. Similarly, it makes recommendations on how states can improve their antiracist policy efforts. ECRI works with NGOs in the region in compiling the report.

CONCLUSION

Overall, the initiatives from the EU and COE will have an important impact on antiracist legislation and practice in Europe and the United Kingdom. European institutions will find themselves under increasing pressure to limit the drive of antiracism as representatives of conservative governments likely have more and more input into these bodies. How the new laws will be implemented and how they will affect current law remains to be seen. The translation of the frameworks into concrete laws and policies will be determined by the degree of collaboration and agreement reached by governments, NGOs, the private sector, and other interested parties. As elsewhere, some will argue that this legislative activity is driven by the devil of political correctness and that racism has effectively been eliminated or curtailed, especially in the modern, advanced states of France,

Germany, and the United Kingdom. After all, black and Asian faces are more prominent than ever in the media, sports, parliaments, and even business spaces of these states. Beneath this surface of inclusion is a very pervasive and persistent regime of discrimination, inequality, and bigotry. Antiracist campaigners will have to continue to make the case for the necessity of the legislation being proposed at the regional level by carefully documenting and presenting a rendering of contemporary racism. Building on the current wave of legislation progress, equality activists in Europe have the opportunity to produce some of the most important antiracist, anti-discrimination, and human rights policies available anywhere.

NOTES

1. The text of Article 13 reads: "Without prejudice to the other provisions of this Treaty, and with the limits of the powers conferred by it upon the Community, the Council, acting unanimously on a proposal from the Commission, and after consulting the European Parliament, may take appropriate action to combat discrimination based on sex, racial or ethnic origin, religion or belief, disability, age, or sexual orientation." The full title of the Race Directive is "Council Directive Implementing the Principle of Equal Treatment Between Persons Irrespective of Racial and Ethnic Origin" (OJ L 180, 19/07/2000), and the full title of the Employment Directive is "Council Directive Establishing a General Framework for Employment Equality" (OJ L 303, 02/12).

2. See, for example, UNESCO's four statements against racisms—a Statement on Race (1950), Statement on the Nature of Race and Race Differences (1951), Statement on the Biological Aspects of Race (1964), and Statement on Race and Racial Prejudice (1967)—as well as the United Nations Declaration on the Elimination of All Forms of Racial Discrimination (November 20, 1963) among others.

3. Prior to World War II, many black activists in the United States, Asia, and Latin America advocated a global thrust against racism, particularly in their colonial manifestations. The leading global powers, however, did not seriously address the issue until Nazism demonstrated the disaster of unchecked racism. See Paul Gordon Lauren, *Power and Prejudice: The Politics and Diplomacy of Racial Discrimination* (Boulder, CO: Westview, 1996).

4. Many problems arise out of the ICERD definition, including the lack of a definition of race and a definition of racial discrimination that is far too broad to be useful. The ICERD definition is a political one, and it fails at the practical policy level, which is why few states or policy makers have actually employed it.

5. The United States government signed the ICERD in 1969 but did not ratify it until 1994. It submitted its first report in 2000.

6. There is considerable debate over whether the ICERD is "self-executing," that is, whether it has the force of domestic law after being signed and ratified by a state. The U.S. government has consistently insisted that the UN conventions and covenants do not have the force of law or authority within the United States or any of its territories.

7. The principle OSCE agency for addressing human rights and discrimination is its Office for Democratic Institutions and Human Rights. In September 2003, the OSCE held a conference specifically focused on racism. See "OSCE Conference on Racism, Xenophobia, and Discrimination," Organization for Security and Co-operation in Europe, PC.DEL/1146/03, October 1, 2003.

8. James Goldston, "Race Discrimination in Europe: Problems and Prospects," *European Human Rights Law Review*, no. 5 (1999): 464.

9. See Anti-Islamic Reactions in the EU After the Terrorist Attacks Against the USA: United Kingdom, Report on Islamophobia, European Monitoring Centre on Racism and Xenophobia (Vienna, May 2002).

10. At one point in May 2002, although without an official title, the person second in leadership of the Pim Fortuyn List party, behind Fortuyn himself, was Joao Valera, an immigrant from Cape Verde. France's right-wing demagogue Jean-Marie Le Pen has also presented black and even Jewish supporters of his cause.

11. See European Race Bulletin (London, January 2002), http://www.irr.org.uk (accessed March 2002).

12. See the European Convention for the Protection of Human Rights and Fundamental Freedoms (1950), popularly known as the European Convention on Human Rights.

13. James Goldston, "Race Discrimination in Europe," 464.

14. Mark Bell, "Meeting the Challenge? A Comparison Between the EU Racial Equality Directive and the Starting Line," in *The Starting Line and the Incorporation of the Racial Directive into the National Laws of the EU Member States and Accession States*, ed. Isabelle Chopin and Jan Niessen (London: Commission for Racial Equality, 2001), 22–49.

15. ENAR General Overview on the Transposition of the COUNCIL DIRECTIVE 2000/43/EC of 29 June 2000" (Public Hearing, July 8, 2003), http://www.enar-eu.org/en (accessed August 2003).

16. See Anne F. Bayefsky, "The Principle of Equality or Non Discrimination in International Law," *Human Rights Law Journal* 11, nos. 1–2 (1990): 1, 5; and Stephen Livingstone, "Article 14 and the Prevention of Discrimination in the European Convention on Human Rights," *European Human Rights Law Review* 1 (1997): 25–34.

17. Livingstone, "Article 14 and the Prevention of Discrimination," 29–30.

18. Karon Monaghan, "Limitations and Opportunities: A Review of the Likely Domestic Impact of Article 14 in the ECHR," *European Human Rights Law Review*, no. 2 (2001): 172.

PART V

CRAFTING RESISTANCE

Identity, Narrative, and Agency

Chapter 16

Salvaging Lives in the African Diaspora

Anthropology, Ethnography, and Women's Narratives

Irma McClaurin

Salvage: the rescue of any property from destruction or waste.
— Webster's New World Dictionary 1979, 423

I am the last girl of a family of seven baby girls, baby sister. My parents were very poor. My father was an alcoholic. He died when I was seven and my mother had to work to raise us, so I grew up among brothers, so I was a very rough little girl. I climbed trees and admired their courage. When I grew up, things slowed down. I started to see what happens to women who live with an alcoholic. I tell myself, I don't want to go through that the pain, the anguish. I met my husband at eighteen. But there was a problem. We were not of the same ethnic group. He was Indian [Maya] and I Garifuna.[1]

—Rose

WITH THESE WORDS, I am invited into the narrative of a young woman who is the same age as I and, like me, has two children: a son and a daughter. We share so much and yet our lives are so different. With these words, I am introduced to a marginal woman (Rose), who lives in a marginal district (Toledo), often referred to as "God's backside" or the "tail end" of a country, which is a marginal place (Belize, formerly British Honduras), of which few people have heard. And even if they have heard of Belize, most people fail to realize that it is uniquely situated in Central America as the only English-speaking enclave in an expanse of Spanish language and culture—a metaphoric Caribbean island in a very real Spanish environment.

Rose's words also mark the beginning of my journey as an "authentic" anthropologist, trained to study the other by people who sometimes, conveniently, forget that I am the other. Her words mark the beginning of my attempts to understand gender as lived experience. They bear witness to my desire as a scholar to salvage (i.e., recover, record, and reveal) the words and memories that constitute women's "life stories," which according to Susan Chase are "narratives about life experience that . . . [are] of deep and abiding interest to the interviewee."[2] Rose's words bear witness to my obsession, "a beautiful obsession," with narrative, which I view as a transformative experience that reveals more often than not who the speaker has become more so than what they were socialized to be. I see narrative as a bridging of representations—the new and transformative (articulated in the telling) with the old and enculturated (out of which the speaker weaves herself anew). In this respect, narrative construction is a form of particularized social knowledge production and a process of identity formation undertaken by the actor as a means of making sense out of her social world. In effect, narrative serves as a way for us to see how personhood, culture, history, and memory are mediated through the telling of a story and the production of a particular life.

So what does it mean when an anthropologist ventures into the heady postmodernist realm to engage these issues of meaning, interpretation, situated identities, and transformative narratives? I think the best way to respond is, of course, through narrative—that is, to create and explicate my own transformative story, to produce my own particularized social knowledge, shaped by my experiences as a black woman (who is simultaneously mother, sister, daughter, friend), a writer, a feminist, and an anthropologist.

Many, many years ago, in a place far removed from where I live now, I set off to explore the unknown, the other, and like every well-trained anthropology graduate student, I was armed with my anthropological toolkit, which included, among other things, an agenda: I arrived in Belize, Central America wanting to study women's grassroots organization, armed with the usual social science paraphernalia: random sample methods, questionnaires, feminist theories, theories of culture, and so forth.

While I adhered to H. Russell Bernard's *Research Methods* like something akin to a religious convert, what I discovered was that some of these methods simply did not work on the ground. For example, in my experience, trying to construct a random sample from an outdated census was nothing short of frustrating, while trying to fit the way women's groups organized themselves in Belize, Central America into the Western feminist paradigms of how women's organizations and support groups ought to work made little sense in the context of what I discovered in Belize.[3]

Ultimately, as I conducted my survey, using "purposive" and "snowball" sampling methods, what intrigued me the most and captured my attention was what women talked about after my survey was completed or, rather, what they said around the corners of my questions. To explicate what I mean by "talking around the comers of my questions," I refer to Karen Sacks's critique of the methods she used to study women involved in a labor strike. Identifying some of the limitations of survey questions in eliciting rich data, Sacks observes: "the questions I posed to the women were sociological, and women responded in that mode, giving me answers that linked sociological variables to personal militance." She concludes, "their answers were as abstract and uninformative as my own thinking."[4] I experienced a similar situation in Belize. I, too, found that when I confined myself to formulaic survey questions or attempted to elicit responses to standard demographic queries, I received formulaic, short, and uninteresting responses in exchange. When I relaxed, however, and sat with women at kitchen tables or spoke with men at bars, worked in the shops with women and helped them with their sewing, cooking, and washing, their narratives disclosed much more about the politics of the "culture of gender"[5] than anything my survey questions could have elicited.

Sacks' point, which is also mine, is that sometimes the conventional methodological forms we use may work to obscure the very aspects of human interaction and experiences that we wish to illuminate. As a consequence of these and other encounters, in my own methodological praxis, I have come to conceptualize narrative as a tool, as a process, as an experience, and as a form of thinking. Understanding narrative in this way is one route out of the methodological conundrum that I have experienced and that Sacks describes.

I have found the elicitation of life histories to be an extremely effective strategy for exacting data for the kind of experience-based, interpretive research I conduct. One of my baseline assumptions is a strong belief that every member of a society has some interesting perspective to offer. In Belize, I conveyed this by simply asking women to "tell me what it was like growing up in Belize as a young girl," to acquire information about gender enculturation and "tell me what your life has been like," to get them to recover their lives through memories and to trace their personal history to the present moment. I think both feminist and indigenous scholars[6] who are interested in documenting and salvaging—that is, rescuing from waste, destruction, and invisibility—the richness of the past and the nuances of the present, have efficaciously used narrative in their diverse configurations of life histories, testimonies, autoethnographies, memories, and memoirs as a precise and rigorous methodological tool.

A concern with which I grapple when applying this narrative method is, how do I effectively capture a life through this research strategy? As an anthropologist committed to producing what Lila Abu-Lughod calls the "ethnography of the particular"[7] or what those of a postmodernist tendency might term "situated" or "subjugated" ethnographies, I continuously grapple with the most originative way to render (depict and translate), without force, lived experiences and write about it in forms that are neither anachronistic nor static. I struggle with what are the best forms to convey all the complexities, richness, layers, and contradictions that inhabit individual lives in social fields of action and that, to fall back upon C. W. Mills, link the "subjective to the political."[8]

One question that may plague you at this juncture is whether these are appropriate questions or the kinds of issues with which an anthropologist and ethnographer should concern herself. You may also be wondering, Are these mere afternoon musings, flights of fancy, fantastical whims of yet another delusional anthropologist who has had one too many field trips in tropical climates or, even more to the point, the rants of a wannabe humanist posing as a social scientist? I assure you to the contrary and strongly assert that issues of method, textuality, and experience-based knowledge production are some of the most critical concerns of contemporary anthropology. These are the issues that speak directly to the nature of anthropological content and epistemology, to anthropological subjects and theory, and to anthropological method and interpretation. Without a doubt these issues touch upon the very soul of anthropology in all its humanistic impulses.

Moreover, these are questions and issues that lie at the crux of current debates about the present and future direction of this discipline. And as George W. Stocking, Jr., the discipline's historiographer, asserts, such tensions and intellectual polarizations are not new to anthropology but rather are endemic, arguing that, "anthropology, in short, has been a discipline pulled between the pole of two radically divergent impulses, one 'scientizing; the other relativizing.'"[9] However, as inspirational as proposing a new approach that leans so heavily upon the tenets of postmodernism may seem, my goal is to follow the lead of Eric Wolf,[10] who argues, quite persuasively, that new approaches and new theories in anthropology, need not be contentious and need not be predicated on the demise of old thoughts and theories. Rather, he asserts, there is sufficient space in the intellectual forest of anthropological theories and paradigms to accommodate new thoughts and approaches.

There is no dearth of neophytes to these new interpretive approaches to anthropology and experimental ethnography. Key scholars like Ruth Behar assert the need for a new kind of anthropology—the kind "that makes you cry." [11] This visceral anthropology is often met with hostility and tremendous resistance from others in the discipline who adhere to a more "scientific," object-oriented anthropology. From the point of view of these latter scholars, often labeled "positivists" or "materialists," anthropology is in crisis. Richard A. Shweder characterizes the debate this way:

> These days in cultural anthropology, the discipline seems palpably and conspicuously divided over three very different conceptions of the field: (1) cultural anthropology as an agora for identity politics and as a platform for moral and political activism in the struggle against racism, sexism, homophobia, capitalism and colonialism; (2) cultural anthropology as an open forum for skeptical postmodern critiques of objective knowledge and ethnographic representation; and (3) cultural anthropology as a positive (i.e., value-neutral and non-moralizing) science designed to develop general explanatory theories and test specific hypotheses about objectively observable regularities in social and mental life, thereby protecting the discipline from identity politics and postmodern critiques.[12]

In my mind, the future of anthropology lies in its capacity to embrace what Stocking calls its "enduring epistemological tension."[13] Moreover, though there may be some who hear the death knell for positivism, I would suggest that the scientizing approach, one that Jerome Bruner[14] describes as "logical-scientific thinking," is not the endangered species postmodernists would like to pretend it is, that is, not as long as grant agencies continue to privilege positivist, materialist, scientific grant proposals over those that are more experimental, experiential, and interpretive and that seek to blur the boundaries between anthropology and the humanities.

One of the chief innovators of this new humanistic and literary approach is Ruth Behar, an anthropologist who challenges those of us who would sink into a kind of disciplinary complacency and who nostalgically, in the midst of postmodernism's antitheoretical posture, support a return to the traditional and to a past that in actuality was not as wonderful as we choose to remember and that we romanticize.[15] Behar characterizes modern anthropology as a discipline caught in a period of indeterminacy, but it is a condition abounding with creative possibilities. Bihar's mission, in this context of unpredictability, is to reconfigure anthropology, interrupt its linear trajectory, and set it upon a new path that leads to the mapping of "an intermediate space we can't quite define yet, a borderland between passion and intellect, analysis and subjectivity, ethnography and autobiography, art and life."[16]

In this new anthropological borderland of interpretive and reflexive anthropology, much more is required—anthropologists must acknowledge that we are nothing more (and nothing less) than "vulnerable observers" whose subjectivity is inextricable from our social observation.[17] If we accept this burden of responsibility, according to Behar, we are on the road toward what she advocates for herself and practices, that is, writing a kind of anthropology that "breaks your heart."[18] In the process, we achieve a qualitative depth to our analysis, reach wider audiences beyond our peers, and, perhaps, just perhaps, in the case of applied anthropology, enhance the potential policy impact of research results. As Behar notes: "When you write vulnerably, others respond vulnerably. A different set of problems and predicaments arise which would never surface in response to more detached writing."[19] It is this humanistic evocation that today is laying claim to its place as a legitimate heir of a contemporary anthropology under the rubric of experimental, new, and postmodern ethnography.

Notwithstanding its persuasiveness, its significance, and its appeal, this heartbreaking anthropology is not unproblematic. The most common critique, which seems to gravitate toward Behar in particular, is that vulnerable anthropology elides issues of power and inequality within the research domain by emphasizing the emotional and aesthetic dimensions of anthropology and ethnography. Yet as Nicole Polier and William Roseberry assert in their essay "Triste Tropes: Postmodern Anthropologists Encounter the Other and Discover Themselves,"[20] "discourses are not self-referential but are instead constructed within social fields of force, power and privilege,"[21] In their critique of "dialogic" ethnography, they also remind us that "the ethnographic context as one of 'cooperative story making' . . . does not address the context in which discourses are situated . . . [for in] social life all 'genres, texts and voices' are not created equal. . . . [I]n the production of ethnographic texts, the ethnographer's privilege is precisely a discourse on the discourse."[22]

Specifically, the forms of power that pervade social life and determine who has the means and resources to produce narratives or to have those narratives made public, are elusive, masked, and ignored as issues in most discussions of a new anthropology. More to the point, writers such as Behar[23] invite critique because they often fail to link their aesthetic and humanistic agenda to a specific set of actions or any critical perspective in anthropology, a fact that leaves their work languishing more in the realm of literature than ethnography and subjects it to denouncements of "solipsism," flawed by a hyper-reflexivity that is more akin to "navel-gazing"[24] and of questionable value to anthropology. The "critical" is an inescapable and defining trait of anthropology; it is what distinguishes it from other disciplines that merely utilize cultural description. I would like to suggest that the engagement of the humanistic and the critical, the positivist and the interpretive, need not be antagonistic. Rather, they can be viewed as equal sides of a Janus-like anthropology.

Toward this end, Jerome Bruner, in his *Culture of Education*, offers a possible bridge between these two tremendously polarized positions.[25] Bruner, who is concerned with how individuals and cultures relate to each other and with the role of education in that process, suggests that the human species organizes and manages knowledge in two ways, of which "one seems more specialized for treating of 'physical' things, the other for treating of people and their plights." He identifies these two ways of thinking as "logical-scientific and narrative." Bruner sees both forms as "universal," rooted in the "human genome," and "givens in the nature of language." He states that logical-scientific thinking and narrative thinking "have varied modes of expression in different cultures, which also cultivate them differently. No culture is without both of them, though different cultures privilege them differently."[26] Bruner's definition of narrative is one that posits it as both "a mode of thought and . . . a vehicle of meaning making."[27] Bruner argues that narrative, in its dualistic form, is the "glue" that produces cultural cohesion and the mental skeleton that structures individual lives.[28] For Bruner, the significance, power, and necessity of narrative are clear and vital: "It seems evident, then, that skill in narrative construction and narrative understanding is crucial to constructing our lives and a 'place' for ourselves in the possible world we will encounter."[29] Moreover, the task of scholars is also clear: "Obviously, if narrative is to be made an instrument of mind on behalf of meaning making, it requires work on our part-reading it, making it, analyzing it, understanding its craft, sensing its uses, discussing it."[30] Along these lines, anthropologists like Behar are well-advanced. Both Behar and Bruner, thus, become starting points for me to explore my own use of narrative, especially in the interest of recouping and developing an African Diaspora perspective. Borrowing from Guy A. M. Widdershoven, I would assert that "from a hermeneutic point of view, life is a process of interpretation in and through stories,"[31] therefore I shall begin this explication by telling another story.

If what Michael J. Fischer asserts in his essay "Ethnicity and the Post-Modern Arts of Memory"[32] is true, which is that "the ethnic, the ethnographer, and the cross-cultural scholar in general often begin with a personal empathetic 'dual tracking,' seeking in the other clarification for processes in the self,"[33] then my primary attraction to anthropology can be seen as an extension of a very personal search for the meaning of cultural identity. Central to this quest has been the use of narrative, which enables me to see culture as more than material conditions, as more than a shared set of beliefs or patterns of social relations. Culture, I argue, is a living entity that is continuously being created individually and collectively and is most frequently represented through individual life narratives.

The use of narrative has been central to the shaping of my life and scholarship as a way of understanding the experiences of the African Diaspora, a place that is part memory and part geography, populated by the descendants of enslaved Africans who were forced through the Middle Passage, dispersed throughout the Americas, and endured and continue to endure patterns of economic and social stratification and exclusion as a consequence of their ancestry and phenotype. The emergence of this "narrative sensibility"[34] is traceable through much of my writing, which began early in life in poetic form. A powerful evocation for me of narrative as a "vehicle of meaning making"[35] is captured in the poem "Pearl's Song," because it shows several efforts: (1) my attempts to recover the life of an ordinary woman (my mother) in the African Diaspora rendered invisible by class, race, and gender; (2) my creation of a narrative structure within which to lend coherency to the bits and pieces that I have gleaned of my mother's life over the years; and (3) my own personal journey of meaning making as I explore my relationship to this woman whom I call mother. The poem, through themes of segregation, migration, black female abandonment, invisibility, and racialized gender continuity, creates its own geography of narrative, and thus identity—mine, hers, ours.

Pearl's Song
My mother, called Pearl
was born in red clay country of Alabama
Sapling planted in April
Fine brown Alabama girl, thrust outward
into an older man's hands to escape red clay, jim crow and home.

2

She is a good singer, this woman. Men have known her as she leaned against Chicago doorways,
Swaying to the movements of day; They touched her
hands, bronzed her face, her eyes, and planted seeds inside her.
Four times she bent to drop shoots:

twice boys, twice girls Fine, brown Alabama girls. Still no one heard her sing,
no one watched her ooze each child out
Knifing the hospital in two octaves.

3

Bronze cast figurine, she chiseled it till often I woke her song in my throat,
Now it settles against the windows,
dusting the curtains.
Lately, turning into my mirror, I glimpse her inside
my face, her dark eyes shut away; her gray hair hidden . . .
Standing over my stove, or seeing myself in the windows of this wallpaper
    I hear her song,
see it dancing in the wild sunlight.[36]

"A Work for the Feminist Imagination"[37] is at the heart of feminist theory and is that which John Gwaltney sees as emanating from the narratives he collected of "core black culture."[38]

From these narratives—these analyses of the heavens, nature, and humanity—it is evident that black people are building theory on every conceivable level. An internally derived, representative impression of core black culture can serve as an anthropological link between private pain, indigenous communal expression, and the national market place of issues and ideas. These people not only know the trouble they have seen but have profound insight into the meaning of those vicissitudes.[39]

In narrative then, history and biography, theory and lived experience, the individual, the local, the national, and the global all intersect and become not atomized elements but rather discourses that are "implicated and intertwined with each other."[40]

The desire to develop a narrative schema devoted to recovering the hidden lives of women of the African Diaspora in other forms besides poetry actually prefigures my encounter with ethnography. It began in 1984, when I started research under my own aegis, on the life and suicide of Leanita McClain, a black middle-class female journalist in Chicago:[41] I had not yet entered anthropology and so lacked the formal terminology to describe my approach to this research; all I had at that moment was a desire to salvage, explicate, and reconstruct as much of this woman's life as I could through interviews with family and friends and through analyses of her journalistic writings, essays, and poetry; the poetry was neatly collected and organized, and left almost as a narrative requiem or a road map by her deathbed to guide us on the journey to reconstruct her life.

As I negotiated the theoretical terrain of anthropology and feminism, I was able to create a space in which I could engage my own views on how to interpret life in the African Diaspora as well as produce a critique of the inability of conventional anthropology and feminist approaches to do the same. In 1990, I reflected that critique in my essay on Leanita's life and death:

When a young black woman dies, it is usually of passing interest, except to those who knew her. When the same black woman has traversed the usual social boundaries and established a place for herself in professional and political spheres traditionally occupied by white males, then her death calls our attention to questions about her life. This is true of Leanita McClain, a gifted black journalist who worked for the Chicago Tribune. By studying her life I hope to bring out the contradictory elements that emerge as a result of the complex interactions of race, class, and gender and the positioning of the individual in the construction of social reality.[42]

Using the life history method, my goal was to employ narrative as a "tool of inquiry,"[43] that is, as a way of deriving meaning from the suicide of a black middle-class professional woman whose life seemed to embody the American Dream paradigm of success and achievement, a paradigm that ultimately offered her little immunity from clashes with sexual, racial, and class barriers. By examining the narratives imbedded in interviews with family and friends, Leanita's own poetry, journalistic writings, and personal letters to friends, I hoped to make sense of how an individual's personal understandings of race, class, and gender inequality informed her decision about survival and about sustaining the will to live.

To some extent, this exploration of Leanita's life is also an exercise in auto-ethnography, which Alice Deck sees as an "introspective personal engagement" wherein "authors rely upon their native ethnographic knowledge to assemble a portrait that is a combination of personal memories (autobiographical) and general cultural descriptions (ethnography)."[44] This conflation of personal experience, memory, and biography is possible because my story and Leanita's are interconnected. We lived in the same city, were of the same generation, attended the same high school, worked on school events together, and had mutual high school friends. Thus however I reconstruct the Chicago of Leanita's past, it is invariably filtered through my own memories of what it was like to grow up in inner-city Chicago in the 1960s. Also, lurking behind each question I posed explicitly to the people who knew Leanita was a silent, implicit question of my own: Why or how had I survived the madness and she had not? Ruthellen Josselson argues that "narrative approaches to understanding bring the researcher more closely into the investigative process than do quantitative and statistical methods,"[45] and feminist scholars like Sandra Harding[46] assert the absolute necessity of feminist researchers locating themselves in the same research plane as the women they study. I was able to bring these two perspectives together in my study of Leanita.

As a result of trying to document and interpret Leanita's life and death, I would venture one step beyond both Josselson and Harding to suggest that the researcher becomes a crucial part of the social action of narration at the moment it occurs by entering into a dialogic relationship with the other person, resulting in what Widdershoven, citing Hans Georg Gadamer, calls a "fusion of horizons."[47] Although Gadamer is referring to texts and readers, Widdershoven argues that this "theory of interpretation can be applied to the relation between experience and story in individual life. In telling stories about past experiences, we try to make clear what these experiences mean."[48] Thus, "interpretation . . . [emerges] as a form of dialogue in which both participants try to come to grips with the truth in a process of mutual understanding."[49] This constitutive sharing was evident during my fieldwork in Belize where I collected narratives in order to understand the "culture of gender"[50] that prevailed in that country. The following excerpt from my field notes gives some indication of my growing awareness of the narrative process as dialogic: "Life history informants are not just found, they are made. In the engagement between interviewer and interviewees, something clicks. In the process of asking survey questions, I looked for a level of self-reflection and thoughtfulness. Even if the informant has little schooling, you can tell that they've given thought to their lives. It is "awareness," 'consciousness' if you will, that lets you know intuitively that there is a story."[51] Of course, the engagement with Leanita is somewhat unusual because she is no longer alive, but I assert that there is still a conversation, a dialogue, and an engagement occurring nonetheless. I am fully aware that when writing a life based on documents, partial letters, and interviews recorded immediately after the trauma of her death, the account is likely to be incomplete in many respects. Moreover, suspicions may abound in the reader: after all, how could I know how to interpret anything I discovered or heard? In an early essay based on Leanita's life, I tried to anticipate and respond to these questions and to address the implicit skepticism that emerges when reconstructing a life. To a large extent, I default to Clifford Geertz's belief that anthropology is at best an imprecise "science," which should have as its main goal not explanation but interpretation. [52]

I have taken Leanita's life and writings and tried to interpret them in a way that is meaningful to her particular situation but that also places her within a given history. Some may disagree over whether I have accurately discerned the "truth" of Leanita's life. To them, I can only respond that in the dialogic process there is no single truth but rather many voices, each telling only what they know.[53]

Moreover, I concur with Geertz that ethnographic descriptions are at best representations and not truths.[54] That is, even when dealing with living people who are predictably unpredictable, we as ethnographers, biographers, and historians can never really know. It is precisely for this reason that I aim in narrative works not to capture truths but rather to explore meaning. In this sense, I am aligned with Widdershoven, who interprets Jacques Derrida as arguing that there is neither "origin nor continuity in the history of interpretation."[55] According to Widdershoven, Derrida rejects the idea of fixed meanings in texts and argues instead for a "principle of intextuality"[56] in which new meanings are constantly produced. As Widdershoven explains: "In our life we are constantly citing ourselves and others, thus creating new patterns of meaning. Life itself is an infinite process of differance, creating ever new texts and contexts."[57]

Derrida's belief that there is no ultimate continuity in individual life stories but ever-emerging instances of divergences as the narrative is brought into "a new web of relations"[58] suggests that the type of certainty and replicability that a "scientific" approach requires is unattainable when studying people.

Much of what I learned during the process of recovering Leanita's life narrative proved to be of immense value to me when I conducted fieldwork in Belize. In this new context I was interested not only in documenting and recording women's voices and listening to the sounds of their lives, but also in the relationship between the individual and her community. What I hoped to achieve was clarity about the relationships between the individual, the local, the national, and global social processes. I also wanted to understand how these processes create and sustain the culture of gender. As was the case with Leanita, I sought to understand by what mechanisms individuals achieved empowerment (a sense of personal autonomy) as they negotiated cultural constraints. Although I focused initially on women of the African Diaspora, which in Belize meant Creole and Garifuna, in recognition of the country's ethnic heterogeneity I cast my net wider and included East Indian women (another group whose voices were frequently invisible in local and national political discourses) in my research.

During the process of writing my dissertation *Women and the Culture of Gender in Belize, Central America*,[59] and eventually the ethnography *Women of Belize*,[60] I struggled to contextualize the corpus of my data (women's narratives) in several dimensions: in relation to my own lived experiences as a black female anthropologist, in relation to the communities in which these women lived with their family and friends, in relation to the national agenda of development and nation-building, and in relation to global concerns about gender inequality in the world. To achieve these ends, I constantly mediated between an emic (insider) and an etic (outsider) perspective. As a reflexive anthropologist, I was aware of my involvement in the narrative process and tried to remain vigilant about what my impact was on the production of the narratives. The conscious negotiation of self, subject, and research objectives are perhaps most evident in the narrative entitled "A Birthday Celebration." It is my "arrival story" to my field site and is a direct transcription of field notes recorded at that moment.

This experience took place less than forty-eight hours after I arrived in Lemongrass. It illustrated for me, in a vivid way, the problems and issues that women face in this country and gave me insights into the ways in which the gender system in Belize is constructed and maintained by both men and women. Listening to Evelyn, Elana, and their friends, I learned that in Lemongrass women's value, either ascribed or self-attributed, comes from the degree to which they conform to social norms of a "good wife or mother" and the degree to which they contribute to the society through biological and social reproduction.

Any other incidents I encountered in Belize make it clear, as many feminist scholars have argued, that gender is far more than an analytical construct or a structural form. Gender, far from being an abstract concept, is a pervasive set of obligations and limitations that saturate the entire being and make up one's identity.[61] What all of this suggests to me is that narrative, as "a mode of thought and . . . a vehicle of meaning making,"[62] is one of the most significant ways in which we acquire insights into understanding the role of culture in structuring the human condition and achieve understanding of the human response to the social facts of class, race, gender, and other forms of oppression. Explication of the social and of human agency through ethnography is one the most significant contributions of anthropological inquiry.

If there was ever a contested domain in anthropology today it is ethnography. Scholars like James Clifford have described ethnographies as "fictions" that are, by definition, "inherently partial," unpredictable, uncontrollable, and imbued with a destiny of their own making.[63] He explains, "Even the best ethnographic texts' serious, true fictions are systems, or economies of truth. Power and history work through them, in ways their authors cannot fully control."[64] Polier and Roseberry also have something to say on the matter. Highly critical of postmodern turns in ethnography, they see the new forms as lacking in "contextualization and repre-sentativeness," which they assert are more important than current emphasis upon "artistic production."[65] "In grafting a literary conceptual framework onto social analysis, a sleight-of-hand is accomplished: it becomes possible to analyze ethnog-raphy just as one would a work of fiction. . . . In the end, using literature as more of an homologue than a limited analogue mystifies more than it illuminates the practice and process of social research."[66] Finally, Polier and Roseberry caution against what they see as the most problematic and extreme aspect of these new approaches to ethnography, "an attitude [that] can lead to an individualistic and self-centered approach in which "ethnography is reduced to personal therapy."[67]

Such critiques of experimental forms are not only legitimate but also neces-sary if we are to create new ways of writing about human experience. Moreover, Polier's and Roseberry's desires for an approach in which there is "a sophisticated consideration of the intellectual, institutional and political forces that shape and constrain the ethnographic encounter and the production and consumption of knowledge in the late twentieth-century academy"[68] is certainly consistent with my own perspective. However, they spend so much time critiquing the new that little attention is given to mapping out a strategy for producing the kind of eth-nographies they want. They do suggest that such an approach should blend post-modernist thought and world-systems theory in order to produce ethnographies that view "anthropological subjects within combined and contradictory historical processes, processes that should be seen as at once determinate (they are estab-lished in particular fields of power) and contingent (the fields of power, as histori-cal products themselves, are subject to change and transformation)."[69] It is toward the production of just such an ethnography that I aim in my salvaging of nar-ratives in the African Diaspora, My own ethnographic strategy borrows heavily from feminist approaches to ethnography that, according to Balsamo, "delineate ways in which we can think anew the simultaneous construction of the personal and the cultural, the one and the many."[70] Balsamo explains, "feminism argues that all interpretive practices, including those inherent in ethnographic research and writing, are political acts that forge links between history and biographies for all participants."[71]

This perspective is quite similar to that articulated by Filomina Chioma Steady in her *African Feminism* in which she posits that "in Africa, as well as in the diaspora, black women engaged in research on the black woman are involved in a process of liberation, as well as in scholarly endeavor, since research, being essentially a product of the power structure, has sometimes been used as a tool of domination."[72] I can envision no worthier task for anthropology, ethnography, or narrative than in the service of decolonizing scholarship and rescuing from destruction and waste the cultural lives, daily experiences, and social action of the people who constitute the African Diaspora.

NOTES

1. Irma McClaurin, *Women of Belize: Gender and Change in Central America* (New Brunswick, NJ: Rutgers University Press, 1996), 47.
2. Susan E. Chase, "Review," *Gender and Society* 9, no. 2 (April 1995): 259–60.
3. McClaurin, *Women of Belize.*
4. Karen Brodkin Sacks, "Towards a Unified Theory of Class, Race, and Gender," *American Ethnologist* 16, no. 3 (August 1989): 534–50.
5. McClaurin, *Women of Belize.*
6. Lila Abu-Lughod, *Remaking Women* (Princeton, NJ: Princeton University Press, 1998); Ruth Behar, "Introduction: Women Writing Culture; Another Telling of the Story of American Anthropology," *Critique of Anthropology* 13, no. 4 (1993): 307–25; Johnnetta B. Cole, "The Education and Endowment of Black Women," *Equity and Excellence in Education* 25, nos. 2–4 (1991): 130–32; Jose Limon, "Representation, Ethnicity, and the Precursory Ethnography: Notes of a Native Anthropologist," in *Recapturing Anthropology: Working in the Present*, ed. R. G. Fox (Santa Fe, NM: School of American Research Press, 1991); Gloria Wekker, "I Am Gold Money: (I Pass through All Hands, but I Do Not Lose My Value)," doctoral thesis, University of California–Los Angeles, 1992.
7. Abu-Lughod, "Writing against Culture," in *Recapturing Anthropology: Working in the Present*, ed. R. G. Fox (Santa Fe, NM: School of American Research Press, 1991).
8. Anne Balsamo, "Rethinking Ethnography: A Work of the Feminist Imagination," *Studies in Symbolic Interactionism* 11 (1990): 75–86.
9. George W. Stocking, *Colonial Situations: Essays on the Contextualization of Ethnographic Knowledge* (Madison: University of Wisconsin Press, 1991), 23.
10. Eric R. Wolf, "Distinguished Lecture: Facing Power—Old Insights, New Questions," *American Anthropologist* 92, no. 3 (1990): 586–96; Behar, *The Vulnerable Observer: Anthropology That Breaks Your Heart* (New York: Beacon Press, 1997).
11. Richard A. Shweder, Richard Jessor, and Anne Colby, *Ethnography and Human Development: Context and Meaning in Social Inquiry* (Chicago: University of Chicago Press, 1996), 1.
12. Stocking, *The Ethnographer's Magic and Other Essays in the History of Anthropology* (Madison: University of Wisconsin Press, 1992), 279.
13. Jerome Bruner, *The Culture of Education* (Cambridge: Harvard University Press, 1996).
14. Behar, *The Vulnerable Observer*; Behar, *Women Writing Culture* (Berkeley: University of California Press, 1995).
15. Behar, *The Vulnerable Observer*, 174.
16. Ibid, 5.
17. Ibid, 161.
18. Ibid, 16.
19. Nicole Polier and William Roseberry, "Tristes Tropes: Post-modern Anthropologists Encounter the Other and Discover Themselves," *Economy and Society* 18, no. 2 (May 1989): 245–64.
20. Ibid., 251.

21. Ibid.
22. Ibid., 251–52; emphasis original.
23. Behar, "Introduction: Women Writing Culture"; Behar, *Women Writing Culture*; Behar, *The Vulnerable Observer*.
24. In *The Vulnerable Observer*, Behar takes an opportunity to respond to critiques of her work by scholars like Daphne Patai ("Sick and Tired of Nouveau Solipsism," *Point of View Essay in Chronicle of Higher Education*, February 23, 1994). Behar defends her form of style in the following way: "The charge that all variants of vulnerable writing that have blossomed in the last two decades are self-serving and superficial, full of unnecessary guilt or excessive bravado, stems from an unwillingness to even consider the possibility that a personal voice, if creatively used, can lead the reader, not into miniature bubbles of navel-gazing, but into the enormous sea of serious social issues" (Behar, *The Vulnerable Observer*, 14).
25. Bruner 1996.
26. Ibid., 39–40; emphasis added.
27. Ibid., 39.
28. Ibid., 40.
29. Ibid.
30. Ibid., 41.
31. Michael M. J. Fischer, "Ethnicity and the Post-modern Arts of Memory," in *Writing Culture: The Poetics and Politics of Ethnography*, ed. James Clifford and George E. Marcus (Santa Fe, NM: School of American Research Press, 1986).
32. Ibid., 199.
33. Bruner 1996.
34. Ibid., 39.
35. McClaurin, *Pearl's Song* (Detroit: Lotus Press, 1988). Reprinted with permission from author.
36. Balsamo, "Rethinking Ethnography."
37. John Langston Gwaltney, *Drylongso: A Self-Portrait of Black America* (New York: Random House,1980).
38. Ibid., xxvi.
39. Balsamo, "Rethinking Ethnography," 46.
40. Irma McClaurin-Allen, "Incongruities: Dissonance and Contradiction in the Life of a Black Middle-Class Woman," in *Uncertain Terms: Negotiating Gender in American Culture*, ed. Faye Ginsburg and Anna Tsing (New York: Beacon Press, 1990).
41. Ibid., 315.
42. Ruthellen Josselson and Amia Lieblich, *Interpreting Experience: The Narrative Study of Lives* (New York: Sage, 1995), x.
43. Ibid.
44. Alice A. Deck, "Autoethnography: Zora Neale Hurston, Noni Jabavu, and Cross-Disciplinary Discourse," *Black American Literature Forum* 24, no. 2 (1990): 237–56.
45. Josselson and Lieblich, *Interpreting Experience*, ix.
46. Sandra G. Harding, *Feminism and Methodology: Social Science Issues* (Bloomington: Indiana University Press, 1987).
47. Guy A. M. Widdershoven, "The Story of Life: Hermeneutic Perspectives on the Relationship between Narrative and Life History," in *The Narrative Study of Lives*, ed. Josselson and Lieblich (New York: Sage, 1993), 13.
48. Ibid.
49. Ibid.
50. Irma McClaurin, *Women and the Culture of Gender in Belize, Central America* (Amherst: University of Massachusetts at Amherst Press, 1993).

51. Clifford Geertz, *The Interpretation of Cultures: Selected Essays* (New York: Basic Books, 1973).
52. McClaurin-Allen, "Incongruities," 330.
53. Geertz, *The Interpretation of Cultures*.
54. Widdershoven, "The Story of Life," 14.
55. Ibid.; emphasis original.
56. Ibid.; emphasis original.
57. Ibid.
58. McClaurin 1993.
59. McClaurin, *Women of Belize*.
60. Ibid.
61. Ibid.,18–21.
62. Bruner 1996.
63. James Clifford and George E. Marcus, eds., *Writing Culture: The Poetics and Politics of Ethnography* (Santa Fe, NM: School of American Research Press, 1986), 6–7.
64. Ibid., 7.
65. Polier and Roseberry, "Tristes Tropes," 255.
66. Ibid., 249.
67. Ibid., 255.
68. Ibid., 246
69. Ibid., 258.
70. Balsamo, "Rethinking Ethnography," 56.
71. Ibid.
72. Filomena Steady, "African Feminism: A Worldwide Perspective," in *Women in Africa and the African Diaspora: A Reader*, ed. Rosalyn Terborg-Penn and Andrea Benton Rushing (Washington, DC: Howard University Press, 1987), 4.

Chapter 17

Going Back to Our Own

Interpreting Malcolm X's Transition from "Black Asiatic" to "Afro-American"

Elizabeth Mazucci

The thing that has made the so-called Negro in America fail, more than any other thing, is your, my, lack of knowledge concerning history.
—Malcolm X, December 1962[1]

The late Imam Benjamin Karim clearly remembered the day in late December 1962 when he introduced Minister Malcolm X to over one thousand Muslims and other curious folk packed into Harlem's Mosque No. 7, anxious to hear him speak on "Black Man's History."[2] In this lecture Malcolm described the disparate genealogies of "Black Asiatics" and white Europeans but focused most of his energy on the devilish nature of whites. Malcolm ended the talk with a summary of the eschatological beliefs held by his organization, the Nation of Islam (NOI), which center on the imminent destruction of the white man. It is implied in this epigraph that Malcolm believed in employing historical knowledge to effect change. Malcolm may have been drawing from his own personal experience: he eventually rose from a young inmate doomed to failure in 1946 to the second most sought-after speaker on college campuses in 1964,[3] propelled by the "true knowledge" he acquired along the way. According to *The Autobiography of Malcolm X*, Malcolm's successful conversion to the Nation was, in part, facilitated by the revelation that black people are systematically denied representation in history books. While in prison, the notion of a suppressed, glorious history of the black man struck Malcolm profoundly and inspired reading marathons in his jail cell. The Nation's focus on history, according to Malcolm, was "one reason why Mr. Muhammad's teachings spread so swiftly all over the United States, among all Negroes, whether or not they became followers."[4] It is not surprising, then, that Karim fondly recalled how Malcolm X engaged his audiences by indicting white America for silencing black history. Malcolm honed his rhetoric during his tenure as the Nation's spokesman and persisted with the same general arguments even when the content of his speeches shifted rather dramatically. Although many scholars have correctly argued that Malcolm's religious transformation while on Hajj in 1964 was largely overdrawn in the *Autobiography*, examining the origin of and apparent shift in Malcolm's self-identity—from "Black Asiatic" to "Afro-American"—helps us understand Malcolm's post-Nation political philosophy. This shift adds another layer of complexity to the increasingly hostile relationship between the Nation and Malcolm in 1964 and 1965.

The issue of ethnic or racial self-identity has been overlooked by most Malcolm X scholars.[5] Ethnic identities are manifestations of experience and interrogating the ways Malcolm X identified himself is one way to access his experience. The scholars that do address Malcolm's abandonment of the Nation of Islam's race theory posit it in terms of his *religious* experience in Asia and Africa. My goal is to frame this transition in terms of his *political* experience and maturation in 1964. While this essay aims to develop a hypothesis regarding the impetus for Malcolm's shifting identities, it does not present a meticulous analysis of how his thought evolved. Such a task would be monumental, since Malcolm's numerous remarks on his heritage varied depending on the context of the speech or discussion.[6] Rather, my objective is to illustrate the likely origins of the Nation's genealogical myth in order to elucidate some of Malcolm X's more cryptic and vitriolic statements regarding Elijah Muhammad's teachings from late 1964 until his death. In doing so, this chapter will also reveal the ironically central place of European anthropology in the Nation's (re)construction of black identity by delineating an etymology of "Black Asiatic." After leaving the Nation, Malcolm X not only rejected Elijah Muhammad's race-based theology but also repudiated his "Black Asiatic" identity. Moreover, I suggest that through his experiences in Africa in 1964, Malcolm X came to realize that the Nation perpetuated a racist, colonialist perspective of Africa and its people. I will begin by situating aspects of early Nation theology within the broader context of African American religion in the early twentieth century.

The industrial age of the early twentieth century marked a period of exponential development in the diversification of African American religion. The emergence of hundreds of small-scale sects was due mainly to the mass migration of rural, southern blacks to the urban North from 1915 to 1930. Several circumstances encouraged millions of people to migrate, the primary reasons being natural disasters that impeded farming productivity (such as the destruction of crops by the Mexican boll weevil and soil erosion) and the increased demand for cheap labor in the period between the wars. Race-based violence in the rural South, particularly a horrific rise in lynching, also compelled blacks to migrate; however, life was not much better once they arrived in the northern cities. This enormous population shift, xenophobic northerners, and the reemergence of the Ku Klux

Klan ignited race riots in city streets. Riots in Chicago and East St. Louis claimed eighty-five lives and injured well over one thousand people. Racial tensions culminated in the Red Summer of 1919, during which twenty-six riots broke out in the overpopulated northern cities. It is within this context of provocation and displacement that certain African American religious sects such as the Moorish Science Temple emerged.[7]

African Americans in the rural South often gathered in small-scale Baptist and Methodist communities that offered temporary refuge from the social and economic inequalities they encountered daily. Upon arrival in the North, many migrants discovered that, despite the variety of religious denominations from which to choose, small-scale community centers and congregations were absent from urban life. Indeed, the population explosion overextended many of the established Baptist and Methodist churches. For some of the migrants, their lively approach to worship, familial connection to the church, and social status clashed with urban sensibilities.[8] Consequently many poor migrants felt that these churches were inadequate for their particular needs. New storefront and house churches were established to meet the needs of southern blacks in the same predicament; they were also places to comfortably reminisce about their former churches in the South. The types of sects that emerged ranged from the integrated, eclectic Father Divine Peace Mission to a separatist black Jewish sect that called itself the Church of God.[9] These congregations were directed by charismatic leaders who, in many cases, were formerly preachers in the South. Arthur Fauset's seminal 1944 urban ethnography, *Black Gods of the Metropolis*, describes and contrasts a motley set of new religious movements. His first-hand experience with these "cults" is useful in explaining their appeal: "Negroes are attracted to the cults for the obvious reason that with few normal outlets of expression . . . the cults offer on one hand the boon of religion . . . and on the other hand they provide for certain Negroes with imagination and other dynamic qualities, in an atmosphere free from embarrassment or apology, a place where they may experiment in activities such as business, politics, social reform, and social expression."[10] These newly established religious sects offered much more to followers than a creed and a space in which to worship. The attempt to build black-owned and -operated businesses reflected a surge in Black Nationalist thought among the masses, engendered most famously by Marcus Garvey's political movement, which took root in the United States in 1916. The Universal Negro Improvement Association (UNIA) provided the social and political blueprint from which most sectarian leaders constructed their ideologies. Garvey's message was directed toward lower-class and working-class people of African descent, especially concerning their economic position and lack of political voice in the cities. Garvey proposed a "Back to Africa" program, which sought to return Diaspora Africans to the motherland to create a unified nation of black people.[11]

Although forging unity among all people of African descent was at the heart of the UNIA, Garvey apparently did make an effort to identify his struggling people with a specific geographic location: Ethiopia.[12] The reasons for Garvey's explicit identification with Ethiopia is undoubtedly due to its remarkable ancient Christian civilization, its role in biblical prophecy, and the Bible's generic use of the term "Ethiopian" when referring to Africans. Not surprisingly, Garvey's Christian beliefs and political philosophy were inextricable. He firmly believed that, with faith in one God, black people would "rise up with their risen Lord and take a firm hold of their heritage as made in God's image, expressed in the soil of Africa, and act courageously to become fully human."[13] Paradoxically, Garvey's mission was for people of African descent to realize self-determination and to "civilize" and Christianize Africans. Given his identification with Ethiopia and emphasis on the beauty of blackness, it is curious that Garvey considered blacks to be blood brothers with "some Indians and even southern Europeans."[14] Garvey focused a great deal of energy on describing the wealth of long-lost kingdoms in Africa as a way to imagine the possibilities for Africa that lie ahead. His talent for using the past to inspire and motivate the masses would be emulated by many—including ministers of the Nation—in the decades to come.

At the same time that Garvey was spreading his Christian-oriented, Black Nationalist message, Timothy Noble Drew Ali developed a unique type of Islamic heterodoxy. Drew Ali established the Moorish Science Temple of America (MSTA) in 1913 in Newark, New Jersey. Arguing that racism is inextricably woven into the fabric of American Christianity, Drew Ali sought to evade the numerous problems confronting poor blacks by presenting a mélange of theologies under the umbrella of "Islam." He selected excerpts of teachings from four texts: the Qur'an, the Bible, The Aquarian Gospels of Jesus Christ, and Unto Thee I Grant but was also influenced by Garveyism, Theosophy, Sufism, and the Shriners.[15]

Through his religious teachings, Noble Drew Ali attempted to replace a race category, "Negro," with an ethnic category. He rejected the terms "Negro," "black," and "colored," maintaining that his African American followers were "olive-skinned."[16] He taught that African Americans are "Asiatics" or "Moors" whose ancestors were forcibly removed from Morocco, enslaved, and stripped of their cultural and religious heritage.[17] The MSTA's genealogical theory centered on a bloc of ethnicities—the descendants of Canaan and Ham—that represented what he considered to be the civilized, "original Asiatic nations." The peoples who fall into this category, according to historian Richard Brent Turner, include "the Egyptians, the Arabians, the Japanese, the Chinese, the Indians, the people of South America and Central America, the Turks, and the African Americans."[18]

Parallel to Malcolm X's contention that a people are doomed to failure without knowledge of history, Drew Ali believed that "the name means everything; by taking the Asiatic's name from him . . . the European stripped the Moor of his power, his authority, his God and every other worth-while possession."[19] While the MSTA may have succeeded in envisioning an alternate ethnicity that symbolically reversed the imposed feeling of inferiority and deflected self-hatred, the movement ultimately did little in the way of actively improving conditions. Moreover, it failed to generate a sense of pride in the great spectrum of dark skin color and in the African roots of black people in America. Despite his reverence for and adoption of portions of Garvey's teachings, Drew Ali's esoteric Orientalist theology hardly dealt with Africa, save the presumed Asiatic presence in Morocco and Egypt. While Garvey was preaching about the need for a physical return to Africa, Drew Ali attempted to psychically (re)discover his brethren's roots while remaining in the United States. The UNIA and the MSTA, with their different approaches to empowering African Americans, together set the stage for the largest, most influential black religious movement to develop.

In 1930 an enigmatic figure, most commonly known as W. D. Fard, began to preach a distinct Muslim message to inner-city blacks in Detroit who were suffering through the Great Depression. His arrival was timely: Marcus Garvey was deported in 1927, while Noble Drew Ali was assassinated in 1929. Fard initially peddled ordinary items like raincoats and "exotic" items such as silk door-to-door. The foreignness of some of these items engendered a conversation with the poor African Americans he would eventually proselytize. According to one follower, Fard told the migrants that "the silks he carried were the same kind that our people used in their home country and that he had come from there. So we all asked him to tell us about our own country."[20]

Despite its sensationalistic title, Erdmann Doane Beynon's 1938 article "The Voodoo Cult among Negro Migrants in Detroit" is a useful source for understanding the early teachings of Fard, especially because Beynon conducted several interviews with Fard's followers. Members claimed that Fard "was born in Mecca, the son of wealthy parents of the tribe of the Koreish, the tribe from which Mohammed the Prophet sprang, and that he was closely related by blood to the dynasty of the Hashimide sheriffs of Mecca who became kings of the Hejaz."[21] Beynon reports that most of Fard's followers, representing the last wave of the Great Migration, were previously exposed to and influenced by the teachings of Marcus Garvey and Drew Ali, albeit indirectly. A few of the early followers were former members of the MSTA, which primed them for Fard's similar heritage claims. "The newer migrants," according to Beynon, "entered a social milieu in which the atmosphere was filled with questions about the origin of their people. Long before their new prophet appeared among them they were wondering who they were and whence they had come."[22] Moreover, some authors assert that in the early days Fard attracted new members by claiming that he embodied Drew Ali's reincarnated spirit.[23]

Fard used his purported Saudi heritage as a paradigm for his conceptualization of African American ethnicity. "The Black men of North America," Fard maintained, "are not Negroes, but members of the lost tribe of Shabazz, stolen by traders from the Holy City of Mecca 379 yrs ago."[24] Similar to Drew Ali's proclamation that African Americans are ethnically Asiatic, Fard also taught his followers that they were Asiatic people but of a presumably fictitious tribe of nobles from Saudi Arabia.[25] While the MSTA situated their preslavery origins in Morocco (Africa), Fard's proto-Nation sect believed that their people merely crossed Africa when they were kidnapped from Mecca (Asia). Unlike Drew Ali's assertion that African Americans are "olive-skinned Asiatic people," Fard used his followers' black skin as a feature to be esteemed, proclaiming that they are "Black Asiatics." In collecting the broken strands of Drew Ali's escapist Asiatic religion and Garvey's inspirational notions of black uplift and self-determination, Fard began to weave together a powerful new vision.

Fard's role quickly progressed from peddler to prophet as his congregation took shape. He founded the first University of Islam (actually a primary school) in Detroit in 1931. That same year, in the former UNIA assembly hall in Detroit, Fard spoke to a few hundred listeners about the tenets of Islam, the origins of blacks in America, and the path to paradise. One of the listeners was Elijah Poole, a migrant from Georgia who was a former corporal of the UNIA's Chicago chapter and would later become Fard's Supreme Minister.[26]

Soon after his conversion, Poole changed his name to Elijah Muhammad, replacing his "slave name" with a surname that identified him with his newfound origin.[27] Together, Fard and Elijah established a temple in Chicago and started to set up one in Milwaukee, but in 1934 Fard disappeared. Despite the FBI's attempts to locate him and testimonies by some members who claim that Fard was in continuous communication with Elijah Muhammad, the truth regarding Fard's obscure origins, his mysterious arrival in Detroit, and his equally mysterious disappearance will probably never be thoroughly understood. The movement became factionalized; Elijah Muhammad vied for the leadership with the claim that Fard was actually Allah in the flesh. He relocated his followers to Chicago and began to teach them that he was the prophet of Allah.[28] Guided by Fard's vision and his own experience, Elijah Muhammad extrapolated elements from Drew Ali's religious teaching and Garvey's political philosophy but ultimately declared that "both men failed to bring about the redemption of the race because they did not possess 'the key' and because the 'time was not ripe.'"[29] The "key" to which Elijah Muhammad refers is the revelation that all white men are devils. Though it was implied in Drew Ali's teachings—he associated the white man with Satan—Fard explicitly claimed that the white man is the devil.[30]

Nation theology centered on two fundamental parables (that were conveyed as truth): the stories of the so-called Negro's Asiatic origin and of the devilish nature of the white man. According to Nation doctrine, over eight thousand years ago a brilliant but malevolent geneticist of the Tribe of Shabazz named Yacub used his scientific knowledge to create the white man out of the original (black) man by gradually mutating the latter's genes. The history of the naturally weak and wicked white man, according to the Nation, only reaches back six thousand years to the moment when the first white man was "grafted." Little is known about the extent to which Fard directly contributed to the canonization of these narratives, especially since during the formative years (1930s to mid-1950s) teachings were mostly transmitted orally.[31]

The appeal of the Nation was largely based on its discourses of identity and black pride borne out of reclaiming and reshaping history. Viewing the Nation as a movement gestating in the context of the Great Migration, Great Depression, headless MSTA, and dismantled UNIA explains how their peculiar genealogical myths could enchant thousands of poor African Americans. The question remains as to *whence* Fard and Elijah Muhammad acquired the particular concepts they preached. The words of Malcolm X during his involvement with the Nation (ca. 1952–March 1964) remains the earliest clear expression of Elijah Muhammad's message.[32] Therefore, it is appropriate to turn to Malcolm's speeches and interviews in order to properly examine the Nation's heritage claims.

In the summer of 1960, Malcolm X met with Nat Hentoff of *The Reporter* to explain the distinctive culture of the Nation, a sect that was around for decades but became known to the world only a year earlier.[33] Malcolm remarked that black people have been miseducated about Africa and fooled into believing the enduring stereotype of the uncivilized African. "Most so-called Negroes," said Malcolm, "know less about their cultural background than a native in deepest Africa. Many, in fact, still think of Africa—despite the growth of African states—as a continent of naked, flesh-eating barbarians. We teach [so-called Negroes] how rich their past is; we teach them there were Black men on earth before the whites, and that we are the chosen of Allah."[34] The juxtaposition of the latter two sentences creates the impression that the Nation teaches that the black man originated in Africa. This sentiment appears to be a far cry from Fard's contention that the Tribe of Shabazz arrived from Mecca some four hundred years ago. In fact, in the overwhelming majority of public speeches that Malcolm and other ministers give in this period, it is strongly stated that the Nation identifies with Africa. By examining speeches given inside the mosques to followers, the seeming contradiction of being both African and Asiatic is clarified.

According to an FBI informant, in 1957 Minister Malcolm X explained to followers that their origins lie in Asia, not Africa:

> Who is the Original Man? The Honorable Elijah Muhammad has taught us the truth. It is the Asiatic Black Man. Why do we say the Asiatic Black Man? Why not the African Black Man or any other Black Man?
>
> Before the white man knew of the planets, they [Black Men] had their true Arabic names but since the White Man has made his culture supreme the whole universe has been altered. Originally this entire planet (that is now called Earth) was called Asia. The first man on it was the Black Man. From him came the brown, red, yellow, and even the white man.[35]

The history of "our own kind," then, had presumably become more complicated and elaborate in the decades after Fard's disappearance. That Malcolm addressed the question as to why they are not African suggests that the Nation may have been distinguishing itself from competing Black Nationalist ideologies. Malcolm offered the most thorough description of the Asiatic black Man in his aptly titled 1962 speech, "Black Man's History," delivered one year before Elijah Muhammad silenced him.

Fifty thousand years ago . . . [a] scientist named Shabazz took his family and wandered down into the jungles of Africa. Prior to that time no one lived in the jungles. Our people were soft; they were black but they were soft and delicate, fine. They had straight hair. Right here on this Earth you find some of them look like that today. They are black as night, but their hair is like silk, and originally *all* our people had that kind of hair. But this scientist took his family down into the jungles of Africa, and living in the open, living a jungle life, eating all kinds of food had an effect on the appearance of our people. Actually living in the rough climate, our hair became stiff, like it is now. We undertook new features that we have now.[36]

Malcolm's teaching differs significantly from that of Fard's, which implied that the Tribe of Shabazz merely crossed over Africa from Mecca en route to the United States 379 years ago. In this altered myth of their lineage, the Tribe of Shabazz settled in Africa long enough to acquire "African" features. Malcolm further explained that the original man had existed in Asia for sixty-six trillion years and migrated to Africa fifty thousand years ago.[37] This dramatic revision may have reflected a consequence of the drastic increase in Nation membership and notoriety among the black masses: members may have sought a precise creation story, one that is (more) compatible with the fact that slaves were taken from Africa, not Asia. The story, at times, may have been even more confusing for new converts; the "true knowledge" explained to Malcolm in prison through letters written by his siblings, was that "Original Man was black, in the continent called Africa where the human race had emerged on the planet Earth."[38]

The most flagrant contradiction in Nation theology, it seems, is that the teachings simultaneously embrace and repel biblical narratives. On the one hand, Elijah and Malcolm often drew from the Bible to illustrate the African American experience. For example, in the last speech Malcolm gave as Elijah Muhammad's chief spokesman, he explained that

Moses' message to the slave master [Pharaoh] was simple and clear: "Let my people go. . . . Let them no longer be *segregated* by you; stop trying to deceive them with false promises of *integration* with you; let them *separate* themselves from you." . . . Moses was trying to restore unto his people their own lost culture, their lost identity, their lost racial dignity . . . the same as The Honorable Elijah Muhammad is trying to do among the twenty-two million "Negro" slaves here in this modern House of Bondage today.[39]

On the other hand, in both early Nation doctrine and in Malcolm's other speeches, it is revealed that "Moses never went down into Egypt. Moses went into the caves of Europe and civilized the white man."[40] To the descendents of Shabazz, the Bible is the devil's history. Thus Adam was the first white man grafted by Yacub's gang. Malcolm explains this belief in his "Black Man's History" speech:

The Honorable Elijah Muhammad teaches us that that man, Adam, was a white man; that before Adam was made the black man was already there. The white man will even tell you that, because *he* refers to Adam as the first one. He refers to the

Adamites as those who came from that first one. He refers to the pre-Adamites as those who were here before Adam. . . . And he always refers to these people as "aborigines," which means what? BLACK FOLK!!![41]

Benjamin 2X explained that Malcolm used biblical narratives as metaphors for black experience, because his followers likely could relate to those stories. "If black people have only one book to their name," Benjamin asserted, "chances are that book will be the Bible, and most of them have at least a nodding acquaintance with portions of it. Much of the time Malcolm would use familiar portions of the Bible to illustrate his point, such as his comparing Moses and the slaves in ancient Egypt to our situation here in present-day America."[42] The notion that black people are pre-Adamite beings reflects an awareness of an intellectual debate from the Enlightenment period engendered by the transatlantic slave trade concerning the nature of humanity.[43] The theological explanation for the creation of humankind was insufficient for these European "men of reason." One school of thought, polygenism, or the theory of a separate creation of the races, held that the Negroid race was subhuman (and therefore pre-Adamite) as compared to their Caucasian counterpart. For, as Edith Sanders points out, "the Western world, which was growing increasingly rich on the institution of slavery, grew increasingly reluctant to look at the Negro slave and see him as a brother under the skin."[44] The Nation's beliefs concerning the disparate genealogies of the races was a deliberate reversal of this race theory that was used to authorize and justify slavery.

What of Malcolm's peculiar claim that Shabazz wanted to toughen his people by leading his straight-haired tribe into the jungles of Africa? It is worth noting that in the *Autobiography's* lengthy passage denouncing the "self-defacing conk," Malcolm/Haley does not raise this aspect of the Nation's genealogy.[45] Beynon provides some insight by merely reporting on some of the books Fard used as study guides—and likely passed on to Elijah Muhammad—during the formative years of the movement. Along with texts on Freemasonry and the Jehovah's Witnesses (unfortunately, Beynon does not specify which ones), in meetings Fard read excerpts from two very popular history books: James Henry Breasted's *Conquest of Civilization* (1926) and Hendrik Van Loon's *Story of Mankind* (1922).[46] Van Loon's book was intended for children; Breasted's work was originally published in 1916 as the standard textbook for young people—so widely read, in fact, that scholars adopted the term "Fertile Crescent" from it.[47] While *The Conquest of Civilization* focuses on prehistory to ancient Roman times, *The Story of Mankind* gives a brief overview of human history, from "Our Earliest Ancestors" to "The Great War" (World War I). Despite their grandiose titles, these historiographies privilege Western civilization and ignore Africa entirely, save Egypt. Both authors concede that while peoples of "the Orient" (comprising Egypt, Babylonia, Assyria, Persia, and the Hebrews) were building nations and developing high civilization, Europeans of the Late Stone Age were struggling with their primitive stone tools in the dark. Breasted, an Egyptologist, wrote: "It was the culmination of that long commingling of ancient Oriental civilizations gathering from Egypt and Asia and forming that Egypto-Asiatic culture nucleus which eventually transformed the life of once savage and barbarous Europe."[48] This historical "fact" was sifted from these (and probably other similar) books only to be exploited and incorporated into the Nation's version of history. "When the black princes of Asia and Africa were wearing silks and plotting the stars," Minister Louis X asserted, "the white man was crawling around on his all-fours in the caves of Europe. The reason why the white man keeps dogs in the house today, and sleeps with them and rides with them about in cars is that he slept with the dogs in the caves of Europe and he has never broken the habit."[49]

However, according to Breasted, because Africans were isolated from "any effective intrusion" by the "Great White Race," "the negro and negroid peoples remained without any influence on the development of early civilization."[50] As for the Egyptians, they belonged to a subgroup of the "Great White Race," "notwithstanding their tanned skins."[51] While there is some suggestion by Van Loon that the Egyptians were part African, his observations are similar to Breasted's: "The fame of the Valley of the Nile must have spread at an early date. From the interior of Africa and from the desert of Arabia and from the western part of Asia people had flocked to Egypt to claim their share of the rich farms. Together these invaders had formed a new race which called itself 'Remi' or 'the Men,' just as we often call the Hebrews 'the Chosen People.'"[52] How is it that this idea of an Asian invasion of Africa, civilizing in its wake, came to be? Why did Van Loon and Breasted consider ancient Egypt to be an "Oriental civilization?" Furthermore, did the Nation merely absorb and perpetuate this blatantly racist view of Africa? Since Elijah Muhammad and his followers believed that Fard was Allah incarnate, it is reasonable to assume that Elijah continued to use these texts after Fard's disappearance. Beynon reports that Fard intended these texts to be understood "symbolically," though he did use them to establish "proofs about themselves."[53] While Elijah Muhammad attempted to Africanize the Shabazz narrative by incorporating a fifty thousand–yearslong sojourn in Africa, what remains in his revision is this myth of a (black) Asiatic migration.

In *Black Messiahs and Uncle Toms*, Wilson Jeremiah Moses compares the black Jews with the Nation and the efforts of Pan-Africanism, Pan-Islamism, and Zionism in general. He explains the ways in which black Jews and Muslims have responded similarly to a racist environment that has thwarted their aspirations. Where they differ, Moses argues, is in the former group's insistence that African culture is Judaic, in "contempt" of "indigenous African culture"—thereby "leav[ing] them with strange bedfellows."[54] He traces the beliefs of the Ethiopian Hebrew Nation and that of other similar Jewish groups to what is now known as the "Hamitic Hypothesis," a theory Moses attributes to the anthropologist C. G. Seligman,[55] which holds that any and every vestige of civilization on the African continent must have been the work of Asian or European settlers (Hamites), not of the people indigenous to Africa. The Ethiopian Hebrew Nation, argues Moses, conflated the Semites with the Hamites, allegedly a racial and ethnic sub-classification of the Caucasian race. The truth is, however, that the black Jews and the Nation have more in common than Moses realized.

The Hamitic Hypothesis was amplified but not created by Seligman; the history of its use is much more complex and expansive than Moses' research suggested. Recent research shows that the Hamitic Hypothesis has influenced historical, philosophical, anthropological, and archaeological study throughout Africa from the late eighteenth century to the modern day.[56] In Edith Sanders' 1969 article entitled "The Hamitic Hypothesis: Its Origin and Functions in Time Perspective," she carefully traces the racialization of Ham to the sixth century CE.

Sanders postulates that there were three different phases of the Hamitic myth. The original, biblical myth from the book of Genesis held that Noah had three sons: Shem, Japhet, and Ham. Noah cursed Ham's son, Canaan, because Ham failed to avert his eyes in shame from his drunken, naked father. The Bible does not describe Ham or his progeny as black; the curse on Canaan only promised that "a servant of servants shall he be." The assertion that Ham was black emerged in the second manifestation of the Hamitic myth, "out of a need of the Israelites to rationalize their subjugation of Canaan."[57] This revision was concretized in the Babylonian Talmud of the sixth century CE.

Napoleon's 1798 invasion of Egypt inspired the third version of this hypothesis. The French rediscovery and popularization of ancient Egypt was obviously problematic in light of slavery. Members of Napoleon's expedition described both the representation of ancient Egyptians on wall paintings and the physical features of living Egyptians as Negroid.[58] The problem was this: "If the Negro was the descendent of Ham, and Ham was cursed, how could he be the creator of a great civilization?"[59] Enlightenment intellectuals deduced that Canaan, son of Ham, was cursed, but Ham and his other sons were not. While the Canaanites had been preordained for slavery, Ham and his other sons were conveniently deemed black but not Negroid but rather Caucasians under a black skin.[60] Subsequent discoveries of civilized African nations in the late nineteenth century compelled explorers to associate every remnant of high civilization and cultural achievement to this mythical Hamitic race. "The confusion surrounding the 'Hamite,'" Sanders argues, "was steadily compounded as the terms of reference became increasingly overlapping and vague. The racial classification of 'Hamites' encompassed a great variety of types from fair-skinned, blonde, blue-eyed [Berbers] to black [Ethiopians]."[61]

Sander's epistemological study explains the classification of ancient Egypt as an Oriental civilization: Breasted's "Egypto-Asiatic culture nucleus" was another way of attributing African achievement to this popular myth of the migratory Hamites. It is hardly surprising that Drew Ali and Elijah Muhammad developed their genealogies in the manner and using the rhetoric that they did—this was the only material on Africa with which they had to work. Similar to teaching that the polygenists were correct (black people are pre-Adamite—and proud!), the Nation's genealogical myth merely represented the inverse of the Hamitic Hypothesis: a "Black Asiatic" invasion of Africa. Rather than dispute the veracity of the tenuous, faith-based claims put forth by historians and anthropologists, Drew Ali, Fard, and Muhammad manipulated European historiographies of Africa to forge a well-documented, "legitimate" history of black people in America.

Whether Malcolm X ever believed in the Nation's genealogy or whether he promoted it only as a propagandistic device is not the issue. Identifying with a particular movement, ethnicity, or group does not necessitate belief. Teaching and promoting these genealogies is equivalent to identifying with them. While scholars such as Louis DeCaro have argued that in the 1950s Malcolm was already using more ambiguous and inclusive phrases such as the "Dark World" to simultaneously embrace African and Asian identities,[62] his promotion of unity among all people of color does not preclude his self-identification as Asiatic. Malcolm declared his "nationality" most emphatically in his 1962 "Black Man's History" speech to fellow Muslims, but he also did not hesitate to divulge an abbreviated version of this history to the public. For example, in the fall of 1963 Malcolm gave a speech entitled "The Old Negro and the New Negro" to a mixed college audience at the University of Pennsylvania, in the question and answer period, Malcolm stated that "the entire earth was once known as Asia and all of the people on it at that time were Asiatic. . . . On my draft card it says Asiatic. And anything that anybody puts in front of me that wants to know what is my race or my nationality, any Muslim will put down Asiatic and that ends it."[63] Malcolm's comment demonstrates that he continued to tout the Tribe of Shabazz narrative in the last year of his ministry for the Nation, however it would be the last time (to my knowledge) that he would publicly comment on being of Asiatic origin.

Moreover, there is no indication that after leaving the Nation Malcolm initially sought to subvert the black Muslim dogma of separation and black supremacy with his new organization. During the press conference on March 12, 1964, at which he announced his defection from the Nation and the founding of the Muslim Mosque, Incorporated (MMI), Malcolm declared, "I am and always will be a Muslim. My religion is Islam. I still believe that Mr. Muhammad's analysis of the problem is the most realistic, and that his solution is the best one. This means that I too believe the best solution is complete separation, with our people going back home, to our own African homeland."[64] In fact, Malcolm originally envisioned the MMI as a more activist (however unaffiliated) branch of the Nation, where he would still teach the tenets of the Nation faith.[65]

While the *Autobiography* may have overemphasized the dramatic changes in Malcolm's religious beliefs as a result of his conversion to Sunni Islam in April 1964, it should be noted that Malcolm did not publicly criticize the tenets of the Nation upon his return to the United States—that is, until he was prodded with threats and slandered in public. By June 1964, when Malcolm X announced the formation of his secular, Pan-Africanist political group, the Organization of Afro-American Unity (OAAU), he was openly criticizing Elijah Muhammad's character in the press.[66] Alex Haley reports that after returning from his second trip to Africa in late November 1964, Malcolm treated the subject of Muhammad's teachings with increasingly "bitter accusations," more "than he ever had."[67] While Haley attributes this "sudden . . . spate of attacks" to a September court decision to evict Malcolm and his family from their East Elmhurst home, it is likely that his virulent criticism of Elijah Muhammad was in response to the culmination of problems he had with Muhammad's organization. Malcolm's initial set of criticisms was centered on the Nation's disinclination to activate its ranks in response to injustices, their mishandling of funds, and Muhammad's infidelities.[68] The looming threat of assault by Nation members coupled with the relentlessly negative coverage of him in *Muhammad Speaks* undoubtedly exacerbated the resentment Malcolm felt toward his former spiritual leader.[69] However, interpreting Malcolm's public denouncements of Elijah Muhammad and the Nation as mere knee-jerk reactions to this hostile atmosphere overlooks the profound insights embedded in his criticisms. Before the Harvard Law School Forum on December 16, 1964, Malcolm remarked:

> Many people will tell you that the black man in this country doesn't identify with Africa. Before 1959, many Negroes didn't. But before 1959, the image of Africa was created by an enemy of Africa, because Africans weren't in a position to create and to project their own images. The image was created by the imperial powers of Europe. Europeans created and popularized the image of Africa as a jungle, a wild place where people were cannibals, naked and savage in a countryside overrun with dangerous animals. Such an image of the Africans was so hateful to Afro-Americans that they refused to identify with Africa. We did not realize that in hating Africa and the Africans we were hating ourselves. . . . We Negroes hated . . . the African nose, the shape of our lips, the color of our skin, the texture of our hair . . . it was not an image created by African or by Afro-Americans, but by an enemy.[70]

The notion that European colonialists were responsible for circulating negative propaganda about Africans would be expressed in a very similar manner and tone in most of the speeches Malcolm gave throughout 1964 until his death on February 21, 1965.[71] At first glance it does not appear that his argument had shifted significantly from what he was describing in the *Reporter* interview of 1960. Malcolm expounded on his views in his January 1965 speech "On Afro-American History:"

> They say mankind is divided up into three categories—Mongoloid, Caucasoid, and Negroid. . . . And all black people aren't Negroid—they've got some jet black ones that they classify as Caucasoid. But if you'll study very closely, all the black ones that they classify as Caucasoid are those that still have great civilizations, or still have the remains of what was once a great civilization. The only ones that they classify as Negroid, are those that they find no evidence that they were ever civilized; then they call them Negroid. But they can't afford to let any black-skinned people who have evidence that they formally occupied a high seat in civilization, they can't afford to let them be called Negroid, so they take them on into the Caucasoid classification.[72]

While Malcolm's rhetoric may appear to be the same, his argument had evolved dramatically from his Nation-sanctioned comments on the etymology of the term "Negro." In those earlier statements, Malcolm deduced that the white man's system of classification was flawed and inconsistent: "If Frenchmen are of France and Germans are of Germany, where is 'Negroland'? I'll tell you: it's in the mind of the white man! . . . You don't call Minnie Minoso a "Negro," and he's blacker than I am. You call him a Cuban!"[73] The important shift lies in his broadening of the focus group—from analyzing the uniqueness of the so-called American Negro's problem ("Nkrumah is an African—a Ghanaian—you don't call him a 'Negro'")[74] to seeing how colonialist typologies have oppressed people of color worldwide. He continues:

> And actually, Caucasoid, Mongoloid, and Negroid—there's no such thing. These are so-called anthropological terms that were put together by anthropologists who were nothing but agents of the colonial powers, and they were purposely given such scientific positions in order that they could come up with definitions that would justify the European domination over the Africans and the Asians. So immediately they invented classifications that would automatically demote these people or put them on a lesser level. All of the Caucasoids are on a high level, the Negroids are kept at a low level. This is just plain trickery that their scientists engage in in order to keep you and me thinking that we never were anything, and therefore he's doing us a favor as he lets us step upward or forward in his particular society or civilization.[75]

Malcolm's argument presents a more mature and informed version of the sentiments implied while on *hajj* in April 1964, namely that race is a social construction—not a scientific fact. During his pilgrimage to Mecca, Malcolm had his Muslim Mosque, Inc. issue a press release that read (in part): "Throughout my travels in the Muslim World, I have met, talked to, and even eaten with people who would have been considered 'white' in America, but the religion of Islam in their hearts has removed the 'white' from their minds."[76] It is important to recognize that Malcolm's views regarding the veracity of race as a scientific truth had by this point permeated both his religious and political agendas. Indeed, Malcolm's experiences after leaving the Nation enabled him to argue what appears to be the antithesis of Elijah Muhammad's program, that "we [black people] were scientifically produced by the white man."[77] Given his astute statements regarding the colonialist discourse on Africans and the spurious taxonomies authored by anthropologists, it is reasonable to deduce that Malcolm had recently become familiar with the deceptions inherent in the Hamitic Hypothesis.

At least twice in February 1965 Malcolm attributed the shift in his views to conversations he had while on his second, longer trip to Africa the previous year, from July through November 1964. "During my conversations with these men [presidents of several countries], and other Africans on that continent, there was much information exchanged that definitely broadened my understanding and, I feel, broadened my scope."[78] Malcolm's shrewd observations regarding the colonialist hand in African historiography was undoubtedly influenced by the contemporaneous African nationalist effort to reconstruct their own histories. The extent to which Malcolm engaged with African intellectuals directly involved in the rewriting of history will likely become known when the diaries of his trips are released to the public.[79]

Juxtaposing Malcolm's perspective on historiography circa February 1965 with some of the passing comments made about Elijah Muhammad's teachings from this period suggests very strongly that the criticism was geared toward Muhammad as much as toward European writers of African history, albeit in a far subtler manner. On two occasions in late February, Malcolm was speaking "off the cuff" about his break with the Nation. On the first occasion, Malcolm was responding to questions after a speech at an OAAU rally held in the Audubon Ballroom on February 15. He remarked, "You cannot read anything that Elijah Muhammad has ever written that's pro-African. I defy you to find one word in his direct writings that's pro-African. You can't find it."[80] Malcolm elaborated on this assertion during a panel discussion on WINS Radio in New York, three days later:

> *Stan Bernard:* (moderator): Let me ask you this, Malcolm. You at one time espoused complete separation of the races.
> *Malcom X:* I must say this concerning what Elijah Muhammad said about separation. He didn't espouse separation. What he said was this: that the government should—if the government can't give complete equality right now, then the government should permit Black people to go back to Africa—He didn't ever say back to Africa. Elijah Muhammad has never made one statement that's pro-African. And he has never, in any of his speeches, written or oral, said anything to his followers about Africa.
> *Bernard* : What about a Black state in the United States?
> *Malcom X:* He was as anti-African as he was anti-white.
> *Bernard:* Did you say a Black state in the United States?
> *Malcom X:* No. So what he said was, "We should go back to our own." And he phrased it like that, because if he spelled it out, he would have to point to some geographic area, and he would have to have the consent of the people in that geographic area, which he knew he couldn't get.[81]

Malcolm's revised understanding of history was not exclusively a recognition and condemnation of the mechanisms of colonialism. Malcolm realized that the Nation's genealogical myth deprived members of the truth about their African origin. Noble Drew Ali and Elijah Muhammad internalized racism to the extent that despite attempting to subvert racist ideology, they appropriated it. Malcolm X sought to revolutionize historiographies of black people rather than merely manipulate the European versions already in place. His comments concerning Elijah Muhammad's treatment of Africa in his teachings show that in his last months Malcolm was also critiquing the Black Nationalist tradition from which he sprang.

NOTES

1. Malcolm X, "Black Man's History," in *The End of White World Supremacy*, ed. Imam Benjamin Karim (New York: Seaver Books, 1971), 26.
2. Ibid., 11, 13. During the period Karim worked with Malcolm X he was known as Benjamin 2X.
3. See Alan Dershowitz's comment in Malcolm X, Archie Epps, ed., *The Speeches of Malcolm X at Harvard* (New York: William Morrow, 1968), 161.
4. Malcolm claimed that "the teachings of Mr. Muhammad stressed how history had been 'whitened'—when white men had written history books, the black man simply had been left out. Mr. Muhammad couldn't have said anything that would have struck me much harder." Malcolm X, *The Autobiography of Malcolm X* (New York: Ballantine Books, 1992), 201.
5. Richard Brent Turner's *Islam in the African-American Experience* (Indianapolis: University of Indiana Press, 1997) is one exception, and I rely heavily on his important work throughout this chapter. However, Turner's work covers the variety of Muslim religious identities that crop up among blacks in America throughout history, sparing relatively little time on Malcolm's own ideas about his origins.
6. Several relationships and moments in his life that could inform that type of inquiry immediately come to mind (and could stand alone as essays unto themselves): analyzing Malcolm's alliances with other Black Nationalists in Harlem, especially Louis Michaux; investigating personal correspondence dating to the period when his sister and confidante Ella renounced the divinity of Fard; and tracking Malcolm's close relationships with Elijah's Sunni-leaning sons, Wallace and Akbar Muhammad. Certainly, the cache of original documents that Malcolm's six daughters have lent to the Schomburg Center for Research in Black Culture in Harlem—but, alas, have not yet released as of this writing—would likely be instrumental in this undertaking.
7. Turner, *Islam*, 73, 75.
8. Hans A. Baer and Merrill Singer, *African-American Religion in the Twentieth Century: Varieties of Protest and Accommodation* (Knoxville: University of Tennessee Press, 1992), 51.
9. Arthur Fauset, *Black Gods of the Metropolis: Negro Religious Cults in the Urban North* (Philadelphia: University of Pennsylvania Press, 1975), 52–68, for Father Divine, 31–41 for the Church of God.
10. Ibid., 107–8. In using the term "cult," Fauset was simply referring to the size of the sect and the strength and unusual charisma of the leader.
11. This, of course, was not the first time the idea of returning to Africa was publicly expressed—church leaders in the mid-nineteenth century first advocated and acted on this idea. See Gayraud S. Wilmore, *Black Religion and Black Radicalism: An Interpretation of the Religious History of African Americans*, 3rd ed. (Maryknoll, NY: Orbis Books, 1998), 125–62. For a comprehensive look at Garvey, see E. David Cronon's, *Black Moses: The Story of Marcus Garvey and the Universal Negro Improvement Association* (Madison: University of Wisconsin Press, 1969).
12. Erdmann Doane Beynon, "The Voodoo Cult among Negro Migrants in Detroit," *American Journal of Sociology* 43, no. 6 (May 1938): 898; and E. U. Essien-Udom, *Black Nationalism: A Search for An Identity in America* (Chicago: University of Chicago Press, 1962), 58.
13. Philip Potter, "The Religious Thought of Marcus Garvey" in *Garvey: His Work and Impact*, ed. Rupert Lewis and Patrick Bryan (Trenton, NJ: Africa World, 1991), 161–62.
14. Arnold Hughes, "Africa and the Garvey Movement in the Interwar Years" in *Garvey: Africa, Europe, the Americas*, ed. Rupert Lewis and Maureen Warner-Lewis (Trenton, NJ: Africa World, 1994), 102.

15. According to Turner, *The Aquarian Gospels of Jesus Christ* is "an occult version of the New Testament" and *Unto Thee I Grant* was published by the Rosicrucians, a Masonic order that is preoccupied with Egyptian mysticism. Turner, *Islam*, 93–94. Curiously, the only white, mainstream Masonic lodge that accepted black members in the early twentieth century was the Alpha Lodge No. 116, located in New Jersey. It is possible that Drew Ali was aware of the Alpha Lodge's mixed membership, though it is highly doubtful that he had direct connections to this lodge. See William J. Whalen, *Christianity and American Freemasonry* (San Francisco: Ignatius, 1998), 29.

16. Fauset, *Black Gods*, 47. Associating black people in America with North Africans (and other people of color who tend to be light-complexioned) perhaps reflects the politics of color and its relationship to status—ideas that were pervasive in this period. For example, Malcolm X remarks on the favoritism his father showed him (ca. late 1920s) and asserts that it was due to him being the lightest child. See Malcolm X, *Autobiography*, 5, 7.

17. Fauset, *Black Gods*, 41; Turner, *Islam*, 92–93, 96.

18. Turner, *Islam*, 93.

19. Fauset, *Black Gods*, 47.

20. Sister Denke Majied, one of Fard's early followers. Quoted in Beynon, "The Voodoo Cult," 895.

21. Ibid., 897.

22. Ibid., 898–99.

23. See Clifton E. Marsh, *From Black Muslims to Muslims: The Transition from Separatism to Islam 1930-1980* (Metuchen, NJ: Scarecrow, 1984), 51; Karl Evanzz, *The Messenger: The Rise and Fall of Elijah Muhammad* (New York: Vintage Books, 1999), 68.

24. Beynon, "The Voodoo Cult," 900. Beynon uses the spelling "Shebazz"; in this essay I will write "Shabazz," because this is how the tribe is spelled in Nation literature. For a compelling study on the etymology of "Shabazz," see Yahyah Monastra, "The Name *Shabazz*: Where Did it Come From?" *Islamic Studies* 32, no. 1 (Spring 1993): 73–76.

25. Benyon, "The Voodoo Cult," 906.

26. Turner, *Islam*, 154.

27. Fard had initially named Poole "Karriem" but later upgraded it to "Muhammad." See C. Eric Lincoln, *The Black Muslims In America*, 3rd ed. (Trenton, NJ: Africa World, 1994), 15. Other members would later substitute an "X" for their slave names. The "X" signified an unknown family lineage, but it also, perhaps, represented Fard's teaching about their lost culture, which included "higher mathematics." See Beynon, "The Voodoo Cult", 900. For a thorough description of the process of discarding slave names, see Lincoln, *Black Muslims*, 105–6.

28. Beynon, "The Voodoo Cult", 907.

29. Essien-Udom, *Black Nationalism*, 63.

30. Marsh, *From Black Muslims*, 56.

31. Beynon cites two key "textbooks" authored by Fard: *Teaching for the Lost Found Nation of Islam in a Mathematical Way* and *Secret Ritual of the Nation of Islam*. The Bible was used in the early period, presumably before Fard's texts were composed. *Teaching . . . in a Mathematical Way* was written in code, requiring Fard's interpretation. At the time of Beynon's writing (1939), few copies of *Secret Ritual* existed; the knowledge was transmitted orally. See Beynon, "The Voodoo Cult," 895, 898; and Lincoln, *The Black Muslims*, 14.

32. Elijah Muhammad's *Message to the Black Man in America* (Chicago: Muhammad Mosque of Islam No. 2, 1965), published after Malcolm X's death, provides the most comprehensive description of the histories of whites and the original people, but this study concerns the Nation's beliefs during Malcolm's lifetime.

33. In July 1959, "The Hate That Hate Produced," an inflammatory news report on the Nation of Islam, aired in five installments on WTNA-TV in New York. Malcolm X, *Autobiography*, 271, 273–76.

34. Nat Hentoff, "Elijah in the Wilderness," *The Reporter* 23 (August 4, 1960): 39-40.

35. FBI Teletype, Malcolm X FBI File, July 11, 1957. Activities of Malcolm Little in New York, New York on June 21, 1957.

36. Malcolm X, "Black Man's History," 48 (emphasis original). Elijah Muhammad offered an abbreviated version of the story at a 1960 Chicago convention. Bontemps and Conroy, *Anyplace*, 226.

37. Malcolm X, "Black Man's History," 44–49.

38. Malcolm X, *Autobiography*, 187. Perhaps his siblings were guided more by the Garveyite elements in Nation doctrine, since this type of thinking was central to how they were raised. See Ted Vincent, "The Garveyite Parents of Malcolm X," *The Black Scholar* (March–April 1989): 10–13.

39. Malcolm X, "God's Judgment of White America" in *The End of White World Supremacy*, ed. Imam Benjamin Karim (New York: Seaver Books, 1971), 126–27 (emphasis original). This was a common theme for Malcolm throughout his tenure with the Nation. An early example can be found in Malcolm X's FBI file. A Chicago-based informant describes a speech given by Malcolm X in 1955 or 1956: "Little cited the ordeals encountered by the children of Israel when they were held captive by the Egyptians in biblical times and likened this to the condition of the Negro in America today." FBI Teletype, Malcolm X FBI File, April 23, 1957. Activities of Little in Chicago, Illinois.

40. Malcolm X, "Black Man's History," 64.

41. Ibid., 42–43; emphasis original.

42. Imam Benjamin Karim, "Introduction," in *The End of White World Supremacy*, ed. Imam Benjamin Karim (New York: Seaver Books, 1971), 16–17. Malcolm himself remarked on his pedagogical style, "I had learned early one important thing, and that was to always teach in terms that the people could understand." Malcolm X, *Autobiography*, 254.

43. See, for example, Mahmood Mamdani, *When Victims Become Killers* (Princeton: Princeton University Press, 2001), 81–82.

44. Edith R. Sanders, "The Hamitic Hypothesis: Its Origin and Functions in Time Perspective", *The Journal of African History* 10, no. 4 (1969): 524.

45. Given Malcolm's emphasis on hair politics while preaching for the Nation, it is curious that he did not reconcile this inconsistency in any of his letters or speeches. See Malcolm X, *Autobiography*, 62–65.

46. Beynon, "The Voodoo Cult," 900; James Henry Breasted, *The Conquest of Civilization* (New York: Harper & Brothers, 1926); Hendrik Van Loon, *The Story of Mankind* (London: George G. Harrap & Co., 1922).

47. Breasted, *Conquest*, 117.

48. Ibid., xiii.

49. Lincoln, *Black Muslims*, 114–15. Minister Louis X later took the name Louis Farrakhan.

50. Ibid., 113. Note that Breasted wrote "Great White Race" in capitals and "Negro" and "Negroid" in lower-case letters.

51. Ibid.

52. Van Loon, *Story of Mankind*, 22.

53. Beynon, "The Voodoo Cult," 900.

54. Wilson Jeremiah Moses, *Black Messiahs and Uncle Toms: Social and Literary Manipulations of a Religious Myth* (University Park: The Pennsylvania State University Press, 1982), 183–84, 189.

55. Moses, *Black Messiahs*, 189. Seligman's major text is *Races of Africa* (London: Oxford University Press, 1930).

56. For a summary of the effects of the Hamitic Hypothesis on black theology in nineteenth-century America, see Wilmore, *Black Religion*, 145–47. For work done on the influence of the Hamitic Hypothesis on the Hutus and Tutsis, see Mamdani, *When Victims*, 79–87. For work done on Nigeria, see Philip S. Zachernuk, "Of Origins and Colonial Order: Southern Nigerian Historians and the 'Hamitic Hypothesis' c. 1870–1970," *Journal of African History* 35 (1994): 427–55. For the role of the Hamitic Hypothesis in justifying slavery, see Robin Blackburn, "The Old World Background to European Colonial Slavery," *William and Mary Quarterly* 54, no. 1 (January 1997): 90–97; and Benjamin Braude, "The Sons of Noah and the Construction of Ethnic and Geographical Identities in the Medieval and Early Modern Periods," *William and Mary Quarterly* 54, no. 1 (January 1997): 103–42. For the role of the Hamitic Hypothesis in African Historiography, see Joseph C. Miller, "History and Africa/Africa and History," *The American Historical Review* 104, no. 1 (February 1999): 1–32. For the use of the Hamitic Hypothesis in interpreting the archaeological remains at Great Zimbabwe, see Martin Hall, "The Legend of the Lost City; Or, the Man with Golden Balls," *Journal of Southern African Studies* 21, no. 2 (June 1995): 179–99.

57. Sanders, "Hamitic Hypothesis," 522.

58. Mamdani, *When Victims*, 82.

59. Sanders, "Hamitic Hypothesis," 526.

60. Mamdani, *When Victims*, 82–83.

61. Sanders, "Hamitic Hypothesis," 529.

62. DeCaro, *On the Side of My People: A Religious Life of Malcolm X* (New York: New York University Press, 1996), 122.

63. Malcolm X, "The Old Negro and the New Negro," *The End of White World Supremacy*, ed. Imam Benjamin Karim (New York: Seaver Books, 1971), 106.

64. Malcolm X, "A Declaration of Independence," *Malcolm X Speaks*, ed. George Breitman (New York: Grove Weidenfeld, 1990), 20. The idea of going back to Africa does not necessarily contradict the Shabazz narrative, since the Tribe of Shabazz presumably called it home for the fifty thousand years prior to their enslavement.

65. See M. S. Handler, "Malcolm X Splits with Muhammad," *New York Times*, March 9, 1964, p. 1; "Malcolm's Brand X," *Newsweek*, March 23, 1964, p. 32.

66. "Malcolm X Flees For Life; Muslim Factions at War," *New York Amsterdam News*, June 20, 1964, pp. 1–2.

67. Malcolm X, *Autobiography*, 484.

68. In his eviction trial appearance before dozens of Elijah Muhammad's followers, Malcolm accused Muhammad of "fathering almost a dozen illegitimate children." See "Malcolm X Flees For His Life," *Pittsburgh Courier*, July 11, 1964, p. 4.

69. See "Malcolm X Flees For Life; Muslim Factions at War," *New York Amsterdam News*, June 20, 1964, pp. 1–2.

70. Malcolm X, *The Speeches . . . at Harvard*, 167–68.

71. See "The Oppressed Masses of the World Cry Out for Action Against the Common Oppressor," in *February 1965: The Final Speeches*, ed. Steve Clark (New York, Pathfinder, 1992), 53; "Educate Our People in Politics," in *February 1965*, 94; "Not Just An American Problem," in *February 1965*, 157, 161; and "A Global Rebellion of the Oppressed Against the Oppressor," in *February 1965*, 178.

72. Malcolm X, *Malcolm X On Afro-American History*, ed. George Breitman (New York: Pathfinder, 1967), 17–18.

73. Malcolm X at the Boston University Human Relations Center, February 15, 1960. Lincoln, *Black Muslims*, 64–65. Minnie Minoso is the baseball legend who broke Chicago's "color barrier" when he signed with the White Sox in 1951.

74. Ibid.

75. Malcolm X, *On Afro-American History*, 17–18.

76. This letter was published internationally. See, for example, "An Open letter from Malcolm X," *Uganda Argus* (Kampala), July 28, 1964, p. 2, in Box 1, Aliya Hassen Papers, Bentley Historical Library, University of Michigan. The collection also contains the press release on Muslim Mosque, Inc. letterhead, dated April 20, 1964.

77. Malcolm X, *On Afro-American History*, 15.

78. Malcolm X, "Not Just an American Problem," 147; also see Malcolm X, WINS Panel Discussion, in *February 1965*, 204.

79. Malcolm X's daughters placed a presumably extensive collection of Malcolm's speeches, letters, photos and diaries on long-term deposit with the New York Public Library's Schomburg Center for Research in Black Culture sometime between late December 2002 and early January 2003. As of this writing, some of the materials have been processed but nothing has been released. Some of the "most significant" items are diaries written during Malcolm's trips to Asia and Africa in 1964. In the epilogue to the *Autobiography*, Alex Haley recalled that Malcolm withheld information from him about his trip, because he was hoping that his "carefully kept diary might be turned into another book." This suggests that a wealth of information about Malcolm's meetings and contacts in Africa can be found in the diaries at the Schomburg Center. See "Malcolm X Papers Will Come to New York Public Library's Schomburg Center for Research in Black Culture," Schomburg Center Press Release, January 7, 2003, http://www.nypl.org/press/malcolmx.cfm (accessed February 2003).

80. Malcolm X, "There's A Worldwide Revolution Going On," in *February 1965*, 132.

81. Malcolm X, WINS Panel Discussion, in *February 1965*, 205.

CHAPTER 18

LINKING AFRICAN AND ASIAN IN PASSING AND PASSAGE

LISA YUN

Ships tekin' Chinese people to Cuba and Sout' America to work in de sugar field because African slavery done. A lot a crimp, dem beat up people and put den pan de ship. . . . I hear de Prince Alexander tekin' Chinese laborers to Jamaica. When I reach Jamaica, I will find a boat gain' Cuba. I don't 'fraid even though I hear some bad story 'bout what happen to indentured people. Indentured Chinese is jus' like African slave, I hear. Me no believe it. . . . Almas' seven hundred Chinese 'pan de boat gain' Jamaica. Right away I see is true: I sign meself ina slavery. We down in de ship and we don't see daylight. Hakka an' Cantonese too. People fightin'. People sick. People hungry an' t'irsty . . . Too much people. Too much story.
—Margaret Cezair-Thompson, *The True History of Paradise*

How many days now, how many nights, had he lain there—the rats walking his face and teasing his veins, waiting; his trousers cruddy with stains and stinking; his tongue swollen with thirst and leathery in his mouth; his head in torment . . . These were not the stupendous journeys his father had outlined. How had his people been swayed like this, fired up by this, when in truth the Chinese he had seen below, during his nocturnal stalkings, were there dying, were there starving and ill with disease. were there chained to one another, chained to iron railings? Chained. An iron gang.
—Patricia Powell, *The Pagoda*

Two RECENT NOVELS BY AFRO-JAMAICAN WRITERS ATTEMPT TO BRING THE Chinese diasporic experience in the Caribbean into creative realization. *The Pagoda* (1998) by Patricia Powell concerns the saga of the Chinese middle passage during the 1860s and life in Jamaica up through the 1890s.[1] *The True History of Paradise* (1999) by Margaret Cezair-Thompson presents a multiracial, multigenerational narrative of Jamaica that includes "Mr. Ho Sing" and his descendants among the main protagonists.[2] The appearance of these two novels complements current work from the Caribbean Diaspora and provides insight regarding the perception of Asians in the narrative of nation-building in the Caribbean. Naturally, one cannot talk of the Asian Diaspora in the Caribbean without examining the East Indian presence (in countries such as Guyana and Trinidad East Indians constitute approximately half of the population). In fact, Cezair-Thompson's *True History of Paradise* begins with the burial of a young Indo-Jamaican woman, an event that casts a shadow for the remainder of the novel. This chapter, though, will focus on the Chinese Diaspora, since it provides a focus for Powell's novel and her conceptualization of passage.

During the past fifteen years Chinese of the Caribbean have provided literary interpretations of creolized experience in the Caribbean. Among others these include Willi Chen's vignettes of comic violence in *King of the Carnival and Other Stories* (1988), set in Trinidad; Easton Lee's poetry collection *From Behind the Counter* (1998) and Victor Chang's short story *"Light in the Shop"* (1997), set in Jamaica; Jan Lo Shinebourne's *The Last English Plantation* (1986) and *Timepiece* (1988), novels about Indo-Chinese women in the context of Guyanese self-determination; Meiling Jin's *Gifts from My Mother* (1985) and *Songs of the Boatwoman* (1996), poetry and stories of Chinese Guyanese Diaspora. More recently, there are Jamaican poets like Staceyann Chin and Lori Tsang, who have performed their work in New York. Much earlier was Severo Sarduy's novel of Cuban cultural multiplicity, which plays on the instability (and absurdity) of language and meaning and the search for a Chinese opera diva in *El Barrio Chino, De donde son los cantantes* (1967).

Each writer offers a markedly different vision of creolized culture and cultural politics. The creative impulse also reveals a desire to inscribe narratives of Chinese Diaspora into an inclusive Caribbean narrative and history. Severo Sarduy, himself a Cuban with Chinese ancestry, expressed this sentiment when he emphasized that Chinese culture, along with the African and Spanish, was at the center of Cuban identity formation. He went so far as to state in an interview: *"Los chinos han sido muy importantes en Cuba porque, aparte de su influencia en el orden cultural, estan en el centro de la concepcion del mundo cubano."*[3] Jamaican writers Powell and Cezaire-Thompson include the Chinese in works that draw on the inheritance of colonialism and the contestations of race and gender. Powell and Cezair-Thompson write about the Chinese in the context of Creole culture, a concept that has been placed on the margins of Chinese diasporic discourses. In their works, creolization emerges as an "historically affected socio-cultural continuum" but does so in terms that expand this continuum to also include Asians in the making of Jamaican history and culture.

Creolization in Chinese diasporic study has remained marginal until recently. The "immigrant narrative" has dominated representations of Chinese diasporas, calling on tropes of immigrant success, "middleman" roles, cultural traditions, and generational conflict. Both Powell and Cezair-Thompson, however, write cultural narratives that emerge from the conflict and cooperation of different racial groups, which is not a benign cultural process but one fraught with contesting claims of raced and gendered survivals. In both novels, the Chinese characters are grocers whose shops are burned down. Cezair-Thompson makes reference to the Anti-Chinese Riots of 1918. Powell makes a shop-burning the central event of her novel. With this narrative she hoped to open dialogue rather than foreclose it: "I wanted to convey the idea of a Chinese history in Jamaica still writing itself, still making itself, still largely undocumented. . . . I wanted to create this open-ended feel to the novel as another way of reflecting on the absence of a conclusive Chinese history and alerting readers to the fact that the story is simply part of an ongoing dialogue about the documentation of Chinese history and culture in the Caribbean."[4] While still working on the novel in 1996, Powell also acknowledged the question of authority in this work, saying, "I can't imagine what space I'll occupy with the publication of *The Pagoda*."[5] After completing the novel, Powell again referred to the imposition of authenticity, "It took me a long time to write this because I felt I didn't know if I had any business writing about this. But again, I wanted to push myself." The act of a black and female author writing from a diasporic location, creating "Mr. Lowe" as the main protagonist—a cross-dressing Chinese woman in nineteenth-century Jamaica who becomes intimately involved with a light-skinned Jamaican woman who is passing as white, no less—becomes immediately vulnerable to marketplace demands for authenticity and authority. Furthermore, Powell suggests a reading of diasporic formation as transgressive and subversive. In the process, the text opens questions of identity as removed from identity politics and more akin to questions of locating power.

Powell focuses on the "coolie" and small shopkeeper history at the turn of the century, whereas Cezair-Thompson focuses on the middle- and upper-class strata of Jamaica of 1980s. Powell's descriptions of Chinese coolies and shopkeepers actually reveal grounding in historical research (not that this should be required of fictional work), while the text situates diasporic culture as a shifting renegotiation in itself, I focus here primarily on Powell's use of passing and passage as doubled-edged tropes that construct a Chinese diasporic narrative as linked to African diasporic experience in the history of colonialism and nation-building.[6] Although passing and passage bring cross-racial linkage in *The Pagoda*, they also destabilize the racial narrative, resulting in a creative interpretation of Diaspora as a continually challenged formation. For this reason—the linkage between Asian and African diasporic experience—a reading of such works might offer interventions in the fields of Asian American, Asian Diaspora, and Caribbean studies. Identifying a linkage in itself is not necessarily of great significance, but rather more interesting are the tropes and methods that both writers have used to construct these linkages (I use the term "linked" to suggest commonality and shared history but not equivalency or homogeneity in experience).

What is unusual about Powell's and Cezair-Thompson's novels is that both implicitly and explicitly link the Chinese to Africans via the middle passage. Powell and Cezair-Thompson (to a lesser degree) depict the Chinese transoceanic passage in terms of the "coolie trade," including details on the more slave-like aspects of the "coolie passage." In actuality, both authors produce a Caribbean story of the passage that hybridizes features of the coolie passage to both the former English and the former Spanish Caribbean and that hybridizes the history of both Chinese indentured laborers and Chinese kidnapped slaves, although it is the latter history that is least known, although it is well-documented. Like the Africans who preceded them, Chinese coolies were brought to the Americas in the holds of European and American ships, "under the hatches."[7] Some Chinese journeyed voluntarily but large numbers were taken by force, The trafficking of Chinese coolies to Cuba was known as the "coolie-slave trade."[8] In this case, labor historians conclude that the majority of the almost quarter-million Chinese of the Spanish colonies of Cuba and Peru were taken against their will, with approximately 50 percent mortality within eight years. The fact that newspapers documented sixty-eight Chinese coolie mutinies on coolie ships, some resulting in total death of the human cargo, suggests that the Chinese were hardly free emigrants at sea.[9] The term "coolie" comes from the Hindi kuli, meaning "bonded labor" and the Chinese ku li, meaning "bitter labor." By including this history, both authors offer Caribbean narratives that provide linkage between Asian and African labor histories in the Caribbean.[10] The creative narrating of the passage allows for the tracing of diasporic emergence and process in profoundly human and sometimes violent terms.

The representation of diasporic culture as mainly predicated on "process" rather than the recuperation of "origin" is illustrated when Jamaican novelists like Powell and Cezair-Thompson offer "memories" of the passage. The question of origin, "Where do I come from?" becomes eclipsed by questions of emergence and process, such as, "How did I come to be?"[11] Powell's novel emphasizes more of the "how" and not just the "where." The protagonist asks in a moment of frustration: "What would it matter whether or not his daughter knew how he'd arrived there on the island exactly and why? What would it matter how he and Cecil had lived down in the gut of that ship? How he and Miss Sylvie lived up there in that house all these years? How the daughter was born and who were the rightful parents?"[12] The protagonist's preoccupations concern remembering the "how." Likewise, Cezair-Thompson's character Mr. Ho Sing arrives aboard the Prince Alexander (an actual ship that arrived in Jamaica in the 1880s). He tells his story but dwells little on a story of origin. Instead, he emphasizes the "how": "Awright, awright, I goin' tell you how I reach Jamaica. But keep in mind what I really want to tell you 'bout is Miss Rema and Marshal Bloom, an' how a 'ooman an' a racehorse did mek an ole man life sweet."[13] Of the nine pages devoted to Mr. Ho Sing's monologue, only one paragraph refers to his village in China. Mr. Ho Sing's "me one dream" is captured instead in a symbol of motion and risk: Marshal Bloom, a racehorse. Around Mr. Ho Sing's deathbed the family awaits his last words. Mumbling, he clutches the shirt of a Chinese translator, Perhaps he will whisper, "Take me back to China." Instead, the translator says, to the surprise of the family, "Him say 'Gimme one adem Benson an' Hedges.' A ripple went through the room, the hilarity of it all, the old man begging a cigarette, not in

Chinese, but in the Jamaican patois that they all spoke" (9). Mr. Ho Sing emerges as a primary Jamaican character and an example of "Asian Creoleness,"[14] speaking in the richness of patois. Cezair-Thompson noted in an interview that "Mr. Ho Sing is the one that I write most purely in the dialect."[15]

It is this passage that makes Powell's and Cezair-Thompson's works of particular interest. Powell described her writing about the passage as a gnawing desire to know what happened on "that other middle passage from China to the Caribbean, where immigrants died from sadness or starvation or suffocation or infectious disease on board ships where they were beaten to death if suspected of revolt, where so many vessels sunk or were gutted by fires and the charred, bloated bodies were swallowed by the jagged jaws of circling sharks."[16] This description of the Chinese coolie passage echoes historical accounts of the coolie traffic, one that profited British, American, and Spanish empiric trades (three of the top five leaders in the coolie traffic and the African slave traffic). The Chinese coolie passage was so rife with inhumanities and brutalities that it was commonly called "the buying and selling of pigs."[17] Powell's novel concurs with numerous published accounts of the time, many being testimonials by Chinese coolies themselves.[18] Her inclusion of this history offers a basis for a shared Chinese and African narrative of forced labor in the Caribbean. The novel points out the deception and false contracts and indicates how Chinese were "captured and taken prisoner, kidnapped and sold to agents with foreign ships" (43). Powell notes the shame experienced by the buying and selling of countrymen: "Then they would become speechless again, broken down and ashamed at how they had been sold like dogs by their own, crimps working for the dirty foreigners" (43). In *The Pagoda*, images of the passage and metaphors of ship voyage repeatedly appear: the ship's hull, the darkness and lurching, shelves and sailcloth, the rats, "the great undulating vessel."

In the end, Lowe attempts to build a community center that is fashioned as a pagoda, a place that marks "arrival." Obviously, the pagoda is symbolic of Lowe's attempts to make space for his creolized descendants in a Jamaican narrative. The novel ends, however, with the pagoda in a precarious state, sporadically built and not finalized. The notion of creolization as an open-ended trope takes on an added dimension of process when Powell's treatment of the body is included as an integral strand of diasporic experience.

MR. LOWE AND THE "TERRAIN OF HIS BODY"

In *The Pagoda*, the experience of the passage is inseparable from experience of the body, gender, sexuality, and power; this inseparability of history, culture, and the body becomes the basis for experience of Diaspora and creolization. Like the African passage, the Chinese passage to the Caribbean was one that involved violation of the body under the maritime state. It is on the ship that Lowe is almost killed and then forced to serve the captain's sexual needs. Lowe is not "sure he would make it to the island alive, or in one piece, whole" (1998, 98). Powell, however, locates Lowe's passing and passage from the beginning. Lowe's past is described as that of a girl who was dressed as a boy by her doting father, a poor coffin maker and former shipbuilder who was sixty by the time Lowe was conceived. Eventually, Lowe is given away as a girl-wife, as a debt payment, to a pedophilic "old cripple" (243). It is from this situation that Lowe breaks away and boards a coolie ship, where again Lowe dresses as a boy.

The Chinese passage, domain of a male migration narrative, becomes unfixed by the insertion of Lowe,[19] Although Chinese migrations to the Americas were dominated by men, Chinese and East Indian women did arrive in smaller numbers in the West Indies during the nineteenth century.[20] "Passage" and "passing" become established as concurrent strands of boundary crossing and transgression in Diaspora. As Lowe's life progresses, he deploys multiple identities; he becomes "father," "mother," "husband," "lover," "whore," and "shopkeeper." Passing, in this case, becomes more than a binary course (the passing from "one" to the "other"). Lowe engages in multiple, simultaneous passings. Powell's work allows for the reading of Diaspora as a passage through cultural formation with deliberate deployment of social and sexual capital. Powell adopts the narrative conceit of referring to the woman Lowe as "he" and "Mr. Lowe" throughout the text, in keeping with "the fabulous masquerade that was his life" (33).

Thus "passing" occurs in the meta-narrative; the references to Lowe as "he" and "Mr." continue long after the early revelation of his sex. The continued reference to Lowe as "he" by the author, reader, and imagined characters replicates the struggle over the terrain and ownership of Lowe's body. At the same time, it allows for transgressive recognition of Lowe's performance as both a "man" and a "woman." This reading of the text raises the necessary question: How should one refer to Lowe? As "Lau A-yin," his/her Chinese name as revealed at the end of the story? As Ms. Lowe? As "she" who is "Mr. Lowe"? As "Mrs. Lowe," since Lowe and Sylvie are living together under the pretense of marriage? Perhaps the question in itself is limiting: "How should one refer to Lowe?" implies the limits of language and the totalizing nature of naming. But for the purposes of consistency with the text's narrative strategy, I refer to Mr. Lowe as "he" while also referring to Lowe's experiences as a woman who performs masculine gender. In this sense, *The Pagoda* offers possibilities for a reading that ruptures standardized narrative treatments of gendered body and mind,

This litany of Lowe's multiple questions leads one to ask: Is this a Caribbean Orientalism? The text invites such a reading—the familiar recitation of exotic Asian, the mysterious Other, feminized native, deceptive "Oriental" on the island of Jamaica.[21] The use of the "pagoda" as a Chinese symbol in the novel also smacks of a touristic orientalizing of Lowe's past. Although these aspects certainly invite an Orientalist critique, another kind of reading exposes contradictions. Lowe is not passive (partly what "Oriental" implies); Lowe is not "feminine" (he resists this categorization); and Lowe disrupts essentialist notions of gender and sexuality (partly what "Orientalism" implies). Lowe's mysteriousness and deceptiveness are juxtaposed with those of Miss Sylvie, the deceptive Jamaican passing as white.

For both of them, their deceptiveness is posed as survival. The contradictions invite a reading that accounts for subversion. Lowe engages in a struggle with racialized and gendered hierarchies that ultimately lead to a web of masking and unmasking by various parties. At times, the novel reads like a postmodern, cross-dressing telenovela[22] with identities destabilized and confounded: Lowe as a woman passing as a man is raped by the ship captain Cecil. Lowe bears the captain's child in secret, a baby girl, and continues to masquerade as a man. Lowe cohabits with a white Jamaican woman named Miss Sylvie, whom Cecil introduces to Lowe. Lowe's daughter is misled to believe that Sylvie is her mother and Lowe is her father. Sylvie's black servant burns down Cecil's shop and Cecil in it. Sylvie later reveals herself to be an "octoroon" who is passing as white, has murdered her white husband, and has given away her three dark-skinned babies for fear of being discovered as black; Lowe engages in sexual relations with men and women and finally falls in love with Sylvie, even though she leaves him.

The plot thickens and it is easy to become engrossed in this (or dismiss this as) high drama.[23] Yet this is drama that breaches all the conventions of class, color, and sex. Lowe thinks of the complex trap that he is in: "Here he was Chinese, and here he was co-habiting with this white-skinned woman, Miss Sylvie, and here he was now living in the biggest house in the district with a dark-skinned maid and a dark-skinned yard boy" (108). The many plot twists speak to the complex codes of racial and class tensions in nineteenth-century Jamaica while implicating gender and sexuality. Lowe's life becomes a power struggle of resistance, negotiation, and strategies of "passing." As a Chinese woman who performs as "he" and "she," Lowe is a survivor but an "outsider" in the hierarchies of race and gender. The regulation of social status becomes directly tied to regulation of the racialized body. Lowe curses "the audacity of these porcelain alabaster people to want to control his life so thoroughly and completely" (107). Regulation of the racialized body is illustrated in Lowe's relationship with the ship captain Cecil. Lowe attempts to kill Cecil twice, yet Lowe also realizes that he needs Cecil to survive. Cecil rapes Lowe but then becomes Lowe's benefactor and sets up Lowe as "Mr. Lowe," the manager of a small shop that Cecil owns. Cecil also introduces Lowe to fair-skinned Miss Sylvie, who provides a "front" of respectability for Lowe while Cecil is mostly away engaging in the coolie trade. Lowe is sponsored yet confined by a white ship captain and a "white" woman of means. She is one of "the porcelain alabaster people who for decades through marriage tried to bleach stains of black Africa from their skins" (91). Ironically, it is the captain's coolie trafficking, the exploitation of other Chinese, that provides the capital for Lowe (and Sylvie) to survive while entrapping Lowe in a lifelong and life-threatening secret. Lowe, as a liminal figure in the making of creolized culture, stands as a symbol of its limitations and boundaries. Powell's novel speaks to codes enforced upon the body and the complicity of the community in constructing the other to reinforce social "order." The chain of blackmail in the novel implies the regulatory power of such codes, with the most successful blackmail entailing life-threatening consequences. Within Sylvie's house are contained the knowledge of "crimes" against normative society: transgressions of race, gender, sexuality, and class. Sylvie as black woman passes as a high-class white woman and wife, and Lowe as a destitute Chinese woman passes as a male shopkeeper and husband. They are sleeping with each other.

Lowe wonders if the town found them out, "would they just descend on the house with their tins of kerosene and catch them on fire? Would they parade both him and Miss Sylvie through the road, stoning them with rocks and lashing them with insults? Would they tie them to posts and leave them there in the scorching sun to die?" (129). As a means of survival, Lowe continues his association with Cecil and Sylvie—as Lowe says, "we all do things to save our lives" (7). These things are uneasy associations based on punitive constructs of knowledge, power, and implicit violence sanctioned by a communalism of heterosexual and racial strictures.[24] Lowe "could detect the precise moment at which innocent conversations verged on violence, when a demure innuendo could leap out of hand, when boundaries were crossed" (118).

As a Chinese person, Lowe is made aware of the clear line between whites and "others." As a result, he resists Sylvie by refusing to respond during sex and yet allows Sylvie's performance of sex to continue. In their most intimate moments, Lowe's experience of sex with Sylvie becomes conflated with racialized and gendered betrayal: "For who is to say she wouldn't fold up her fantasies into him and turn him further into something he wasn't, as his father had done and then Cecil? And who is to say she wouldn't abandon him once her mission was accomplished. Who is to say?" (113–14), Sylvie's need for Lowe contains scripted roles of power. Sylvie needs "Mr. Lowe" as a "husband" publicly (particularly as she continues her relationship with a "Miss Whitley" on the side) but also desires that Lowe privately wear corsets and be a "wife." Lowe resists being feminized as the wife, not because he rejects femaleness but because he rejects the disempowerment associated with Sylvie's fantasy. The instance of being Chinese and wife and lesbian lover to Sylvie, in this moment, becomes tripled in its potentials for Lowe's disempowerment.[25] Likewise, Captain Cecil's need for Lowe contains scripted roles of power and fantasy. It is ambiguous whether Cecil fantasizes Lowe as a "girl" or a "boy," particularly as he cuts Lowe's hair and dresses him, again, as a boy. The captain's assaults on Lowe are from behind: "his callused hands rough on Lowe's shoulders, his breathing quick and sharp in Lowe's ears; Cecil's face buried in the back of Lowe's head" (96). With "everyone else having claims to his body" (179), the struggle over "cultural" expression is the struggle over the body and the mind. Lowe's transgendered actions reveal a certain agency that Lowe decides to exercise even at great risk. Thus Powell creates a character of the Diaspora who breaks social taboos upheld by normative conventions.

LOWE AND "SO HE OPERATE"

Powell's linking of Lowe and Sylvie in passage and passing, Chinese and African, becomes the novel's beginning marker for understanding the island's multiracial

past. Lowe and Sylvie both contend with systems of color and capital that structure the terms of self and community. Both Lowe and Sylvie are blackmailed
and held under debt by Cecil, who is described as coming from a family of slave
traders (67). In fact, Lowe's passage is aboard a former slaver ship that Cecil
inherits. Slave ships were the models for coolie ships, and the same slave-trading
nations, shippers, investors, and plantations reaped profits by turning to the coolie
trade.[26] Powell also links the African and coolie passages by indicating the transatlantic route (25, 45). Coolies were taken to the Caribbean via both transpacific
and transatlantic routes. The novel's connection of coolie ships to slave ships, and
their routes, is reflective of the maritime history and global economy of the nineteenth century. Whereas Lowe is indebted to a coolie trader, Sylvie lives under the
shadow of her first husband, a slave trader, a "big government man with money
and clout and a face to hold up and a lot of people working for him and capturing
black people and still selling them" (143, also 219). Coolie trade and slave trade
cast shadows over Lowe's and Sylvie's lives and order the social structure in which
they live. Sylvie reveals her own coercion as a black woman under colonial regimes
of color and class. Sylvie's husband, a powerful white man, discovers that she is a
"nigger" (146) and plans to kill her "for shaming him like that" (146). As Sylvie
reveals, she kills him first: "'I kill a man, Lowe . . . I kill a man, a white man. My
husband'" (142). When Sylvie reveals herself, Lowe feels for the first time that he
understood "her silences and her double life," and he saw "that their fates were
linked together" (144). It is only after this revelation that Lowe begins to love
Sylvie (178, 222). Yet Lowe's secret euphoria becomes short-lived as he thinks,
"There was still the daughter to contend with, his marriage to Miss Sylvie, and the
fabulous masquerade that was his life" (33). The ramifications of the social and
racial configurations that the colonial system created still remain.

Powell's vision of Jamaica's, history of color, and creolization reveals the struggle and the fissures. She stated in an interview:

> I've been thinking about the racial/ethnic/political/social spaces that those Jamai-
> cans who are neither of African nor European descent occupy. About the space
> they occupy in the minds of the dominant racial group. In researching the history
> of the Chinese in Jamaica, I've had to reflect on the images of the Chinese that are
> portrayed in Caribbean literature and culture. and the racial fears and stereotypes
> of the Chinese I absorbed while growing up in Jamaica. It's been making me think,
> too, about the East Indians and the other non-white/black groups and their experi-
> ences of exile and displacement, their experiences of otherness and of home, there
> on the island.[27]

Powell illustrates the operations of colonialism that foment friction between
Africans and Asians. Lowe comes to a painful awareness of being despised and
misunderstood, Lowe refers to the black violence against Chinese as "common
accord for them to burn down the Chinese people's shops. Common accord for
them to loot. The more militant types intending to clear his [Lowe's] people out
of the country" (32). In public, Lowe is even avoided by an old Chinese restau-
rant proprietor, whose face tightens when Lowe attempts to speak a few words of

Chinese (56). The Chinese warn each other not to amass publicly, speak Chinese, or draw attention to themselves. The country is composed of "men of a million assortments of brown," but it is "erased of its original inhabitants" and "composed of a transplanted citizenry" (55). Lowe's experience of "transplantation" and the island's creolization is an uneven and conflicted one in a nation and political process that is not automatically inclusive in itself. The novel also indicates a new kind of presence; Lowe notices "America's growing involvement" and "increasing unemployment" (168). As Lowe struggles with the realization "they had been brought there only to supply cheap labor and keep down wages. They had been brought there only to keep the Negro population in check" (45). The burning of the shop hangs over Lowe as he attempts to rebuild his own trust in the people around him and grapples with his own anger. Lowe begins to resent the villagers for their resentment of him—a circle of suspicion and hate. Yet Lowe knows little of their lives and history and they know little of him.

Social relationships become fractured due to interests of the ruling class: "the few Europeans who controlled the country, ever since they arrived" (6). This divisive history erodes Lowe's relations with both the villagers and his daughter. Although the term "benefactor" imparts positive connotations, Cecil aids Lowe's survival but only on Cecil's terms—terms that isolate Lowe and divide Chinese populations from black ones. Lowe's attempts to build relations with black Jamaicans are thwarted by the self-serving actions of Cecil, "for every time Cecil came, it disrupted the relationship he [Lowe] had built up with the villagers. . . . Every time Cecil came, he [Lowe] was assaulted with the memories of the ship" (95–96). Although Lowe recognizes the logic of survival, he barely contains his anger, wanting Cecil dead for a life of entrapment, "shackled to the shop," the target of unwanted envy and hostility. Lowe hides this history from the villagers and also from his daughter, thus resulting in familial estrangement: "For what else could he have said: that her father was a trickster, a rapist, a thief, a smuggler of illegal Chinese, a kidnapper, a madman, a demagogue?" (66).

It is a shop burning that destroys Cecil, who is sleeping in the shop. Here Powell rewrites history, as the burnings of Chinese shops in Jamaica resulted in the deaths of Chinese, not white men. The displacement of the Chinese shopkeepers' death with Cecil's death skews the history of anti-Chinese violence and shop burnings. As a result, Powell's novel sidesteps the ramifications of racialized "othering" and the loss of Chinese lives. But at the same time, the creative decision points to something deeper. The novel effectively narrates the conflict between Africans and Chinese but also shifts this conflict to African with Chinese against European. The creative decision displaces the localized violence and points the finger at the macro-structure of colonial politics. Cecil, the white man-capitalist-trader-owner, dies as a consequence of his profits and ownership of the people. Lowe gains a modicum of freedom upon Cecil's death: "strangely enough he was relieved. . . . He felt clean and unburdened from the shop and from Cecil's plans . . . somewhere deep inside he knew that for the first time he could sort out what it was he wanted to do with his life. . . . He could rethink again those reasons that had brought him to the island and try to live out some of his dreams" (32). The burning of the shop fulfills one of Lowe's deepest desires, as he had wished Cecil dead before.

LOWE'S HISTORY "FROM THE EDGES"

What the minority voice cannot tell becomes history unrecorded: "He would have no history for his daughter; he had told her nothing, taught her nothing" (213). The narrative depicts Lowe's isolation as cultural isolation driven by colonial politics rather than simply "cultural differences" and in this way anchors the culturalist argument in terms of power formation. Lowe's and his daughter's entry into Jamaican society are accompanied by the resentment of white merchants and black villagers and the burning of the shop (57). Lowe manages to place his daughter in a "convent school which only as an exception admitted the children of the Chinese and those of the porcelain alabasters" (114) and manages to make a meager living in a shop while local villagers watch with simmering envy. But it is the telling of history that becomes critical for Lowe, something that implies the sharing and knowing of history as critical to the building of a creole culture, that is, the inclusion of multiple histories. By Lowe telling his story, Powell's narrative writes the story of the Chinese passage into a Jamaican narrative. Lowe attempts to share the passage with his grandson, couched in "storytelling," calling Cecil (the boy's grandfather) a man who did "try hard" (70) yet also revealing him as a man who, "two, three times a day he bring them [Chinese] out for air, walk them up and down the deck, a chain gang of them clanking as they drag the iron. He hose them down with jaze and then water to kill crab lice and then leave them there little to dry off, to yabber at one another. . . . Half of them were already dead' and thrown overboard or just leaning up next to one another, stiff with rigor mortis" (71). Lowe's telling also leads to a lasting relationship with Afro-Jamaican Omar, who stays with Lowe in the end. It is Lowe's telling that help him (and Omar) to make sense of the circumstances under which Lowe came to and entered Jamaica. It is repeatedly reinscribed as the memory enters his mind over and over, as "the shelves of his mind tumbled again, down into the rusty hull of Cecil's ship" (47).

As Omar tells Lowe that the way to tell the story is "from the edges" (228). Images and sounds of slapping waves, decaying maps, tied wrists, the ship's hum, steps of men, and a dark cabin repeatedly enter "the edges" of Lowe's consciousness. At moments, the shop, the ship, and the house overlap as metaphors of containment and with this Lowe's identities as shopkeeper, whore, and wife. At the shop, he paces between the "barred windows of the shop that looked out at the squalling rain, and back again to the locked door that faced the empty square. And back again to the barred window that looked out at the silvery slant of persistent squalls" (97). This echoes Lowe's ship passage and confinement and his earlier circumstance as a girl-wife waiting behind the "barred windows of a house" (179). Ship, house, and island become conflated; the dynamics of passage and passing, and the politics of class, color, and gender are accentuated in the power structures of a coolie ship, ship-as-island, house as society. These become a combined metaphor for a life story that Lowe must negotiate.

The term "whore" also becomes a metaphor that links the colonial underclass for Lowe and Omar. Repeatedly, "whore" enters Lowe's telling of passage and passing as a figurative and literal term that speaks to Lowe being prostituted by Cecil on several levels. When Lowe charges that Cecil "turned me into his whore," Omar replies, "He turn everybody into whore, sir, man and woman. Young or old. He blackmail everybody, sir. So he operate" (227). Lowe puts it to Omar:

> And then there was the pregnancy, and all through that he was so good. I didn't know 'bout the man markets, I didn't know bout the hellish plantations that kill people with work. You know how many Chinese die on the ship with me? I didn't know which life was better, the one underneath him whenever he want, the one tie up, shackle underneath the ship. And then he sew the clothes so people wouldn't know, in the heart of night he smuggle me out. Put me up at some friend or other. Pay some woman to help me deliver the baby, strengthen me again. Then he give me the shop. Give me the money, but every time coming back for more of me, wanting to humiliate me, remind me. (227)

The allegory of race and labor and their costs is summed up by Omar, who repeats, "Yes, sir, so he operate."

The ownership of Lowe as underclass Chinese is marked by his elevation to the "whore" of the captain. Lowe becomes the captain's "mistress," while below are "eight hundred of us [Chinese] pack up down there. Eight hundred. And you should see how little the ship, . . . [f]ive hundred of them kidnap. . . . Yes, Kidnap. Drag out of bed deep into the night. Tie up and beat up and bundled off"(68). Lowe's limited ability to escape is compounded with fear that hordes of men might descend on him if he should resist the captain. Lowe also fears "being thrown down below into that sewer of human waste with the other Chinese" (100). This passage thus reinforces the forced art of passing: the art of consciousness and of controlling one's movement, being, and expression. It becomes a technique of survival for Lowe on board, as it would be for the remainder of his life, as woman and immigrant made "other" and "illegal." Lowe "learned how to stifle sneezes and hold back coughs, how to lock the muscles in his rectum and tighten his kidneys, how to squat for long long hours behind crates of shimmering silk and handcrafted fans, dodge between cargoes of fuel tea and silver dishes wrapped with ropes, dart between towering columns that divided the ship into small, neat boroughs. And always there was the fear clanging in his chest, thudding in the veins by his head" (16–17). As the captain approaches, Lowe lunges out and stabs him in an unsuccessful defense. (This marks one of several times when Lowe initiates force—he later chokes his son-in-law and pulls a knife on Sylvie's manservant. Lowe's behaviors confound the stereotypical categorization of so-called passive-female or active-male behaviors.) The combination of sex, gender, class, and race evolve as the factors that characterize Lowe's passage to

Jamaica. Lowe's telling of this continues to the very end, when he writes a letter to his daughter, an act of passing down a "history." The writing of the letter becomes a device for Lowe to address "past and future" to his daughter, who becomes the next generation in a creolized nation, marked by the legacy to which she was born as a biracial daughter, conceived by rape on a coolie ship and now married to an Afro-Jamaican.

LOWE AND THE WRITING OF "SURPRISE" HISTORY

Powell's novel suggests a definition of belonging for the Chinese in the commitment to stay and build community. Lowe becomes vested in maintaining a claim to a "narrative" of Jamaica, even after the blackmail, the antagonism of the local villagers, and the destruction of his shop. Lowe remains committed to his life in Jamaica with dreams of opening a community center or school. When Sylvie suggests that they leave Jamaica, he is incensed: "Jesus Christ, where? We belong here, this where we live" (142). For him, there is no such thing as a "fresh history," something that Sylvie offers by proposing that she take him to another island. She repeatedly talks to Lowe about moving to Barbados or Panama (149), Lowe rejects the idea as simply undesirable and does not "want to live out anybody else's fantasy. . . . He wanted to have the Pagoda right here on the island" (145). Lowe's commitment to community-building and to Jamaica makes him more a part of Jamaica's Creole identity. Still there is cultural negotiation and creolization emerges as a loss of language. Among other things, Lowe writes of the pagoda to his daughter, suggesting that the daughter might learn about Chinese-Jamaican history and hear some of the Chinese language that was erased. Importantly, it is fellow Jamaicans who aid in the building of the pagoda and who eventually become Lowe's closest friends, Jake and Omar. The diasporic story, as presented here, resists the standardized understanding of Diaspora as the extended practice of maintaining heterosexual structures and reproducing patriarchal traditions. In the last version of the letter Lowe writes to his daughter, he begins with a queering of his history. The sharing of his body as politic becomes an act of claiming and writing his history and body. His body had become "a familiar source of commentary" (172) for Cecil, Sylvie, and members of the community but now also for himself. Although Lowe's "passing" is constructed as a strategic method of survival, it is also an act of queering and transgendering Lowe's diasporic history such that it is not easily commodified or homogenized.[28] Yet Lowe's letters are never sent during the course of the book. Lowe's history is tenuous and contingent, along with Miss Sylvie's "decaying correspondence" and Cecil's moldy, disintegrating travel journal "with pages that had been ripped out"(65). Lowe's own writing of history is only viewed by himself. Thus Lowe's own desires are simultaneously intimated, "presented," yet still unvoiced and unshared history. He discovers that "People don't like surprises. They don't like the truth" (240). Moments of Lowe's struggle are presented as grim yet dryly humorous. When Lowe tells Omar point-blank, "You know I'm a woman," and begins to strip, confirming Omar's suspicions,

Omar panics and says, "Oh God Mr. Lowe": and runs out the room. The surprise is an historical act outside the narratives of normative society. Lowe's truth is one of several truths or perhaps the resistance to fixing one "truth." The novel implies that "official" history is maintained yet subverted by the people on another level. Those in Sylvie's house, along with people in the village, have known of Lowe's passing for a long time. As Omar once said to him, "These eyes don't miss anything at all, Mr. Lowe. I see everything. I know what is going on in that house. I know the show" (126). As well, a village woman named Joyce Fine informs Lowe that his laugh is what gave him away as a woman and "sometimes is a walk, a look, sometimes is a silence, a dis-ease" (153). She adds: "But you know, Lowe, everybody seducible. Man or woman" (153). The "disease" is one of fear, as Lowe apprehensively thinks of Joyce's policeman husband. (The brief story also alludes to the rumored cause of the Anti-Chinese Riot in 1918: a relationship between a Chinese shopkeeper's wife and a policeman.) Joyce's "wink" at Lowe, her complicit knowledge of Lowe's passing, and her philosophy that everyone is sexual, whether "man or woman," is also a "wink" shared by the villagers. The complicity of the villagers and Lowe is one of recognition but also one of boundaries that divide the unspoken from spoken and the official history from the "surprise" history. Lowe's efforts become a constant struggle to write history or to claim his body. Lowe writes his surprise history in letters that seem to never get sent, and then he writes more. He signs the last one, "Lau A-yin," his birth name. Still, throughout the novel Lowe has been unable to shake the name "Mr. Lowe." Lowe has discovered that the power to name and represent becomes a struggle between society, state, and individual. Creole culture, with the need to name and represent the history of all its subjects, from its subjects, becomes an ongoing creolization that relies on continual challenge to its constitution, something that the novel implies by the writing of this letter and the writing of Lowe's "body."

NOTES

1. Patricia Powell, *The Pagoda* (New York: Knopf, 1998).
2. Margaret Cezair-Thompson, *The True History of Paradise* (1999; repr. New York: Plume, 2000).
3. Emir Rodriguez Monegat, *El Arte de Narrar* (Caracas: Monte Avila Editores,1998), 277.
4. Faith Smith, "An Interview with Patricia Powell," *Callaloo* 19, no. 2 (1996): 324–29.
5. Edward Brathwaite, *The Development of Creole Society in Jamaica 1770–1820* (Oxford: Clarendon, 1977), 309–10. I also thank Donette Francis for suggestions regarding Creole identity during my revision of this chapter.
6. Patricia Powell, "The Dynamics of Power and Desire in The Pagoda," in *Winds of Change*, ed. Adele Newson and Linda Strong Leek (New York: Peter Lang, t 998), 194.
7. Faith Smith, "An Interview with Patricia Powell," *Callaloo* 19, no. 2 (1996), 324–29.
8. Craig S. Semon, review of *The Pagoda*, by Patricial Powell, *Mantachusett Telegram & Gazette*, February 6, 2000, p. 6.
9. I do not suggest a trope that is emptied of historical significance and experience. In the case of Chinese coolies, the trope is one of historical grounding. See Joan Dayan's concerns

regarding the middle passage as metaphor in "Paul Gilroy's Slaves, Ships, and Routes: The Middle Passage As Metaphor," *Research in African Literatures* 27, no. 4:7–9.

10. See *Harper's Monthly* and the *New York Herald Sun* for articles during the 1850s to 1870s on American involvement in the coolie trade. See Juan Perez de la Riva, Juan Jimenez Pastrana, Jose Baltar, Basil Lubbock (maritime history), Evelyn Hu DeHart, Denise Helly, Gary Okihiro, and Rebecca Scott. Also see travel literature of the era. See Lisa Yun, "Under the Hatches," *Amerasia Journal* (2001); and Lisa Yun and Ricardo Laremont, "Chinese Coolies and African Slaves," *Journal of Asian American Studies* 4, no. 2 (June 2001).

11. At the same time, the Chinese passage has been exploited and appropriated as propaganda for white labor politics. Conflating "coolie" passage with all Chinese immigration, some American politicians argued to halt Chinese immigration in the nineteenth century. See Stuart Lyman, "The Chinese Question and American Labor Historians," *New Politics* 7, no. 4 (Winter 2000); and John Tchen, *New York Before Chinatown: Orientalism and the Shaping of American Culture, 1776-1882* (Baltimore: Johns Hopkins University Press, 1999). Similarly, the "illegal" passages of Chinese and Africans in the twentieth century have been strategically used by the American media and politicians for bordering and consolidating an "American nation."

12. Powell, *The Pagoda*, 24.

13. Cezair-Thompson, *The True History of Paradise*, 77.

14. Also see Jean Bernabé, Patrick Chamoiseau, Raphael Confiant, and Mohamed B. Taleb Khyar, "In Praise of Creoleness," *Callaloo* 13, no. 4 (Autumn 1990): 894, 886–909.

15. Randall Kenan's interview, "Margaret Cezair-Thompson," *BOMB*, no. 69 (Fall 1999): 55–59.

16. Powell, *The Pagoda*, 190.

17. Also see Loni Ding's well-researched and award-winning documentary film *Ancestors: Coolies, Sailors, Settlers*, which narrates coolie trade to the Americas, http://www.answers.com/topic/ancestors-in-the-americas-coolies-sailors-settlers-tv-episode?cat=entertainment.

18. See Eldon Griffin, *Clippers and Consuls* (Ann Arbor, MI: Edwards Brothers, 1938; repr. Taipei).

19. Powell, *Pagoda*, 172.

20. Also see Cezair-Thompson's novel, in which Ho Sign's second wife, Lim Su, arrives from the passage and never quite recovers. Cherry, the Ho Sing family matriarch, observes: "You don't know what go on in dem faraway places, what kind a meanness go on 'pan dem ship" (254).

21. See Lai, *The Chinese in the West Indies*, 149–53, 166–67; Crawford, *Scenes from the History of the Chinese in Guyana*, 32–36, 112; and Kirkpatrick, *From the Middle Kingdom to the New World*, 65–20.

22. See also various critiques of David Henry Hwang's play *M Butterfly* (which includes a cross-dressing character who performs gender and racial fantasy) that interrogate the politics of body trace, orientalism, and subject performativity, such as Dorinne Kondo's *About Face* (New York: Routledge, 1997); and David Eng's essay "Heterosexuality in the Face of Whiteness," in *Q & A: Queer in Asian America*, ed. David Eng and Alice Hom (Philadelphia: Temple University Press, 1998).

23. The popular genre of televised Latin novellas is well-known for its particular brand of soap opera that often reinforces middle-class norms of color, class, and sexuality. I suggest the term "cross-dressing novella" to mean one that flip-flops norms while maintaining the genre of high drama.

24. One *New York Times* reviewer dismissed the novel as a misconceived modem romance: "It can be a glorious thing to make a modem romance out of a remote time and person, to heighten the colors as you go along, painting pictures for the sake of having pictures.

The trouble is that *The Pagoda* feels more like painting for the sake of painting, a technical exercise from an astonishingly fluent writer who may be suffering just a temporary passion deficiency (Michael Pye, review of *The Pagoda*, by Patricia Powell, *New York Times*, November 1, 1998, Sect. 7, p. T5).

25. See Thomas Glave, "Toward a Nobility of the Imagination: Jamaica's Shame," *Small Axe* (March 2000): 122–25; Timothy Chin, "Jamaican Popular Culture, Caribbean Literature, and the Representation of Gay and Lesbian Sexuality in the Discourses of Race and Nation," *Small Axe* (March 1999): 14–33; Timothy Chin, "'Bullers' and 'Battymen': Contesting Homophobia in Black Popular Culture and Contemporary Caribbean Literature," *Callaloo* (Winter 1997): 127–41; Lawson Williams, "Homophobia and Gay Rights Activism in Jamaica," *Small Axe* (March 2000): 107–11; and "Boom Bye-Bye Batty-Boy," *The Economist* (February 17, 2001): 42.

26. See also Richard Fung's essay "Looking for My Penis," in *Asian American Sexualities*, ed. Russell Leong (New York: Routledge, 1996).

27. Faith Smith, "An Interview with Patricia Powell," *Callaloo* 19, no. 2 (1996): 324–29.

28. Smith, "An Interview with Patricia Powell," 328.

CHAPTER 19

OUT OF CHAOS

AFRO-COLOMBIAN PEACE COMMUNITIES AND THE REALITIES OF WAR

ASALE ANGEL-AJANI

> Queremos vivir en paz
> Como veníamos viviendo,
> Que nos respeten la vida
> Al pueblo indígena y negro.[1]

DISPLACEMENT'S LAMENT

TWILIGHT FINALLY MANAGED TO BREAK THROUGH NIGHT'S BRUTAL EXCHANGE BETWEEN the Revolutionary Armed Forces of Colombia (FARC) and the Colombian military. In Andalucia, a small river community along a cuenca (a tributary) off the Río Atrato, we rattled our way through our morning rituals, bodies still sleeping, minds alert, our ears filtering out the sounds of the jungle for man-made chaos. Weeks of nocturnal combat piled on months of the presence of armed actors joined with years of threats, displacements, forced recruitment, and death and more death renders this a war zone. But this plot of land, this carved out community on top of a hill surrounded by sugarcane is more than just the space between warring armies. In this soil lives memory's ghost. First words were spoken here. The trees contain the maps to dreams carefully crafted in the once quiet evenings. On the walls, sounds of laughter and first love are recorded.

This morning we move like mute ants, busy with the work of defiance. Older girl children gather up toddlers and corral young boys into order. Women crowd around an open flame to cook the last bit of food, and the men, the few that remain, tend to the boats down at the river. Visiting human rights activists and academics from Bogotá and the United States dissect time's significance and catalogue the history of what cannot be taken on our journey. Together we defy the universal classifications that are the spoils of war: civilians, noncombatants, and collateral damage. Collectively, the women, men, and children of Andalucia, like many of the families who live up and down the river, resist the proselytizing of violence's disciples. This is a Peace Community, one of fifty-nine that struggles to maintain a life of dignity in the midst of the reigning chaos of war.

We have little so we take everything, leaving only the shells of small houses and our souls that live in them. When a military helicopter hovers overhead, the sound of the blade slicing through the sky momentarily disorients us. We congregate at the base of the Peace Community *bandera*, a white flag attached to a pole cut from a long tree branch.

The flag is a symbol of all the incalculable ways war has changed life's routine. Widows and mothers of the disappeared have gone through the corridors of Hades' knowledge and earned doctorates in steel-plated truths. Children have learned to restrain their laughter with thick ropes and have developed the habit of reading the invisible script of fear. These are dubious honors.

Forty years ago, peace was driven out of this country. Since then, Colombians were left with the corpse of soldiers, hollowed-out men and women who have tasted the power to give and take life. Their righteousness is sanctioned by the state or by their ideologies, all of which are fueled by greed. The North and South Americans, the Europeans, Israelis, and the Japanese all have a hand in the kidnapping of peace. But here, on the margins of the nation, in a place that is unseen by the eye of the cartographer, we have dared to envision a world of nonviolence. Hope in tomorrow and in our children's tomorrow; faith in the beauty of love and humanity is what makes us subversive. Our will has made a home of our conviction.

We begin our descent to the river. We carry our future on our backs. Our eyes are braver than our mouths. They cast us forward into the unknown. Our feet compel us to stay for this will be the last time they touch the smooth earth of Andalucia. We are heading down river to the Peace Community of Costa de Oro where we will, for now, be safe. Our souls heave with excruciating sobs. Our memories of home drop from our bodies like birth. We are silent. The leaves look on as we pull our boats down the river and whisper their good-byes. The wind dances off the water, taking with it our secrets.

The dead speak in Colombia. So do the living, if one cares to listen. They confirm that Colombia has, in recent years, held the record for the worst human rights violations in the hemisphere. It has been the record keeper long before George W. Bush promised to extend the Clinton Administration's 1.3 billion dollar aid package known as Plan Colombia. Well before 2000, the year Plan Colombia came into effect, the country was drowning in violence and had been for four decades. The violence displaced over two million people.[2] Colombia shares the long sad history of political violence that plagues Latin America and can count the United

States among its most fervent accomplices. Since 1986, Mathew Knoester notes that more Colombians have died at the hands of the military and right-wing paramilitary forces each year than "throughout the entire seventeen years of political repression in Chile under the Pinochet dictatorship."[3] The Andean Commission of Jurist reported that in 1988 and 1989 there were eleven political killings a day. Between 1988 and 1992 there were 9,500 politically motivated assassinations, 830 disappearances, and, in a period of just two years, 313 massacres. The victims were mostly poor rural farmers. As Noam Chomsky writes, "Throughout these years, as usual, the primary victims of state terror were peasants. In 1988, grassroots organizations in one southern department reported a 'campaign of total annihilation and scorched earth, Vietnam-style,' conducted by the military forces 'in a most criminal manner, with assassinations of men, women, elderly and children. Homes and crops were burned, obligating the peasants to leave their lands."[4] Critical observers of Colombia's conflict have found that the rhetoric from U.S. officials about the continued financial support for the "war on drugs," a partnership that was the brainchild of George H. W. Bush in 1989[5] and the "war on terror," the brainchild of George W. Bush, has never amounted to anything more than a counterinsurgency campaign similar to the other U.S. campaigns waged in other parts of Latin America—but perhaps far worse. For example, nearly half of all of the Latin American graduates of the notorious School of the Americas, a combat training school for Latin American soldiers, were Colombian officers. In 1998, it was reported that a least fifty of these officers trained in this Fort Benning, Georgia school were "involved in ten civilian massacres, totaling over 521 victims."[6] Between the U.S.-led military training and the weapons that are either given or bought with the funding from the United States (for example, the promised thirty UH-1 "Huey" and sixty Blackhawk helicopters—sixteen of which are already in the country, or the sixty super Huey helicopters) makes living in Colombia a very dangerous proposition, if you are among the targeted populations: the poor, community leaders, trade unionists, journalists, teachers, nuns, priests, women, human rights activists, lawyers, and children.

The United Nations High Commissioner has deemed Colombia one of the most unstable countries in the world. In 2001, a Colombia Commission of Jurists report found that there were twenty deaths a day directly connected to the political situation; about six thousand people were kidnapped a year, and three hundred thousand people were forcibly displaced a year. Edward S. Herman and Cecilia Zarate-Laun despair that some in the international community may regard Colombians as "unworthy victims."[7] This may well be true if we consider which classes of people mostly face the brutal consequences of the forty-year conflict.

In 2001, Carlos Rosero, an Afro-Colombian activist from the Proceso de Comunidades Negras (PNC) based in Buenaventura, told me that the war has disproportionately impacted the lives of Afro-Colombians with alarming results. Indeed 49 percent of the internally displaced are Afro-Colombians. At the same time, countless other Afro-Colombians have fled into neighboring countries seeking safe-havens, sometimes to little avail.[8] Sadly, Afro-Colombians are among the most marginalized and disfranchised of any community struggling to carve out an existence in Colombia today. In August of 2001, the major daily newspaper, *El Tiempo*, published a long two-page report establishing the fact that Afro-Colombians do exist in large numbers—"Son 10 millones y medio!"—and discussed the abysmal social and political climate for 26 percent of Colombia's population.[9] The paper's two-page story publicly corrected the official oversight of the Administrative Department of National Statistics (DANE), which reported in 1993 that there were only six hundred thousand Afro-Colombians. However what was perhaps most shocking for this nation, which has the third highest population of African descendants in the Americas (after the United States and Brazil), were the realities confronting black people. An astounding 80 percent of all black people live below the poverty line. When there is paid work, 74 percent receive salaries less than the minimum wage. Only two out of every one hundred blacks who finish high school in all of Colombia go on to university. Fifty percent of blacks who live in rural areas are illiterate—three times more than the national average.

Afro-Colombians have historically faced rampant discrimination and the state has actively and unabashedly undermined the community. In 1959, for example, a law was passed that designated the Pacific Basin as public lands. The 1959 law placed regulations on these public lands to ostensibly conserve natural resources. Bettina Ngweno notes that, "this dispossession by the state included lands that had been previously titled to Afro-Colombian families . . . making them illegal occupants of their lands."[10] Afro-Colombian Senator Piedad Cordoba Ruiz states the case more firmly for black people when she writes of the 1959 law: "As a result they lost forty percent of their traditional territory, which was divided among diverse institutions, mostly private businesses. Since then, legal efforts to grant Afrocolombians' autonomy in their territories have been realized on paper only."[11] The Constitution of 1991 attempted to rectify the inequalities that were created by the 1959 law. Ironically, while the Constitution of 1991 authorized state protection and recognition of ethnic diversity and territorial, cultural, and political rights, Marino Córdoba, originally a community activist from Riosucio and later a founder of AFRODES (an association for displaced Colombians) writes, "Blacks were not represented in that . . . [a]ssembly, but we asked the indigenous representatives to take up the defense of our culture and land rights. They won some recognition of our rights that were small, but important."[12] While there was some reluctance to grant protective measures for black communities, the Transitory Article 55 was included. This paved the way for what Senator Piedad Cordoba Ruiz considers to be the most significant piece of legislation for black communities, Ley 70 (Law 70), which was passed in 1993. In theory, Law 70 opened the space for black communities to apply for collective land titles to their ancestral lands. Additionally, the law sought to improve the material conditions for black communities by outlawing discrimination and improving access to credit, education, and training.[13] It was mandated that education in the communities had to foreground black cultural traditions. In practice, Law 70 has been a whole other matter.

Colombia is a country that is swimming in the riches of its natural resources. Given the long-standing tradition of wholesale corporate extraction of these resources (for example, the 1959 law), the requirement that Colombia increase exports of primary goods under structural adjustment policies, and privatization and "free" trade (to widen under the Free Trade Area of the Americas), black communities, the indigenous, and mestizo campesinos are in the way of corporate profit. For instance, the department of Chocó, which has the highest population of blacks (roughly 88 percent), is considered to be among the world's richest regions in biodiversity. Journalists Edward Herman and Cecilia Zarate-Laun: "The Chocó area in the Northwest part of Colombia below Panama is rich in minerals and oil and contains one of the world's last pristine rain forests. It has been rapidly opened up to mining, oil and timber exploitation, and pipelines, ports, railroads, a canal, and the last 65 miles of the Pan American Highway are being pushed forward to bring this region into the global market. Free trade zones are in the planning stages for the Chocó area."[14] Many Afro-Colombians and others view Colombia's conflict as not being a "drug war" but rather a war over resources.[15] Marino Córdoba and other black activists, journalists and policy makers, have all argued that the very communities that have sought to legally obtain land titles under Law 70 have been victimized by armed actors. Spanish journalist Paco Gómez Nadal, who wrote about the largest massacre in the country's history that occurred in the Afro-Colombian community of Bellavista (medio Atrato, located in the Bojayá municipality of Chocó), has asked why armed groups use so much force in the expulsion of these communities, especially the ones that have advanced in the process of collective land titling?[16] Gómez Nadal shows that massacres in Chocó, which directly caused the displacement of hundreds of thousands of people, coincide with land titling. On December 13, 1996, Riosucio, located along the Atrato River (bajo Atrato), received the first six collective titles for sixty-one thousand hectares of land for two hundred and fifty families. Four days later, on December 17, paramilitaries began fighting in and around Riosucio, eventually taking over the community two days later on the December 20, 1996.[17] With assistance from the Colombian Army, the paramilitaries (considered to be illegal right-wing terrorist forces in official discourse) violently undermined the legal rights of black communities. Marino Córdoba, then a community leader in Riosucio before he was forced to flee the area, describes the violent events of December 1996, which began at 5 o'clock in the morning:

> Paramilitary groups arrived in my town . . . intent on murdering the leaders and their families. Many were taken from their beds and paraded naked through the streets. Anyone who resisted was killed. The shouts woke me up. I ran to take refuge in the swamp along with many others. At 8:00 am, army helicopters started patrolling. The paramilitaries radioed the pilots to attack the swamp, claiming the people were guerrillas. The army attacked us with bombs and rifles, killing many people.

Those who survived stayed in the water for three days until hunger and desperation forced us out.[18]

In Riosucio alone, the paramilitaries killed 750 people and nearly 50,000 were displaced between 1996 and 2000. In 2001, the year I was first in Riosucio with writer Victoria Sanford, and again a year later in 2002, the paramilitaries were still in the community. In fact, they occupied much of the bajo Atrato area by 2002 and still worked in collaboration with the Colombian military. In addition to the massacres in Riosucio in 1996 and Bellavista in 2002, there have also been massacres in Bagadó in 1997—in which the entire community of more than twenty thousand people fled, also in 2000 in Anchicaya, Vigía del Fuerte, and Bellavista (again), in the communities of Naya and Yurumangui in 2001, and most recently in Zabaletas, a municipality of Buenaventura in June of 2003.[19]

Perhaps Gómez Nadal is right when he says that, "except in dramatic cases, Chocó doesn't exist. Only the dead are visible."[20] Or that indeed "Chocó is a specimen in the laboratory of the forgotten."[21] But massacres in the name of profit are only half of a very troubling reality.

In the wake of violence, communities in Urabá-Chocó and Antioquia have organized themselves into peace communities as a viable way to nonviolently resist the armed conflict that has dominated their lives. At the end of 1995 and the beginning of 1996, communities along varying points of the Río Atrato endured threats, bombardments, selective homicides, massacres, and blockades on their food supply. Several thousand people were forcibly displaced from their homes; many traveled by land and by river for nearly a month to reach safer locations. By the end of 1996, many communities in the Riosucio municipality living along the river basins of Curbaradó, Jiguamiandó, Salaquí, Truandó, Quiparadó, and others committed their lives to peace. Thousands of people, largely Afro-Colombians, indigenous, and mestizos have declared themselves to be part of a large network of communities seeking peace, autonomy, justice, dignity, and the right to raise their children outside of the context of war. However, even under the banner of the San Francisco de Asís Peace Communities, families still endure pressure and the threat of death by armed groups who do not respect the will of communities who are pursuing a life of nonviolence as a means to end the cycle of war. As Sanford shows, the Urabá-Chocó peace communities are currently the only effective alternatives to war.[22]

Men, women, and children in these small rural communities have gone beyond what the Colombian state claims to want to provide for their citizens. As poor, mostly black and indigenous campesinos who are marginalized by the state and brutalized by the guerillas, paramilitaries, and the army, they have asserted their right to not participate in the conflict at great personal risk. Writing about her time in 2000 with the Afro-Colombian peace community of Cacarica—a community that had been displaced by the paramilitaries and the army during 1997 and 1999—Sanford tells of the principles that guide all peace communities: "At the entrance to Cacarica, a hand-painted sign in the rainbow colors of the peace community states: 'We are a Peace Community. We are special because we do not carry any weapons. No armed actors, whether legal or illegal are permitted in our community.'"[23]

The mandate of peace communities is that they will in no way voluntarily support or collaborate with armed actors. Often when communities—mainly women—are forced to provide food or shelter to one group, they become targets for the next group that moves into the area. This could be a never-ending cycle of terror if communities do not actively resist armed actors through peaceful means.

In 2002, I returned to the Urabá-Chocó region with Sanford to join a celebration that commemorated the five-year anniversary of the formation of the peace communities. In the village of Curbaradó, where we had been a year a earlier, families and representatives of several peace communities, international non-governmental organizations (NGOs) and Colombian-based NGOs gathered to remember the dead and disappeared and honor the remarkable efforts of the living. As community members stood in line with crosses baring the names of victims, it was clear that, over the past five years in particular, violence shook the foundation upon which peace communities are built. Despite the presence of the international NGOs and the military (who were brought in to "protect" them), it was clear that members of these peace communities struggled to remain alive, despite the fact that Colombian and international NGOs (for example, the United Nations, Red Cross, and Doctors without Borders) have abandoned them. As a result, these communities are not able to receive ongoing medical care, educational support (i.e., teachers and school supplies), or food. Only a handful of local NGOs have offered any real support to peace communities.

On a very limited budget, members of the Pastoral Social of the Diócesis de Apartadó and Quibdó and CINEP (Research and Popular Education Center) and Justicia y Paz have been the only organizations providing consistent assistance to peace communities in the Urabá-Chocó region.[24] In addition to humanitarian aid, one of the many ways that these organizations support the survival of peace communities is through accompaniments. Small teams of people working for the Pastoral Social, the social and humanitarian wing of the *diocesis*, travel to rural communities providing support. Increasingly, communities are asking the Pastoral Social to accompany their communities out of combat zones to safer locations, and several communities have requested those accompaniment teams to stay on in their community, hoping that their presence will provide a measure of security. Often, accompaniment team members, who are extraordinary Colombians—many of whom have themselves been displaced and have families of their own—work over twenty days out of each month moving from one conflict zone to another to reach local communities, sometimes working without pay. The ability for these organizations to do accompaniments effectively means that peace communities do not fall prey to violence at the hands of the military, paramilitaries, and guerilla.

From October 13 to October 20, 2003, Urabá-Chocó peace communities met in Costa de Oro and Domingodó to commemorate their continued existence and collectively organize about the best ways to assert their right to live in peace. To be sure, their political project illuminates the very ways in which the Colombian state and the international community have failed the average citizen. These small rural peace communities exist in a context where more than half the country of thirty-three million people live between "absolute poverty" and a poverty that is called "absolute misery." The top 3 percent, the elite, own more than 70 percent of the arable land; the campesinos, who make up 63 percent of the rural population and who own less than 5 percent of the land, are fast becoming an extinct population because of war related causes: displacement, murder, disease, and malnutrition. Peace community members are defying long-standing social orders that have placed their communities in feudal-like relationships to the state. Sanford further argues the case: "The construction of the peace communities becomes a new site from which the international community can judge the Colombian state and put pressure upon it for the way in which the state exercises power. In this way, river communities, while still geographically isolated, are no longer simply sites of surveillance and violence, but also new sites of state legibility."[25]

Audre Lorde once wrote that "survival is not an academic skill. It is learning to stand alone, unpopular and sometimes reviled, and how to make common cause . . . and seek a world in which we can all flourish."[26] The need to recognize the realities confronting black communities and support the necessary life work of their peace communities must be part of black America's political agenda. Why not? As Father Honelio Mercado, a young Afro-Colombian priest who cites the teachings of Martin Luther King, Jr., to his black constituents, has asked me on many occasions: "Do you think black Americans would care what they are doing to us here?" Once he added, "We [Afro-Colombians] know of the struggles of blacks there. After all, we arrived to the Americas the same way, as slaves. Isn't it only a matter of where our boats landed that separates our struggles?"

NOTES

1. Paco Gómez Nadal, *Los muertos no hablan* (Bogotá, Colombia: Editora Aguilar, 2002), 30.

2. The March 14, 2001, Bureau of Western Hemisphere Affairs notes that Plan Colombia is actually a 7.5 billion dollar multilateral aid effort. Plan Colombia is the largest military assistance package to any Latin American country from the United States since El Salvador in the 1980s.

3. *Justicia y Paz* estimated that the Colombian military and their allies, the paramilitaries, are responsible for 70 percent of killings there. Further, in 1997, the U.S. State Department found that of the twenty thousand political murders since 1986, the paramilitaries were responsible for 59 percent of them. In 1996, the State Department found that the paramilitary killings, "increased significantly, often with the alleged complicity of individual soldiers or of the entire military units and with the knowledge and tacit approval of senior military officials." Cited in Mathew Knoester 1998, "Washington's Role in Colombian Repression," http://www.zmag.org/zmag/articles/jan98knoes.htm (accessed April 2001).

4. Noam Chomsky, "The Culture of Fear," *Z Magazine* online, http://www.zmag.org/chomsky/other/ culture-of-fear.html (accessed May 21, 2001).

5. In 1969, Nixon first officially declared a "war on drugs and crime." In 1988, the office of the Drug Czar was formed. Under these auspices, Bush Sr. officially invited Colombia to join U.S. drug interdiction activities in 1989.

6. Mathew Knoester, "Washington's Role in Colombian Repression,"*Z Magazine*, http://www.zmag.org/zmag/articles/ jan98knoes.htm (accessed May 21, 2001).

7. Edward S. Herman and Cecilia Zarate-Luan, "Globalization and Instability: The Case of Colombia," *Z Magazine Online*, http://www.zmag.org/articles/herman%202.htm (accessed May 21, 2001).

8. For example, the Panamanians illegally and involuntarily repatriated over one hundred people who fled to Panama in the 1990s.

9. Nadal, *Los muertos no hablan*, 103.

10. Bettina Ng'weno, "Displacement, Violence and Afro-Colombians" *Against the Current* 104 (May–June 2003), http://solidarity.igc.org/atc/ngweno104.html (accessed August 16, 2003).

11. Piedad Cordoba Ruiz, "Black in Colombia: Organizing to Promote Afrocolombian Territorial Rights," *Revista: Harvard Review of Latin America* (Spring 2003): 84–85, http://www.drclas.harvard.edu/revista/issues/view/16.

12. Marino Córdoba, "The Afro-Colombian Struggle for Land and Justice," *News and Letters*, (May 2002), http://www.newsandletters.org (accessed August 1, 2003).

13. Peter Wade, "Law 70 (Law of Black Communities): Legislation Recognizing and Benefiting Afro-Colombians," http://diaspora.northwestern.edu/cgi-bin/WebObjects/DiasporsX.woa/wa/displayArticle? atomid=691 (accessed August 25, 2003).

14. Edward S. Herman and Cecilia Zarate-Laun, "Globalization and Instability: The Case of Colombia," *Z Magazine*, http://www.zmag.org/articles/herman%202.htm (accessed May 21, 2001).

15. Mary Jo McConahay, "For Afro-Colombians, the War is Not About Drugs," *NCM New Online*, June 2002, http://news.ncmonline.com/news/view_article.html?article_id=504 (accessed August 16, 2003).

16. On May 2, 2002, 119 people where killed in Bellavista—forty-five of whom were children.

17. Nadal, *Los muertos no hablan*, 35.

18. Ibid., 34.

19. Marino Córdoba, community leader in Riosucio.

20. "Massacre in Territory of the Black Communities, Colombia," http://colhrnet.igc.org/newitems/jun03/ afsc.615.htm (accessed August 13, 2003.

21. Nadal, *Los muertos no hablan*, 14.

22. Ibid., 20.

23. Victoria Sanford, "Peacebuilding in a War Zone: The Case of Colombian Peace Communities," *International Peacekeeping*10, no. 2 (Summer 2003): 107–18.

24. Ibid., 109.

25. Victoria Sanford writes, "The Catholic-based Justicia y Paz and the Diocese of Apartado have worked with some forty-five thousand internally displaced people (IDPs) in the Uraba-Choco region since their displacement in the 1990s. They have also worked with the 12,000 IDPs who decided to return to their lands and who have organized themselves into 59 peace communities," "Peacebuilding in a War Zone," 113.

26. Audre Lorde, "The Master's Tools Will Never Dismantle the Master's House," *This Bridge Called My Back: Radical Writings by Women of Color* (New York: Kitchen Table Press, 1983), 98.

PART VI

RACE, POWER, AND POLITICS IN AFRICA

Chapter 20

African American Expatriates in Ghana and the Black Radical Tradition

Kevin K. Gaines

THE AFRICAN AMERICAN RADICAL TRADITION, particularly in the post–World War II era, is a subject shrouded in historical amnesia and misunderstanding. One of the most striking misunderstandings surrounding black radicalism—and the left more generally—concerns their interaction and, at critical times, convergence with mainstream politics. It has become reflexive for many commentators to dismiss the black left, asserting its marginality or claiming that it is "out of touch" with black communities and interests. On this point, white liberals and black nationalists alike have found themselves in unexpected agreement.

Such a refusal to note the affinities and tensions between radical and centrist politics is fatal to historical analysis. To reflect on the African American radical tradition, then, is to recall the conditions that made personal and social transformation possible. There was, first, a moment in which African American radicals emphasized the mutuality of the U.S.-based civil rights movement and African liberation struggles and confronted the postwar emergence of U.S. global hegemony. African American radicals' (and liberals') indictment of the contradiction between American "free world" ideals and the denial of civil and voting rights was steadily approaching a consensus in American (and international) youth cultures, over against the cold war's suppression of dissent. In addition, possibilities for freedom were manifested in the vitality of black popular culture, including rhythm and blues, the church origins and resonances of which eventually symbolized the moral authority of desegregation. In addition, after World War II, modern jazz was a site for cross-cultural African diasporic collaboration and global political awareness (despite U.S. attempts to claim jazz musicians as "goodwill ambassadors"). Finally, the emergence of African states from colonial rule, most famously the republic of Ghana under the leadership of Kwame Nkrumah, further lent a sense of historical momentum to U.S.-based freedom struggles and inspired black diaspora solidarities. For many coming of age at this moment, these developments were hardly marginal but deeply and ineluctably formative. At such moments, for many such persons, aspects of cultural radicalism define the mainstream; they foster new hopes, aspirations, and identities and constitute a sense of participation in the making of history.

African American radicals, in tandem with the masses of southern blacks mobilized in revolt against segregation, were among the earliest exponents of an emerging consensus for freedom and social transformation in the late 1950s and early 1960s, shaped by the movements for civil rights and African liberation. This is not to deny tensions or fissures among and between black radicals and their allies. Nor would it be appropriate to minimize the conservative influence of cold war anticommunism in mass journalism, educational films, newsreels, and the like. Nor, in addition, would it be accurate to deny that there were fundamental disagreements between African American radicals and liberals on the crucial matter of the relationship of the civil rights movement and African liberation to the cold war. But we should not underestimate the fact that, for many, oppositional perspectives were undermining the legitimacy of official pronouncements and positions.

The hold of Ghana and African liberation on the political imagination of so many is demonstrated by the phenomenon of African American expatriates, whose political outlook was representative of that moment's sense of radical possibility.[1] At the height of the civil rights movement—between the late 1950s and 1966—hundreds of African Americans, including intellectuals, technicians, teachers, artists, and trade unionists, left the United States for Ghana, the first sub-Saharan African nation to gain its independence from colonial rule. There was nothing accidental about this extraordinary migration. Kwame Nkrumah, Ghana's first president, studied in the United States during the 1930s. In several visits to the United States during the 1950s, Nkrumah strengthened his ties to the African American intelligentsia, recruiting its members to contribute their skills to Ghana in the name of Pan-African solidarity. W. E. B. Du Bois, for example, spent his last years as a citizen of Ghana and the director of the Encyclopedia Africana project. Ghana was a magnet for African Americans, whose support for Nkrumah's politics of nonalignment, African continental unity, and revolutionary Pan-Africanism was reinforced by their frustration with the racial inequities and cold war constraints of U.S. society.

The independence of Ghana and emerging African states had an impact on black Americans' consciousness and worldview. Black expatriates in Ghana represented an independent black radical critique of cold war liberalism. During the early civil rights era, Ghana was, for the expatriates and other supporters, an inspirational symbol of Black Power (which I distinguish from the later articulation of "Black Power," namely the militant slogan that arose among black activists in the United States). Emblematic of this prior incarnation of Black Power, blacks throughout the Diaspora celebrated Ghana and Nkrumah for their leadership in the cause of Pan-African liberation. The example of Ghana reaffirmed for many civil rights activists in the United States the conviction that history was on their side. For others, Ghana stood as the realization of traditional African American aspirations for African nationhood dating back to the origins of the Pan-African movement, Garveyism, and the Italian invasion of Ethiopia during the 1930s. With such black Diaspora intellectuals as C. L. R. James, George Padmore, Richard Wright, and others inclined to look beyond European and Soviet Marxisms, nonaligned Ghana served as an exemplary expression of their socialist politics. Indeed, Wright's 1954 account of the independence struggle in Ghana was titled Black Power.

As Ghana attempted to carve out a position of nonalignment in the cold war that was independent from both the Soviet Union and the United States, the black American expatriates in Ghana and their allies in the United States waged a similar struggle for independence against cold war ideology and the U.S. government's attempts to impose limits on the political language and tactics of black activists and movements. The radical promise of Ghana's first republic and its potential influence on the nature and terms of struggle in the United States beyond campaigns for civil rights and formal equality led to concerted official strategies of surveillance and containment. In fact, during the early civil rights era, black radical militancy was expressed in moral critiques of cold war liberalism, support for the abolition of the House Un-American Activities Committee, an advocacy of armed self-defense against what was perceived as official tolerance of segregationist violence, support for the Cuban revolution, and advocacy of African liberation. This was what black militancy looked like during the late 1950s and early 1960s, and this radical outlook reflected an important trend in black politics, however foreclosed or forgotten. The legacy of black politics during the 1960s cannot be fully understood without reference to Ghana and the international radicalism embodied by the expatriates.

Black radicals, emboldened by the radicalism of Ghana and Nkrumah, were engaged in an ongoing debate over Pan-African identity during the cold war. During and immediately after World War II, claims of black solidarity connecting African Americans activism with the independence struggles of African peoples was central to black radical politics. But as Penny Von Eschen has shown in *Race Against Empire: Black Americans and Anticolonialism, 1937–1957*,[2] with the cold war persecution of Paul Robeson, Du Bois, and others, black Americans' advocacy for anticolonial struggles on the African continent risked official censure as intolerable criticisms of the U.S. government's policies at home and abroad.

With the convergence of the civil rights movement and African liberation struggles, African American jazz musicians, performing artists, and intellectuals challenged the dominant cold war strictures that attempted to circumscribe black Americans' political identities and, indeed, to airbrush the political activism of the post–World War II era from black collective memory. The well-publicized independence of Ghana from colonial rule in 1957, coinciding with violent racial unrest in the South, sparked resurgence in black protest and solidarities.

The late 1950s also saw the founding of the American Society of African Culture (AMSAC) by the political scientist and civil rights activist John A. Davis. AMSAC sought to promote cultural exchange, collaboration, and heightened mutual awareness between African and African American intellectuals. Although its leading intellectuals were cold war liberals, AMSAC provided a space for radical and liberal black musicians, artists, and writers to independently enact their international visions of solidarity. The jazz musicians Abbey Lincoln and Max Roach; members of the Harlem Writers Guild, including John Oliver Killens and the future luminaries Rosa Guy, Maya Angelou, Audre Lorde, and Paule Marshall; contributors to the radical Harlem journal Freedomways, including Julian Mayfield and John Henrik Clarke; and the poet laureate of black America, Langston Hughes, were all associated with AMSAC.

African American journalists were also crucial to this network of activist intellectuals, including the foreign correspondents Marguerite Cartwright, Charles Howard, and William Worthy. The actors Ossie Davis, Ruby Dee, Sidney Poitier, and Harry Belafonte shared the radical internationalism of the playwright Lorraine Hansberry. As members of the Cultural Association for Women of African Heritage, Angelou, Guy, and Lincoln organized the demonstration against the United Nations and its role in the arrest and assassination of Patrice Lumumba, the revolutionary Pan-Africanist and prime minister of the independent Congo. Such prominent writers as Lorraine Hansberry and James Baldwin defended the demonstration and excoriated cold war liberals' attempts to Red-bait and dismiss independent black dissent.

African American expatriates in Ghana shared the outlook of the radical dissenters in the United States who were sharply critical of a narrowly conceived civil rights agenda, cold war anticommunism, federal indifference to segregationist violence, and interventionist U.S. foreign policy. This group placed the highest priority on black and African struggles for equality and refused to subordinate these causes to cold war anticommunism. The group revered Robeson for his refusal to mute his criticisms of U.S. racism by capitulating to anticommunist witch-hunts. The examples of Robeson, Du Bois, and others taught this transnational cohort of black radicals that cold war anticommunism was a barrier to antiracist struggles at home and abroad. From the outset, these radicals were sharply critical of the U.S. government, which seemed throughout the early 1960s more concerned with African states' perceptions of a United States rent with racial strife than with responding directly and immediately to the demands of civil rights protesters. For American policy makers who tended to see racial discord as a propaganda windfall for the Soviets, it was crucial that African American leadership, Ghana, other African nations, and the domestic civil rights movement refrain from openly challenging U.S. domestic and foreign policies.

For the expatriates, there were many paths to Ghana, ranging from voluntary relocation to forced exile. Nevertheless, they were united in their abhorrence of American racism and their advocacy of the cause of African liberation. From their unique vantage point, such figures as the novelist and actor Julian Mayfield, the social scientists St. Clair and Elizabeth Drake, W. E. B. and Shirley Graham Du Bois, the writer Maya Angelou, the art historian Sylvia Arden Boone, and their stateside allies articulated during the early 1960s critiques of U.S. racism and empire that would become commonplace by 1968. As thousands marched in the nation's capital for jobs and freedom in August 1963, a delegation of the Ghana expatriates picketed the U.S. embassy in Accra, condemning Kennedy's interventions in Cuba and Vietnam, the administration's appeasement of the apartheid regime in South Africa, and its foot-dragging on civil rights. Such forceful criticism of the Kennedy administration was carefully censored from the officially managed March on Washington. The expatriates' parallel demonstration in Ghana, arguably the most radical of several international demonstrations, was in sympathy with the March.[3]

Washington, DC, accordingly attracted far more U.S. government scrutiny than the others (including those held in Paris, Oslo, Munich, and Tel Aviv). This was evident in the detailed State Department memorandum describing the protest, which included the expatriates' original petition to President Kennedy. The demonstration received extensive coverage in the *Liberator*, the magazine of the Liberation Committee on Africa, a radical black organization based in New York.

As the most politically active intellectual among the Ghana expatriates, Julian Mayfield contributed greatly to such demonstrations. Mayfield reached Ghana in 1961, fleeing federal agents investigating his role in the armed self-defense movement led by Robert Williams in Monroe, North Carolina (Williams himself was a fugitive living in exile in Cuba). In Ghana, Mayfield worked for Nkrumah as a speech writer and edited the *African Review*, an independent left journal on Pan-African politics. Expatriates gathered frequently at the home of Mayfield and his wife, Ana Livia Cordero, a physician from Puerto Rico who ran a public clinic for women in Accra. Mayfield's journalistic writings promoted Ghana's policies of continental unity and socialist development and were widely circulated throughout the United States (in Freedomways, Muhammed Speaks, and other black newspapers) and on the African continent. His columns in the Ghanaian press often carried graphic exposes of white southerners' violent resistance to desegregation and voting rights for blacks. Such efforts made Mayfield the most notorious member of the American expatriate community in the eyes of U.S. embassy personnel.

The circumstances of Mayfield's exile to Ghana and his activities there reflect the uneasy relationship between the expatriates and U.S. officialdom, and even with the mainstream civil rights movement. Indeed, several expatriates, including Du Bois, Alphaeus Hunton, and the trade unionist Vicki Garvin, were political refugees from McCarthyism. Mayfield and others questioned the movement's emphasis on nonviolent protest and demanded broadening the struggle to address the economic plight of African Americans in northern ghettos. From their standpoint, nonviolent protest lacked credibility when the U.S. government itself seemed unwilling to protect civil rights organizers or punish segregationist vigilantes. The expatriates's skepticism regarding nonviolence was shaped not only by the reign of terror in the Jim Crow South but also by the violent repression of African peoples and their liberation movements, including the Sharpeville massacre in South Africa and the assassination of the independent Congo's Prime Minister Patrice Lumumba in 1961. With the bombing two years later of Birmingham's Sixteenth Street Baptist Church (which killed four children), the commitment to nonviolence among many activists did not extend beyond lip service.

The Ghana expatriates and their allies in the United States were highly skeptical of orthodox views of domestic race relations. Indeed, their international outlook was a product of the gamut of local and global experiences of white racism in the cosmopolitan black community of Harlem (which extended downtown to the United Nations). For black radicals, this internationalism informed their struggles against police brutality and school and housing segregation in the North. All these issues, domestic and international, received thorough coverage in the Nation of Islam weekly, *Muhammed Speaks*. This deep awareness of the global dimensions of black struggle and U.S. resistance also informed the politics of participants in the Student Nonviolent Coordinating Committee's (SNCC) struggle for voting rights in Mississippi and the Deep South, many of whom, including Robert Moses, John Lewis, and Fannie Lou Hamer, toured Africa during 1965. Moses and his then wife, Dona Edwards, planned to join the expatriate community in Ghana. With the military overthrow of Nkrumah's government in 1966, Moses and Edwards eventually relocated to the next destination for radical expatriates: Tanzania.

For African American expatriates, Ghana's Black Power was multifaceted. They experienced Ghana and Africa variously as a political sanctuary, as a haven for professional and technical opportunities unfettered by racism, and, more important, as the last best hope for democracy and human freedom.

More romantically, some of them regarded Ghana and Africa as a homeland. The crucial point is that for the expatriates, "home" was where they identified the vanguard of black struggle, and during the early 1960s this was understood as Ghana. Relatively unencumbered there by the repressive cold war climate that branded antiracist dissent "un-American" and that scorned attempted linkages of domestic and international struggles for democracy, the expatriates reveled in the expanded horizons in Ghana for black statehood and for black identity. From Ghana, Mayfield penned a critical appreciation of James Baldwin, which appeared in *Freedomways* in 1963, defending him in the face of a rising chorus of dogmatic and homophobic attacks by African American militants.[4]

One important aspect of the transformative significance of Ghana for black identity is the extent to which black feminist consciousness was shaped by the expatriate experience. During the early 1960s, Ghana and Mali hosted international conferences on the status of women of Africa and African descent. It was axiomatic for many of the expatriate women that gender equality was integral to their revolutionary agenda. Such ideology was tested and ultimately reinforced by the profoundly gendered nature of the expatriate experience in Africa, contrasting male sexual adventurism with obstacles to women's independence. This confrontation between feminist ideals and patriarchal realities contributed in part to several women expatriates' early formulations of black feminism by the decade's end. Indeed African American male expatriates, as well, were compelled to reflect on tensions arising from the differently gendered experiences of the expatriates. Mayfield noted that African American women enjoyed an advantage in gaining access to the higher echelons of Ghanaian politics, an access that was not available to most African American men. He also confessed that men like himself would visit Maya Angelou, Alice Windom, and Vicki Garvin for home cooking, which would invariably be served with a scathing critique of their exploitation of Ghanaian women. In a lengthy analysis of the Ghana coup, Sylvia Boone identified sexism and the exclusion of black women from leadership as the downfall of black progressive movements. Interestingly, Boone exempted the late Malcolm X from her indictment, identifying him as the only black male leader who, in her view, granted women equal status in his fledgling Organization of Afro-American Unity. The Ghana expatriates were unique for their gender diversity and egalitarianism, which emerged out of their vision of racial solidarity.

The coup that exiled Nkrumah to Guinea, where he spent the rest of his days, also occasioned the dispersal of most of the black expatriate community. Although the coup extinguished the independent radical project embodied by the expatriates, the Ghana experience was too deeply formative in their lives to end with Nkrumah's overthrow. Back in the United States, the expatriates remained a self-constituted, tightly knit group of colleagues, though it should be added that some remained in exile, unwilling or unable to return to their country of origin. For most, the expatriate interlude in Ghana informed their subsequent activities, as they supported and promoted each other's intellectual endeavors.

They continued to champion African and Third World liberation, yet their focus was on the United States, as they sought to nurture the evolving African American political consciousness. In 1968 Mayfield, along with attorney Conrad Lynn, publicly took issue with Harold Cruse's scorched earth polemic against black left intellectuals in *The Crisis of the Negro Intellectual*. In 1970, Sylvia Boone convened a path-breaking conference at Yale titled "The black Woman," at which Maya Angelou, Shirley Graham Du Bois, and John Henrik Clarke spoke. Against formidable odds, Mayfield, St. Clair Drake, and other returned expatriates offered their perspectives and experiences in hopes of effectively channeling the anger of young militants who, in Mayfield's estimate, had taken Cruse's one-sided account to heart and mistakenly believed that history began with themselves.

What was the legacy of the radical internationalist vision of the Ghana expatriates and their allies? In *No Name in the Street*,[5] James Baldwin produced a scathing indictment of the crimes of Western colonialism and cold war liberalism. The spectacle of anti-Stalinist intellectuals remorselessly betraying each other and principles of civil liberties had begotten the bloody betrayals of the civil rights movement. Baldwin furiously denounced what he described as an international conspiracy of white supremacy determined to crush black dissent. Powerless to obtain the release of an assistant, a black man falsely imprisoned for murder by corrupt New York City police and courts, Baldwin concluded that white America's self-congratulatory vision of civil rights was limited to the matter of transforming hearts and was ultimately blind to the necessity of fighting inequality in the streets. Angrily renouncing the place offered him at the "welcome table" as darling of the American literary establishment, Baldwin devoted himself to affirming in his subsequent fiction the distinctive ethical sensibilities and outlook of black communities, a tradition refined by the community's historical struggle to survive the destructive forces of racism.

Baldwin's indictment of police brutality and the criminal justice and prison systems as instruments of racial discrimination is sadly prescient for understanding the contemporary plight of African Americans. And tragically, with the killing by NYPD officers in early 1999 of Amadou Diallo, an unarmed African immigrant, internationalism is no less relevant for black consciousness. As African American militants, elected officials, professionals, and business leaders joined with other citizens of conscience to protest former mayor Rudolph Giuliani's arrogant attempt to stonewall justice in the Diallo case, we are reminded once again that African American radicalism, as a response to racial oppression and state violence, is central to ongoing struggles for justice and a more democratic society.

NOTES

1. Richard Wright, *Black Power* (New York: Harper and Bros., 1954).
2. Penny Von Eschen. *Race Against Empire: Black Americans and Anticolonialism. 1937–1957* (Ithaca, NY: Cornell University Press, 1997).
3. Harold Cruse, *The Crisis of the Negro Intellectual* (New York: Quill, 1984).
4. Julian Mayfield, "And Then Came Baldwin," *Freedomways* 3 (Spring 1963): 148–49.
5. James Baldwin, *No Name in the Street* (New York: Dial Press, 1972).

CHAPTER 21

"CRIMES OF HISTORY"

SENEGALESE SOCCER AND THE FORENSICS OF SLAVERY

MICHAEL RALPH

> One of the largest migrations of history was also one of the greatest crimes of history.
> —George W. Bush

LOCKED UP

SENEGALESE AMERICAN HIP-HOP RHYTHM AND BLUES SINGER AKON SOARED TO the top of U.S. music charts in 2004 with his autobiographical gangsta ballad, "Locked Up." The track chronicles Akon's frustration with being profiled, harassed, and ultimately imprisoned by law enforcement officials. Though Akon's narrative centers on his experience in the United States, the sense of captivity outlined in his hit song—with its chorus, "I'm locked up, they won't let me out"—parallels events that took place in his native Senegal the previous year.

According to those interviewed, U.S. military personnel arrived on Gorée Island on the morning of July 8, 2003, around 4 o'clock in the morning, accompanied by bomb-sniffing dogs. Soldiers evacuated and then searched homes; afterward, they moved most of the island's residents onto a local soccer field. The field, once packed to capacity, was barricaded. The small satchels of water distributed haphazardly provided little relief for the crowd during the eight hours—from nearly 6:00 a.m. to 2:00 p.m.—some of them spent in the same spot, though the visit, which lasted less than two hours, took place between 11:00 a.m. and 1:00 p.m. All cell phone communication was disabled during this time.

The reaction of the local population can be distilled into a Wolof phrase that appears frequently in transcripts from my interviews: *Da fa mélni Diaam mo gna watt* [It was like slavery had returned].

How should we understand this discourse on slavery more than 150 years after the institution was officially abolished in French territories, andwhy during a visit from the United States?[1] This scenario provokes more questions than answers in my mind: Who designed the security measures taken for Bush's visit? Why did the Senegalese government comply andunder what conditions? This repressive treatment contradicts the political congeniality that exists between George W. Bush and Abdoulaye Wade: for their July 8 trip to Gorée, I was told, the two men used the Senegalese president's personal yacht. There is but one way to enter and exit the island. What exactly concerned U.S. security personnel? Did this community of people on a small island off the coast of Senegal really constitute a threat to the U.S. Commander-in-Chief?

What does it mean that a soccer field was the setting where African bodies faced the coercive presence of U.S. power on Senegalese soil? In recent years, Senegal has gained international notoriety as a soccer powerhouse, especially after its upset victory over France in the 2002 World Cup. For several years now, soccer has been one of the most effective ways for Senegal to distinguish itself—even assert its potential—to the world. The priority of sport, in other words, is not simply evident in local competition but also part of an idea that has folded back into the Senegalese political imagination with startling implications. This essay is concerned with understanding the way this perspective on sport operates with regard to these two distinct but interconnected pursuits.

These differential aims are embedded in ethnographic events and historical developments that this chapter seeks to untangle by highlighting the shifting shape of Senegalese geopolitics and the country's fascination with representing itself through sporting successes. To do this, I dribble back and forth between a series of overlapping events tied to Senegal's response to the World Trade Center attacks of September 11, 2001, on the one hand, and the country's aggressive 2002 World Cup bid and its Cinderella-esque victory over defending champions France in the opening round, on the other. After exploring these two distinct trajectories, I show the way they yield political anxieties that collapse clumsily into President George W. Bush's 2003 visit and become intensified through the way that moment is represented.

Given that both the U.S. and Senegalese governments were complicit in the ritual of captivity that marked Bush's 2003 visit, one is tempted to ask not simply how we grapple with "crimes of history" but also what they are. Who is guilty of what? Who is the judge? Who makes up the jury? So that I might develop a theoretical strategy for tracking the agents and events that form the subject of this chapter, I pursue a brief methodological detour through which I develop what I call a "forensic" approach to anthropology: in the effort to understand historical events, we have only their traces—in conversations, material objects, and media imprints—from which to construct some sense of the action.

More empirical than methodological, my second reason for studying Senegal by means of a "forensic" inquiry is simple: postcolonial Africa is often viewed as a criminal context.

EVIDENCE OF CORRUPTION

Lawlessness and criminal violence have become integral to depictions of postcolonial societies. . . .
—Jean and John L. Comaroff, *Law and Disorder in the Postcolony* (2006, 6)

The idea of democracy is nowhere more suspicious than in the African postcolony. Military regimes in Nigeria, coups in Ghana and Uganda, and chronic political instability in Angola, the Democratic Republic of the Congo, and Sudan are often taken as evidence of inadequate, if not illegitimate, institutions of authority. Senegal, by contrast, is considered "one of Africa's model democracies,"[2] as one of only five African countries not run by dictatorships (Taylor 2005). But despite its pristine reputation there is reason to question the dominant perception of Senegalese republicanism.

Leopold Sédar Senghor, who led the country to independence, steered the course of a nation that went more than a decade without holding competitive elections so that, for much of his presidency, there was only one legal political party. Though Senegal officially achieved independence in 1960, it was 1976 before Senghor passed legislation permitting the presence of three political parties—one Marxist-Leninist, one Democratic Socialist, and the last, Liberal—which, he argued, represented the dominant political inclinations of Senegalese people (Fatton 1987). Perhaps more democratic than a one-party state, this development created the pretense of much improved electoral participation but nevertheless placed constraints on political involvement.

Known throughout the world as one of the greatest poets in the French language and as co-founder of the Négritude movement Senghor, for Nobel laureate Wole Soyinka, evidenced "the transcendence of the humanist over the trappings of" the postcolonial presidency. This was a "lesson in power" Soyinka (2002, 2) "wished so desperately other African Heads of State would heed." Still, Senghor is known as much for "the misappropriation of funds" as for instituting a civil code that protected individual rights. The same man who defended the political and cultural integrity of the Senegalese nation is also known for agricultural collapse and industrial ruin (Diagne 2002, 13).

And when Senghor finally left office in 1981, he did so in a nonelection year so that the presidency was, constitutionally, transferred to his prime minister and protégé, Abdou Diouf, which means that, for the first forty years of independence Senegal was ruled by Senghor's party and the political course he had established. Despite pioneering changes in the national constitution in order to institute competitive elections, Diouf simultaneously faced accusations of vote-tampering and electoral fraud.

Though Diouf served as the Senegalese Head of State for two decades after Senghor's departure in 1981, the dominance of the *Parti Socialiste* (Socialist Party) was officially shattered when long-time opposition leader Abdoulaye Wade became president in 2000. But Wade's politics are suspicious for other reasons. A presidential candidate as far back as 1993, he was then implicated in the assassination of constitutional court judge Babacar Seye, even though he insisted the *Parti Socialiste* was responsible (Sy 2005).[3]

It is a true testament to Senghor's political genius that, despite a dubious commitment to democracy and policies that failed to stave off economic crisis, schol-ar-activists like Soyinka and specialists of African politics have taken what could be understood as a rather ambivalent relationship to democracy as a success story for the Senegalese nation-state in a narrative that, according to this reading, sets it apart from criminal states (Bayart, Hibou, and Ellis 1999) elsewhere in Africa whose claims to authority are considered spurious. In these alternate contexts, the suspension of democracy structures governmental politics and, paradoxically, protocols of protest. The Nigerian government, for instance, has accused Wole Soyinka of treason for condoning violent acts of rebellion against the state. Meanwhile Soyinka, instead of defending himself against this assertion, contends his challenge to the Nigerian state is justified, since it aims to topple a government that is illegitimates anyway (Pietz 1997b).

Soyinka and his adversaries tend to insist that each side has violated well-recognized democratic standards (usually, the national constitution, or legislated forms of due process). Yet the way transnational oil conglomerates that wreak environmental devastation (Apter 2005; Smith 2005)—and which have been implicated in the assassinations of student activists who protest against them (Pietz 1997b)—shape domestic policy could, from a different angle, be viewed as the most sinister aspect of Nigerian politics rather than the state's tendency to violate juridical standards.[4] Thus while scholars concerned with theorizing "corruption" have made it possible to see the limits of statecraft and to trace the contours of social exclusion (Smith 2001, Chabal and Daloz 1999; Bayart, Ellis, and Hibou 1999; Gupta 1995; Bayart 1993; Achebe 1983), this designation inadvertently encourages undue confidence in republicanism as a mode of governance. To insist that democracy fails in Nigeria simply because the reigning government is "corrupt" (Soyinka 1996) is to imply that the political instruments in question are pristine and that if a few culprits can be dethroned and, perhaps, imprisoned, the problem would be solved. But what if—as the foregoing analysis suggests—the concept of political (il)legitimacy is rather more complex? What if, indeed, there is an intimate relationship between law and lawlessness in the African postcolony, and not only there (Comaroff and Comaroff 2006a)? What implications might this have for political theory?

After all Soyinka has admittedly (and, one could argue, justifiably) resorted to extra-legal violence as a means to convey his own political position, "In 1965, he took over a radio station in Ibadan at gunpoint and persuaded the engineers to play a tape denouncing the fraudulent federal election results that the government [had] planned to air" (Masiki 2006). That he was subsequently imprisoned for three months but later acquitted—as much as the fact that two years later he was jailed on spurious charges—indicates the idea of "corruption" in Nigeria is, at best, an imprecise and unstable signifier.

These issues weigh heavily on a Senegalese context where democracy is considered crucial to the national political project but simultaneously undermined. In contrast to countries like Ethiopia and the Gambia, long criticized for imprisoning journalists, Senegalese government officials promote free press (they even permit the easy circulation of international press), yet the state censures critique by charging dissenting media with libel. And Abdoulaye Wade, who is himself a lawyer by training, has routinely used untenable treason accusations and other forms of political malfeasance to silence potential rivals, most famously in the case of Idrissa Seck, a man instrumental in Wade's ascent to the nation's top office in 2000[5] who—in 2005, after widespread speculation that he might run for president—was held in police custody on dubious "corruption allegations" stemming from his alleged mismanagement of state funds. Subsequently, there were even more serious charges, and Seck was transferred to a prison although the new charges were never made public.[6] That Wade secures his political vitality through a crime defined as an attempt to sabotage and violate state sovereignty is especially ironic given that his economic strategy implicitly hinges on the out-migration of Senegalese citizens whose remittances account for one of the nation's largest incomes. The challenge for specialists of Senegal to reckon with the charge of treason tendered by a nation-state that envisions itself as increasingly border-less is not the expressed aim of this chapter but still something to be considered in what follows, though the state's power to reconstitute itself, paradoxically, on the very grounds through which it displaces its primary responsibility (of, for instance, providing its citizens with opportunities for employment) is precisely what gives Senegalese politics the criminal quality I hope a forensic inquiry might help us theorize.[7]

As a method specifically designed to help analysts reconstruct historical events—as a theoretical toolkit concerned with tracking agency by ascertaining culpability (which is surely this method's greatest strength and limitation)—the discipline of forensics might prove a useful point of engagement regarding the crimes that characterize Senegalese postcoloniality.

THE VALUE OF FORENSICS

Forensic anthropology has, historically, been concerned with themes one now finds even in the popular television dramas its techniques have inspired (*CSI*, *Bones*, and *Forensic Files*): the way human remains figure in ideas about life and death and evidence related to criminal cases. In the latter they help investigators determine "what really happened." But given that all narratives are structured by certain assumptions, these methods—instead of leading us to the truth—are better served helping us consider the politics and assumptions of alterity that structure competing discourses.[8] In this way, forensic insights might help us refine tools of inquiry in the social sciences. Apart from continuously being infused with technologies for deciphering human difference in criminal cases, forensics has simultaneously provided a format for enacting compacts that assuage the injured party to a civil dispute, though it clearly, in both instances, has been concerned with making sense of the historical events that structure legal episodes.

One prevalent story about the origins of forensic science traces it to a famous murder in 1905. When an elderly British couple was discovered murdered in the shop they owned, the stage was set for a "scientific breakthrough that solved one of England's most brutal murders and forever changed the criminal justice system."[9] But besides this specific concern with tracking the material residue of human activity, I would like to emphasize a second, different, argument for the origins of forensic science relevant to theoretical issues I raise during the course of this chapter.

In this alternate account, the science of forensics was born some time between the 1830s and 1840s, when the birth of the railroad industry resulted in sudden, unexpected, and tragic deaths in the English countryside. Trying to determine how much families and heirs should be compensated, insurance companies were "faced with the brand new challenge of trying to assess the money value of a human life" (Pietz 1997b, 97).[10] This case, thus, marks a historical shift in civil litigation as understood in the Anglo-American legal tradition.

"Criminal laws," says Rudolph Peters (2005, 1) "give an insight into what a society and its rulers regard as its core values." But since criminal *and* civil suits are often linked to emotional and commercial concerns, forensic insights might cue us to emergent forms of symbolic and economic value. As a theoretical framework concerned with the intersection between history, alterity, technology, social belonging, and value—as configured in social contexts and cultural artifacts—forensics might provide a reliable way to excavate the present.[11] In that spirit, this discussion draws from several different sources for evidence: speeches, interviews, and newspaper articles, such as those I will now discuss, which return us to the event sketched at the outset so we might better understand how Senegal became the stage for one of Bush's most significant campaigns against Islamic terrorism.

Before the tragic events of September 11, 2001, were even ten days old, Senegalese President Abdoulaye Wade had already formed an African coalition to fight terrorism. What the nation lacked in firepower it made up in enthusiasm. Positioning himself as a representative for the African continent, Wade declared that his coalition condemned the terrorist acts committed against the United States and made it clear that, following the lead of the Western coalition fighting terrorism, he would create a committee of seven African nations dedicated to the same cause. This was a collective he expected to include, at least, the leaders of Nigeria, South Africa, and Algeria, as Wade[12] assured the world he would do his part to ensure no African nation financed or otherwise enabled terrorist activities.[13] Complicating Bush's idea of a "crusade" against Islamic terrorism, Wade remained convinced there were no clandestine activities of this sort underway in Senegal despite leading a country that is more than 90 percent Muslim.[14]

By the time Wade's summit was convened on October 17 in Dakar, it boasted fifteen African heads of state.[15] The countries assembled created a "Déclaration de Dakar" ultimately signed by twenty-seven nations, which expressed a conviction that Africa remain free of all terrorist activity whether motivated by political, philosophical, ideological, racial, ethnic, or religious concerns. The signatories believed the guidelines of the United Nations and the Organization of African Unity offered the best blueprints for fighting terrorism and recommended that the "Pact Against Terrorism," proposed and coordinated by Senegal, be ratified through official protocol.[16]

Thirteen days later, President Bush would publicly affirm his appreciation for Senegal's newest foreign policy objective at the "Forum for African Economic and Commercial Cooperation." There, Bush announced that the United States would create a $200 million fund to stimulate private investment for countries in sub-Saharan Africa and would offer American corporations certain protections against risk to encourage interest in the continent.[17]

African economic recovery was a prominent issue in the world that year. In January 2001, Wade had unveiled his *Omega Plan*, which stressed the need to improve physical and human capital in Africa and to bring continent-wide development plans under the aegis of a single international authority (Nabudere 2002, 8). The project was formerly introduced to the public during an international conference six months later. Eventually the *Omega Plan* was combined with South African President Thabo Mbeki's *Millennium Partnership for African Recovery* program. The result was NEPAD, the New Partnership for African Development, an initiative the United States has endorsed, alongside other donor nations.[18]

The United States has so quickly become one of Senegal's most important financiers and political allies, in fact, that French President Jacques Chirac held talks with President Wade in 2004 to "remind" him which country has historically been his greatest ally.[19] This alliance, more meaningful for both countries since 2001, started to assume its present contours during the cold war. While Senegal has operated differently from many socialist nations in its historic reluctance to align itself with other socialist or communist countries, it was never intimate with the United States either. But this trend started to change in the 1980s, "With its adamant opposition to any Soviet influence in Africa, the Reagan administration" began to see Senegal, which had "sharply criticized Cuban and Soviet involvement in Angola and Ethiopia, as a friend and potential ally" (Gellar 1982, 82–83).[20]

The "war on terror" (Kraxberger 2005) offered the perfect occasion to bolster this alliance: Bush's $200 million investment fund sent a message that the United States was prepared to support Senegal's foreign policy both rhetorically and financially. As African nations are frequently expected to demonstrate their commitment to democracy as a precondition for receiving foreign direct assistance, Senegal's fight against terrorism, from Bush's perspective, indicated that the country was committed to sustaining and enhancing democratic political institutions.

From Bush's perspective, one either supported or sabotaged the fight against terror. He aspired to recruit allies in various international arenas—and these domains did not even have to be explicitly political: he threatened to boycott U.S. participation in the 2002 World Cup, for instance, if other nations did not demonstrate a satisfactory commitment to fighting against the encroachment of global terrorist networks.

Ultimately, however, Bush must have been pleased that he decided not to do so because the U.S. national team reached the quarterfinals. This was the subject of a friendly conversation with another world leader whose team had also played exceptionally well during tournament competition. As they met in 2002 to discuss each country's newest foreign policy initiatives, George W. Bush and Abdoulaye Wade joked that if both teams continued to excel in their respective divisions the United States and Senegal would face off in the World Cup Championship.[21] At the same time, soccer was becoming increasingly significant for Senegalese subjects locally.

Thus it seems that even as Senegal worked to bolster public faith in the appearance of its commitment to democracy through its own African-led war against terrorism, it was pursuing another course of action partially aimed at improving the international perception of Senegalese potential.

This leads us back to a question posed at the outset: why did residents of Gorée Island experience slavery again for the first time on the island's only soccer terrain? To explore that question, I draw from certain key events surrounding the political campaign of President Abdoulaye Wade, on the one hand, and Senegal's participation in the 2002 World Cup, on the other. Soccer, here, is key to the way neoliberal aspirations play out.

In 2000, Abdoulaye Wade became Senegal's third president, because he managed to secure the support of Senegal's most powerful and disaffected constituency: the youth. In a nation where more than 50 percent of the population is under the age of twenty (Diouf 1996) and where "urban youth" [22] constitute more than 40 percent of the total 48 percent unemployed population,[23] the youth labor problem is the nation's most significant economic obstacle.

Wade intended for Senegalese youth to become the engine of national productivity—and production, more specifically. His inaugural speech, in which he referred to youth as the country's most valuable "capital resource," builds to a crescendo that closes with the motto that immediately made him famous: "Il n'y a pas de secret: il faut travailler, encore travailler, beaucoup travailler, toujours travailler" [There is no secret [to success]: you should work, work some more, work a lot—always work]. Wade's words were memorialized and resurface in a number of popular *mbalaax* songs throughout Senegal. *"Il faut travailler, beaucoup travailler, encore travailler, toujours travailler"* is always the refrain.

The most popular song was made by the group Pape et Cheikh. But as the song ends, the chorus changes, the most significant word in the mantra being transformed by the writers' efforts to index what they surely hoped would be the outcome of all this hard work, "Il faut gagner, encore gagner, beaucoup gagner, toujours gagner" [You should win, win some more, win a lot, always win].

As Senegal marched through its qualifying matches for the 2002 World Cup, national enthusiasm increased with every victory. News that the team was officially selected occasioned an impromptu parade as the populace flooded downtown Dakar in celebration of this momentous achievement. For his part, President Wade cut short an official trip to France so he could party with the national team at home: "At this time, it's the most important thing that could happen to any country and I will join the team and the nation in celebrating by reducing the amount of time I was expected to stay in Paris." He offered, as well, his sentiments about the importance of this moment, "My deepest congratulations go to the courageous Lions who have *made history* for Senegal."[24]

The president resurfaced in Senegal wearing the jersey of striker El-Hadji Diouf to join the "madness" that characterized local celebrations, according to one spectator. A few days later, Wade held a special ceremony and concert at the presidential palace, where each team member was presented a bonus of 10 million FCFA (then $15,000),[25] this was in addition to cash rewards offered by the nation's wealthiest residents and families.[26] From the perspective of El-Hadji Diouf, the team's best but also its most controversial player, "People" was acting like Senegal had already "won the World Cup."[27]

The drama was heightened once fans realized Senegal, in its first World Cup match, was slotted to battle France. Supporters stressed the geopolitical dimensions of this occasion, "Senegal-France is an historic match. . . . This is the European country that colonized us. And, God willing, we will beat them."[28]

The government, in its effort to galvanize support for the team, promoted the slogan that the *Lions de Teranga*—as the team is affectionately called—hail from "Le Senegal qui gagne" [The Senegal that wins]. Where the slogan came from is not altogether clear, but in the days, weeks, and months leading up to the match against France, the motto littered fliers, posters, and signs across the country. *Le Sénégal qui gagne*. Once emblazoned across the nation, the phrase stuck.

El-Hadji Diouf seemed profoundly to understand the political consequences of this postcolonial drama. As the team swept through qualifying competition on the strength of his eight goals, he found himself in the position to command the Senegalese forces for this important battle. "It's like being the leader of a country," he explained when asked to describe his feelings about the ensuing match against France.[29]

So when Senegal pulled off the 0-1 upset victory, fans poured into the streets of Dakar gravitating, significantly, around *Le Place de l'Independance* and the presidential palace. Red, yellow, and green Senegalese flags, hats, scarves, T-shirts, and African-style boubous, were the only acceptable attire to commemorate the occasion.

One supporter, Omar Ly, extended the meaningfulness of this event beyond Senegal's national boundaries. In his words, it was "a victory for black people everywhere." As he elaborated, "I'm Senegalese, but I've lived in the U.S. and France. I've experienced racism. I'm so happy to be here for this."[30] Whether in terms of the old or new superpower ally, this win communicated an important message.

Yet the victory camouflaged a complicated relationship between Senegal and its former colonizer. In the months leading up to the match, Khalilou Fadiga, who actually scored the game-winning goal, was quoted as saying it would be difficult for him to play against France, "The truth is that I know the streets of Paris better than the streets of Dakar." Fadiga left Senegal when he was six. He grew up in France. These factors would make it tough for him to line up against the representatives of a place that was so special to him, "but," he was careful to add "if I play in the World Cup finals, I've got to try my best to beat them."

Fadiga's loyalties ultimately rested with his ancestral land. "I feel Senegalese," he eventually concluded. "When I was home, everybody would speak our language and we listened to Senegalese radio and music. We ate Senegalese food." The significance of his heritage went even beyond his acculturation, "I share both [French and Senegalese] cultures but I have a lot of family over in Senegal and my *color* is Senegalese."[31]

Fadiga's affectionate testimony highlights an important paradox. Many of the Senegalese players spend at least as much time in Europe as they do in Africa, if not more. To the extent that they spend most of the year playing for club teams—in France, Switzerland, and England, for instance—they are seldom in Senegal.[32] Yet they are its emissaries on important diplomatic missions, such as this one.

This seeming contradiction exposes a significant feature of Senegalese social life, especially during the postcolonial period: whether in terms of professional athletes, musicians, students, politicians, merchants, or professors, the persons occupying the highest ranks of power and wealth are those who have spent some period of time absent from the nation (Diouf 2000, 700). Socioeconomic mobility in this context means moving out to move up.

Still, this team of citizens with multiple national allegiances was constituted, in the World Cup moment, as a coalition of the nation's best talent. Suddenly *Le Sénégal qui gagne* referred not simply to a nation with the ability to win but also to one that had proved it could and was destined to do so. Supporters delivered the chant when welcoming the national team back home. It was a slogan that, when offered, immediately invoked the euphoria attached to this victory. The motto followed the national team through World Cup competition.

The president was quick to associate himself with this turn of events. Immediately declaring a national holiday in honor of the team's win, Wade appeared at the national parade in a vehicle with the top open so everyone could see him juggling a soccer ball to commemorate this important event. Having been in office only two years at that point, Wade's presidency had coincided with Senegal's eruption onto the world scene as a soccer team of renown. His undying emotional and financial support for the squad made him into a national hero of sorts, even as Wade's opposition criticized what they considered to be vulgar opportunism. "Our president is trying to capture this performance of the Senegalese boys, but I think I it is very childish . . . because [this victory] is not the result of some football policy," said Amath Dansokho, leader of the Independence and Labor Party.[33] For some people, the president had done little—in terms of sports or politics—to influence the national athletic successes for which he credited himself.

Senegal would win again before tying a match and losing another to finally exit World Cup competition. But they had already made history affirming a place in the international spotlight for the country and its president.

What are we to make of the team's success and of its ability to cast a positive spin on Abdoulaye Wade's tenure at the nation's helm? This athletic spectacle eclipsed the government's inability to cope with recurring energy shortages or the countless jobs lost to agricultural stagnation amidst the escalating numbers of youth crowding urban areas for the past few decades in search of work as a result.

Without even offering an elaborate treatment of the gender issues under consideration, Dansokho provokes us to consider them by complaining about the way Wade appropriates the hard work of the Senegalese boys who achieved international acclaim in athletic competition. Indeed, one of the most prominent features of Senegalese neoliberalism has been the way it valorizes masculinity in aspirational narratives. From overseas traders to athletes, Senegalese success stories frequently privilege male subjects with extensive access to foreign capital. This tendency is tied to the way Senegalese politics was characterized during the past several decades by increased reliance on structural adjustment and donations from wealthier nations. The state, instead of building a national infrastructure, hopes private investment will engender economic revitalization.[34] Taking my cue from the way this ideology is entangled with the nation's commitment to highlighting sporting successes, I have tentatively called this set of political commitments *gagnism*.

Gagner is, of course, the French verb meaning "to win." Concerning Senegalese politics, the government is presently proceeding as if it believes neoliberal capitalism furnishes a set of rules that, when followed, will automatically yield the desired results. This is the same idea promoted in athletic competition, which a proper disciplinary regime automatically translates into victory. For that reason, the parallels are startling.

Yet the ideology of gagnism is rather more elaborate than Senegal's commitment to winning by pursuing this particular formula.[35] Athletic competition always presumes an idealized subject imagined to be best suited for certain sporting contests. Just as teams—as firms—try their best to select the athletes most capable of attaining desired results, a key aspect of Senegalese neoliberalism involves convincing potential donors that it has achieved a form worthy of their investment.[36] Every country is expected to fit a particular profile: democracy according to a particular definition, for instance. Otherwise, it is considered unfit for sponsorship and is, in these instances, disqualified from competition altogether (Taylor 2005; Pender 2001).[37] NEPAD, after all, "calls on African leaders *to put their houses in order* in exchange for foreign direct investment" (Owusu 2003: 1660, emphasis added; cf. African Development Report 2004: 28–29).[38]

The male-focused image of success and emphasis on foreign investment that structure Senegal's pursuit of prosperity help explain why a soccer field was the site for a neoliberal spectacle that provoked comparison with this region's most inhumane traffic in human commodities historically. All of my interviewees insisted that, during George W. Bush's visit to Gorée, the entire population of the island was taken to a soccer stadium and locked inside. I later learned it was a sandlot soccer field, barricaded to prevent escape. Yet by referring to the field as a solid fortification, interviewees suggested the makeshift barriers had, for them, become concrete enclosures.

Though U.S. foreign policy in Senegal has crystallized through the successive visits of different presidents, interviewees contrasted the repressive treatment that accompanied Bush's arrival with the enriching experience of hosting Bill and Hillary Clinton in 1998. "Bill would be in the countryside playing with little children," a thirty-something-year-old female vendor insisted. Hillary, meanwhile, participated in a debate on female genital mutilation.[39] Bush's 2003 visit, by contrast, was considerably more coercive: reporters noticed that, even before the U.S. president arrived, security personnel had placed his anticipated hotel under strict surveillance,[40] besides cutting down centuries-old baobab trees to enhance visibility and, ultimately, imprisoning much of the island population on a soccer terrain. And few people missed the profound irony that all this took place a few feet from the historic *Maison des Esclaves* (slave house or slave "dungeon") (Hartman 2002, 766), which is responsible for the tourism that usually provides this island economy's primary revenue.

Some scholars have, in recent years, challenged Gorée's historical legitimacy as a central port in the transatlantic slave trade, suggesting the current site of the *Maison des Esclaves* served primarily as a private residence for one Anne Pépin (Hinchman 2004a, 49). Although Pépin sometimes engaged in various minor forms of overseas exchange and occasionally held enslaved Africans in the basement of the residence (Hinchman 2004b), the *Maison* was apparently not the pivotal site of Atlantic dispersion it is often imagined to be (Austen 2001). Given all this, the slavery discourses that emerged in the aftermath of July 8, 2003, seem at best exaggerated, at worst, unjustified.[41]

The Bush visit, though, ultimately hinges on a profound irony: the technique of coercion used to subdue Senegalese peoples in that moment reproduced the way enslavement historically occurred on Gorée Island more faithfully than any event that has ever occurred at the *Maison des Esclaves*, for slavery in this locale typically did not proceed through dungeons and warehousing. Instead, enslaved Africans were usually hoarded together and bound in open-air *captiveries* (Samb, ed. 1997),[42] awaiting placement onto ships that would send them across the Atlantic. The residents of Gorée Island conveyed to me that, besides a few invited guests, they were treated similarly on July 8, 2003.[43]

Though Bush's speech was ostensibly directed toward the people of Senegal, many were too busy fighting heat exhaustion and too far removed from the podium to hear his address. The speech nevertheless had a purpose *and* an audience; for these reasons, we might consider how this address could be read from the vantage of Senegal's new relationship with the United States.

With the power and resources given to us, the United States seeks to bring peace where there is conflict, hope where there is suffering, and liberty where there is tyranny. And these commitments bring me and other distinguished leaders of my government across the Atlantic to Africa.

—George W. Bush, in his address at Gorée Island

Beginning some time around 11: 45 a.m., the U.S. Commander-in-Chief situated his remarks in the context of transatlantic slavery:

For hundreds of years on this island, peoples of different continents met in fear and cruelty. Today we gather in respect and friendship, mindful of past wrongs and dedicated to the advance of human liberty.

At this place, liberty and life were stolen and sold. Human beings were delivered and sorted, and weighed, and branded with the marks of commercial enterprises, and loaded as cargo on a voyage without return. One of the largest migrations in history was also one of the greatest crimes of history.

Below the decks, the middle passage was a hot, narrow, sunless nightmare; weeks and months of confinement and abuse and confusion on a strange and lonely sea. Some refused to eat, preferring death to any future their captors might prepare for them. Some who were sick were thrown over the side. Some rose up in violent rebellion, delivering the closest thing to justice on a slave ship. Many acts of defiance and bravery are recorded. Countless others, we will never know.

Here Bush endorses resistance as a feasible strategy for Africans who refuse to accept their own captivity. Perhaps Bush believes that, when faced with a "criminal," or tyrannical regime people ought to pursue their freedom by any means necessary.

And yet, Senegal became an important U.S. ally at this historical juncture because of President Wade's expressed disdain for Islamic jihads waged by Muslims who see the United States as an imperial regime. But Bush certainly could not be speaking about them. The "War on Terror" entailed an effort to identify rogue militants associated with an international axis of evil, not commending people who declare themselves revolutionaries in the face of a political superpower. This phraseology, then, applauds a nebulous sense of resistance that does not correspond to any particular historical actors or events. Perhaps this explains why no specific personages or sites of struggle are named, though Bush would be more specific at other moments in his speech.

Those who lived to see land again were displayed, examined, and sold at auctions across nations in the Western Hemisphere. Because families were often separated, many were denied even the comfort of suffering together. . . .

In America, enslaved Africans learned the story of the exodus from Egypt and set their own hearts on a promised land of freedom. Enslaved Africans discovered a suffering Savior and found he was more like themselves than their masters. Enslaved Africans heard the ringing promises of the Declaration of Independence and asked the self-evident question, "then why not me?"

In this passage, Bush establishes a parallel between the people of Senegal and the descendants of "enslaved Africans" who now live in the United States. Such a move is rhetorically strategic, since in 2001 the U.S. government was criticized for leaving a United Nations conference on racism in Durban, South Africa where many believed it would have been asked to issue an apology for its participation in the transatlantic slave trade and to deliver a formal statement concerning its position on the issue of reparations for African Americans. By establishing a correlation between African Americans and Africans in Senegal, Bush commends the latter while paying homage to the former, yet he avoids detailing a specific commitment to either population.

In this same phrasing, Bush speaks of the Church's role in the slave trade but does not condemn it. Instead of addressing the contradictions that structured Christianity in the context of plantation enslavement in the United States, Bush submits that European Christians were not Christian enough. Thus, Africans who discerned spiritual lessons in the Bible taught Christians what they really needed to learn about humanity, so all those "generations of oppression" could not "defeat the purposes of God," intentions that perhaps align with Bush's tendency, in the aftermath of September 11, 2001's devastating World Trade Center attacks, to use "religious arguments that encourage[d]" the U.S. military's "aggressive tendencies" in his speeches (Lincoln 2007).

> In the struggle of the centuries, America learned that freedom is not the possession of one race. We know with equal certainty that freedom is not the possession of one nation. This belief in the natural rights of man, this conviction that justice should reach wherever the sun passes leads America into the world. . . .
>
> African peoples are now writing your own story of liberty. Africans have overcome the arrogance of colonial powers, overturned the cruelties of apartheid, and made it clear that dictatorship is not the future of any nation on this continent. In the process, Africa has produced heroes of liberation—leaders like Mandela, Senghor, Nkrumah, Kenyatta, Selassie and Sadat. And many visionary African leaders, such as my friend, have grasped the power of economic and political freedom to lift whole nations and put forth bold plans for Africa's development.[44]

The narrative of "liberty" that figures prominently in this speech glorifies African "fathers" of independence: a roll call that includes Senghor and someone Bush refers to as his "friend," who remains unnamed. Most likely he is referring to Wade, who he commends for having "grasped the power of economic and political freedom to lift whole nations and put forth bold plans for Africa's development." Here, Bush affirms his support for NEPAD, an initiative that, in the words of Francis Owusu (2003, 1655), "support[s] neoliberalism and sees global integration as the key to Africa's development" (see also Nabudere 2002; Taylor 2005). Wade, as I have indicated, was one of the proposal's main architects and is one of its biggest advocates in Africa.

> Because Africans and Americans share a belief in the values of liberty and dignity, we must share in the labor of advancing those values. In a time of growing

commerce across the globe, we will ensure that the nations of Africa are full partners in the trade and prosperity of the world. Against the waste and violence of civil war, we will stand together for peace. Against the merciless terrorists who threaten every nation, we will wage an unrelenting campaign of justice. Confronted with desperate hunger, we will answer with human compassion and the tools of human technology. In the face of spreading disease, we will join with you in turning the tide against AIDS in Africa.

We know that these challenges can be overcome, because history moves in the direction of justice. There is a voice of conscience and hope in every man and woman that will not be silenced—what Martin Luther King called a certain kind of fire that no water could put out. That flame could not be extinguished at the Birmingham jail. It could not be stamped out at Robben Island Prison. It was seen in the darkness here at Gorée Island, where no chain could bind the soul. This untamed fire of justice continues to burn in the affairs of man, and it lights the way before us.

May God bless you all.[45]

And with that benediction Bush ends his speech, though even then it is not clear who he was addressing. Given the way Senegalese citizens in the immediate vicinity were, by and large, locked away, it seems the speech was aimed not at them but at an audience located elsewhere who would apprehend the speech as it would be mediated by print journalism, snapshots, and televisual snippets.

What was he saying to *them*? Bush's initial concern with the African American protest tradition blends easily, in the latter part of his address, with the political inclinations of all "Americans," typified by the utopian impulse Martin Luther King, Jr., embodied. African Americans, as far as Bush was concerned, had joined his "unrelenting campaign for justice." But in speaking of a shared commitment to "dignity . . . [i]n a time of growing commerce across the globe," Bush implicitly references the legacy of Senegalese traders in the United States, by now so prevalent that a section of Harlem's renowned 125th street marketplace has become known locally as *Le Petit Sénégal*. In that regard, the epigraph to this chapter offers a relationship between "migrations" and "crime" we should perhaps consider at greater length, especially in terms of the distinctions the president draws between "Africans" and "Americans": "We are never so steeped in history as when we pretend not to be" (Trouillot 1997, xix)

While theorizing the activities of Senegalese actors in U.S. commercial spheres, some scholars have been complicit in reproducing the dubious notion that their success grows from a work ethic that sets them apart from their African American peers operating in shared contexts. In some of the literature, the African American traders who work alongside Senegalese merchants are, understandably, beyond the scholar's analytic focus (Buggenhagen 2003). Elsewhere, though, some writers uncritically recycle the ethno-racial stereotype that, "Entering legally, working furtively, leaving harmlessly," the Senegalese trader is a model citizen (Millman 1997, 180). The anthropologist Paul Stoller (2002, 88–90), in his discussion of New York City's Senegalese traders, troubles the assumption they are all law-abiding citizens but nevertheless imbues them with a devotedness to their work ethic, which implies that other people occupying the African American communities in which they tend to operate fail to achieve comparable levels of prosperity because they do not apply themselves to the same principles (cf. Millman 1997, 177). When these merchants arrived in the Big Apple for commercial opportunities in the early 1980s, Stoller tells us, they immediately applied to the Consumer Affairs Board for vending licenses. As a result of the excessive harassment state agents rained upon them, the traders quickly racked up thousands of dollars worth of fines. As a result, these vendors quietly allowed their licenses to expire then "continued their operations outside the regulatory aegis of New York City" (Stoller 2000, 88). This, from Stoller's perspective, is a success story even though these merchants had willingly entered the unregulated—and illicit—informal economic sector (90). Not, in their view, an indication of the harassment traders racialized as black are likely to experience, Stoller and others consider this one more example of the way "West African merchants in New York City use their familial traditions to construct long-distance trade networks in North America." Stoller is especially interested in the way African actors use the perceived cultural impoverishment of African Americans as an economic advantage—as a strategy for marketing "Afrocentricity" (Stoller 2000, 90). Little regard is shown here for the way African Americans manage to develop economic opportunities despite battling state-sponsored forms of social exclusion (Venkatesh 2000; Wacquant 1999) or the moments when Senegalese traders and African Americans embrace shared categories of racial or cultural identification (Ralph 2007), at times in the context of illicit enterprise.

By contrast when Akon released the hit single "Locked Up" in 2004, he was joined by the African American rapper Styles P. Styles who, to date, has garnered more critical attention than commercial sales and is well-known for gritty tales about drug trafficking and critique of hypocrisy in the criminal justice system. Appropriately, Akon summoned Styles for a song that discussed his own trials with law enforcement officials. The album, tellingly entitled, *Trouble*, hit shelves soon after Akon was released from prison for an auto theft conviction. Subsequently, he formed the label "Konvict Musik," and showed up in songs from a range of rap artists—from Snoop to Young Jeezy to Eminem—between 2004 and 2006. In 2005 Styles P joined Akon in Dakar for a concert; during their stay the Senegalese artist was declared a "youth ambassador," (Checkoway 2007, 92) besides being received at Wade's presidential palace. In this moment, Akon embodied the Senegalese emigrant's success story: son of the famed percussionist Mior Thiam, Akon was raised in the United States where he eventually became a top-selling recording artist. Yet what separates this story from others of a similar ilk is the sense that Akon came of age in unmonitored economic enclaves alongside African American actors likewise interested in seizing available economic opportunities. For Akon this entailed being the "leader of a national car-theft operation" (Checkoway 2007, 94).[46]

To his credit Stoller, in moments, seems concerned to highlight the sociological dimensions of Senegal's "astute entrepreneurs" who seize the "economic advantage" to be gained from the new forces of "global restructuring" (Stoller 2000, 88–90), yet what is meant to be a flattering portrayal of Senegalese vendors unfortunately—in overstating the role of "familial traditions"—reproduces a primitive sense of African genius, which is ultimately a dehumanizing gesture. Besides that, his sense that Senegalese immigrants to the United States invariably understand themselves to occupy a different ethno-racial category is empirically wrong.[47] Akon, when questioned by law enforcement officials about the stolen luxury cars he regularly resold, strategically used American stereotypes about Africans to his advantage, "[I would speak] with an accent as if my English ain't no good . . . like I didn't know what they're talking about. . . . [The police] bought it. And I stuck with that story forever" (Checkoway 2007, 94–95).

Stoller's assessment ultimately relies too heavily on an absolute irreconcilability of racial categories one that, in quite a few instances, sabotages his analytic. In place of this ethnic reification, I would suggest greater recognition of the indeterminacy of identity that arises in moments as part of the African diasporic experience and which sometimes structures commercial and affective engagements. While conducting research among Senegalese basketball players—in Dakar and in New York City—who aspired to play professional basketball in the United States, I regularly encountered players who insisted they had a natural advantage over competitors from other racial categories simply because they were "black, just like African Americans" (Ralph 2006b).

The African American encounter not simply with Senegalese people but also with the nation of Senegal forces one, in moments, to confront the tangled web of Diaspora.[48] I first visited Gorée Island with Mark, a graduate student in psychology from Howard University who was, like me, in Senegal to conduct research for a doctoral thesis. Mark was interested in seeing Gorée Island for what he hoped would enable him to connect with a vital part of his personal legacy, one which geographical distance and alternate financial obligations had previously prevented him from experiencing. Mark once asked some of our Senegalese friends whether they might be interested in accompanying him to Gorée and was subsequently outraged when they responded with enthusiastic offers to join him for a swim in Gorée's beautiful beaches. Mark was disgusted by the thought of lounging in the same body of water where some of his ancestors had drowned centuries before, though the U.S. dollars in his possession still eventually contributed to Gorée's primary industry, a significant component of which is "heritage" tourism of the kind he typified. Thus Gorée achieves a symbolic resonance for vacationers, merchants, and visitors, evidencing a complicated set of relationships too easily elided in some of the scholarly literature and in Bush's Gorée Island address.

The U.S. Commander-in-Chief's extended political meditation on the affective ties between Africans and African Americans—as well as his implicit theorization of the way people inhabiting the former position come to occupy the latter—invites comparison with one recent accomplished, albeit anecdotal, discussion of these processes.

Saidiya's Hartman's acclaimed memoir, *Lose Your Mother*, evidences one writer's journey through scholarly discourse, into and then beyond archives, across the Atlantic to the slave castles (or dungeons) of Cape Coast and Elmina, Ghana. Breathtaking in its prose Hartman's account is, from its opening pages, framed as a case of unrequited longing for an ancestral connection. Brilliant as an exposition of diasporic angst, Hartman, like Stoller, unfortunately treats categories of identification with undue veracity.

Hartman (2007, 4) in her account is, from the moment she arrives in Accra, treated as a "foreigner from across the sea." A "stranger," in her words, as "*confirmed* by her appearance": a "vinyl" jacket, a "gait best suited to navigating the streets of Manhattan," and "German walking shoes." This sense that appearance provides a reliable way to track Afrodiasporic difference includes, for Hartman, phenotypic distinctions. In her "jumble of features," Hartman writes, "no certain line of origin could be traced," as if to contrast her own hybrid heritage with the ethnic affiliations of her Ghanaian counterparts: "Clearly, I was not Fanti, or Ashanti, or Ewe or Ga" (Hartman 2007, 5).

I do not mean to suggest that Hartman invents the social exclusion she experiences. In her account she is, immediately upon her arrival, understood to be *obruni* (a designation variously understood as "white, foreign, or stranger," depending on the context). Curious, though, for a scholar of the Afrodiasporic world, she credits this term with scientific precision ("I didn't relish this label. . . . But then I learned to accept it. After all, I *was* a stranger from across the sea" (Hartman 2007, 4, emphasis added).[49]

The certainty with which Hartman understands herself to be a stranger undoes the complicated indeterminacy of diasporic belonging theorized in much of the scholarly literature (Edwards 2003, 2001, 1998; Clifford 1997; Tölölyan 1996; Hall 1990) especially, for instance, in Kenneth Warren's "Appeals for (Mis)recognition: Theorizing the Diaspora." Langston Hughes' frustration at not being received as an African in Liberia despite wanting to be seen in that light underscores a "comedy of misrecognition where Hughes who appeals for misrecognition as an African is misrecognized as white," meanwhile his comrade, "George who appeals for misrecognition as a Kentuckian is misrecognized as African." These necessarily imprecise performances of geo-racial positioning highlight, for Warren, "the condition of the diasporan subject" (Warren 1993, 400), which is marked by identities that crystallize in moments but never in any absolute manner. Hartman, by contrast, confines the African American to a very specific "slot" (Trouillot 2003) in the diasporic imagination, "We all avoided the word 'slave,' but we all knew who was who." Yet the complicated history of the African Diaspora always threatens to unravel the certainty with which Hartman's discussion proceeds.[50]

Given that the transatlantic slave trade was structured in part through historical contingencies, even actors who occupy fixed positions at one moment might find themselves in other roles at another point in the evolution of this economic and social matrix. In other words, if even Africans who were complicit in the sale of people from elsewhere on the continent could themselves later be enslaved, as we know, "who," indeed, "is who"? Hartman is quite correct to note that a slave is always, to some extent, an "outsider." But given that outsiders are produced from the inside—and vice versa—these seem more like historical gradations than ontological (or even epistemological) categories.[51] They certainly are not categorical differences, as they appear too often in Hartman's work.[52]

The Senegalese case makes it especially difficult to retain such a view since, despite the presence of distinct ethnic affiliations, Wolof language and culture constitute the shared social currency of the country.[53] And Senegalese political aspirations retain a fluid quality based on the preponderance of pragmatic Islamic traditions that, in moments, blend with Christianity but in other ways accommodate different political sensibilities as well as one historical consequence of a nineteenth-century French colonial experiment that granted Senegalese inhabitants of the *Quatre Communes* citizenship rights (Diouf 1998, 671–94; Johnson 1971). The cosmopolitan quality of Senegalese social life has fed transnational labor migrations that intensified during recent decades of economic distress, so that Senegalese people of various social classes benefit from the circulation of European luxury goods through informal domestic markets and as part of remittance packages sent home from overseas (Buggenhagen 2003). Although Hartman felt the sting of social dislocatedness as a result of the "German" shoes she wore upon her arrival in Accra, commodities from wealthy nations alone do not render one a cultural outsider in Dakar. I support Hartman's (2007, 234) ultimate endeavor to seek diasporic solidarity in a shared commitment to ongoing political transformation instead of *simply* in a sense of kinship or descent. But I think the promising critical project she outlines requires greater compassion for the ingenuity of historical narratives that manage to override ritualized violence and closer attention to the contradictory and inconsistent ways identities become institutionalized, since social actors discover their political inclinations, in part, by developing forms of historical consciousness that prepare them to negotiate different subject positions simultaneously.

The complicated relationship between race, ethnicity, gender, sexuality, and nation that shapes diasporic relationships also structures political performances as evidenced in Bush's Gorée Island speech, to which I now return.

The U.S. president emphasized in his speech, the need to make Senegalese people "full partners in the trade and prosperity of the world." The alternative—not being "full partners . . . in . . . trade"—is linked, in his view, to all manner of social problems, including the spread of AIDS and civil wars—both infectious diseases that, apparently, run rampant in Africa. Not to "move in [this] direction," too, would fuel the efforts of "merciless terrorists." Senegal's historic mission, in Bush's view, would not allow it to do that, given the way its present geopolitical projects dovetail with the spirit that sustained Martin Luther King, Jr., in his Birmingham jail and Nelson Mandela in his Robben Island cell. These leaders endured imprisonment to awaken a sense of "hope" that dwells in the "human heart" of every "man and woman."

Bush's message was apparently directed at all segments of the population, which explains its gender-inclusive phraseology. Yet, the leaders mentioned were all men. How are Senegalese women to understand the part they play in the "full partnership" that the country's historical arc is leading it to embrace in the world? Will this "partnership" be realized as well in the social spaces with which they are familiar? Or does their exclusion here correspond to the way the enslavement of Senegalese people is hidden and silenced as a seemingly necessary part of this new economic enterprise?

What of the specific images used to anchor Bush's address, in particular the sense communicated toward the end of his speech that Martin Luther King, Jr., and Nelson Mandela dramatize and communicate to the world a powerful message of "hope"? Who, after all, was charged with maintaining "hope" for the Senegalese people incarcerated during that same moment in the center of the island where they reside? This question is not mine alone: the French word for "hope" (*espoir*) was scrawled on the wall of the Gorée Island site where Senegalese peoples were detained.

This word has a special significance for me, based on my researches and travails in Senegal. As I moved back and forth between my base in Chicago and my research site, friends would invariably ask what I could do to facilitate their career pursuits. The word "espoir" was always used to articulate their aspirations.

My friend Lamine, a sculptor, once asked if I might be interested in joining him for a new commercial venture. If I could front the money for a major purchase of wood from Mali, Lamine suggested, he could produce a number of sculptures in bulk, then split the profits with me. He recognized this was a long-term investment but was sure we both stood to gain: "I know it's a lot to ask, Michael," he conceded, "But you're my last hope [*dernier espoir*]."

Some months after a research trip to Senegal in 2002, I received a letter from Pierre, a security guard in the home where I had stayed. He was writing to ask if I knew of any security or law enforcement opportunities in the United States for which he might be suited. As the letter drew to a close he, too, made sure to indicate that I was his *dernier espoir* (last hope).

Significantly, I think, the phrase of note was *dernier espoir*—always in French, even for my friends who usually only spoke Wolof and conceded they were barely literate in the European language. I still struggle to grasp the semantic significance of communicating the concept in French, but I suspect it has do in part with the sense that opportunities exist in a context connected to the elsewhere. For Senegalese people usually start to learn *Wolof* from the moment they are born. French then, as the language of official business and formal education, dwells in a register of expanded possibility.

This term "espoir" as an index of desperation stood apposite another, which was used to articulate the prospect of prosperity: *gagner*.

Lamine and Pierre (who don't know each other), used to both say they needed my assistance, because in Senegal it was *difficile à gagner quelque chose* (difficult to find something). Among others, a former basketball player I once interviewed likewise insisted that, "In Senegal, it is *difficile à gagner quelque chose* [difficult to find a job] unless you are well-connected [*bien placé*]." The phrase *difficile à gagner quelque chose*, though relatively simple, resists translation; the sense being communicated is that it is "hard to find something" or "difficult to find work." But the word *gagner* was usually deployed by interviewees to speak about prospects for *earning* money. They hoped an opportunity would enable them to "earn" an income. *Gagner*, then, means at once to "find," "win," or "earn," revenue.

In this context, being imprisoned on Gorée Island during a visit from the "leader of the free world" reminded Senegalese citizens of the extent to which they are "trapped" in a marginal economic and political position, just as it led other Senegalese citizens I interviewed to the conclusion that, although the treatment they received was indeed "unjust," the Bush visit was something the country "needed" to improve its stature in the world of nations:

What is ghastly and really almost hopeless in our racial situation now is that the crimes we have committed are so great and so unspeakable. (Baldwin 1965, 10)

As important were the silences and evasions. (Hartman 2007, 16)

Why would the Senegalese state enable its population to be "enslaved"? Why did the trope of enslavement figure so prominently in Bush's visit and why such emphasis on the criminal nature of this cruel traffic in human beings? The precise motivations driving this political performance are difficult to discern, but this sort of narrative has become more frequent in recent years (cf. Brooks, ed. 1999). Michel-Rolph Trouillot (2000, 171) for instance, has noted that, "collective historical apologies are increasing worldwide." Not simply evidence of but also factors in historical transformation, "these rituals of apology create pastness by connecting existing collectivities to past ones that either perpetrated wrongs or were victimized," even though, frequently, the events to which these "apologies" refer took place several centuries before. It is not always clear, then, who is even being addressed in many of these speech events.

Still, given that the point about such claims may be less what they assert than the fact of their assertion" (Trouillot 1995, xvii), we might look for the true significance of a performance like this one in the structure of the message and its relationship to the historical context the utterance helps to construct, "As transformative rituals, apologies always involve time . . . [t]hey mark a temporal transition: [a] wrong done in the past is recognized as such, and this acknowledgement itself creates or verifies a new temporal plane, a present oriented toward the future" (Trouillot 2000, 174). Indeed, from the evidence marshaled in this chapter, it seems people from both Senegal and the United States were interested in using evidence of a "wrong done in the past"—in this case, the transatlantic traffic in human beings—to erect "a present oriented toward" a mutually beneficial future.

In a time of economic desperation, moments like July 8, 2003, emerge as an opportunity for the nation to attract a capital commitment that could, potentially, reverse its economic course. The victims of the "enslavement" that characterized Bush's Gorée Island visit might be skeptical of the message he promoted, but a message need not be well-received to transform a social context. For at least the second time, delivering a speech at Gorée Island helped a U.S. president cultivate foreign policy in Africa while strategically evading the social consequences of the United States' historic participation in the transatlantic slave trade. As a way to meditate on the way alterity structures the multiple mediations this sort of political performance requires, I quote this passage from *U.S. News & World Report* about Clinton's visit to Africa in 1998:

> In stopovers in Africa last week, President Clinton was careful not to issue a formal apology for America's slave past, but rather to express regret and contrition. One . . . factor—rarely discussed by the White House—is concern over the *legal implications of an apology.* If Clinton, as head of the U.S. government, issues such a statement, it could increase legal, as well as moral, pressure for reparations to the descendants of slaves . . . That's why the White House is particularly grateful for the Rev. Jesse Jackson's defense of Clinton's handling of the issue. White House officials say privately that Jackson, who accompanied Clinton to Africa, has been especially effective in giving Clinton credibility on the apology question within the press corps and, they believe, with many African-Americans.[54]

As evident as much in the speech Bush delivered as the one Clinton avoided, political performances frequently rely on "abortive rituals"—infelicitous speech acts that nevertheless resonate with audience members, because of what they superficially signify (Trouillot 2000)—to help people "erase" past actions that would otherwise undermine present pursuits perhaps, in this case, by indicating the escape route the perpetrators of historical "crimes" have taken.

But though Trouillot's erudition helps us to understand what is at stake in these speech acts, his term for them—"historical apologies"—is something of a misnomer in this case. Bush, despite the lyrical gymnastics he undertook to elaborate the horrors of transatlantic enslavement, did not actually offer the apology for slavery many feel is long overdue.[55] Trouillot's suggestion that apologies are used to create "pastness" between the perpetrator and recipient of a "wrong" perhaps explains why the United States government has never heeded that request, neither at the 2001 United Nations World Conference on Racism where the U.S. delegation left early nor on this fateful day at Gorée. Maybe there is something about the historical trajectory of U.S. political and economic ambitions that, from the perspective of the state, renders slavery inhumane but prevents it from being considered "wrong." Maybe even beyond the widely acknowledged point that slavery and freedom are conditions of each other's possibility (Palmié 1985, ix, Morgan 1975) slavery, as a form of economic extraction, remains difficult to disentangle

from the structure of capitalism (Graeber 2006), although it is imagined in popular discourses as part of a rather different historical epoch. This is a sorry state of
affairs, although apologies of the kind we see here (even when they *are* felicitous)
serve only to undermine and eclipse present wrongs—"crimes," even.[56] In that
regard, can we consider Bush's meditation on slavery as, if not exactly an apology,
a kind of confession—though one that, in line with the Fifth Amendment to the
U.S. Bill of Rights, lacks the power to incriminate its issuer?

   With an eye to forensics, we might conclude that "slavery" emerged as a way to
categorize the injustices perpetrated on Gorée Island, because it captures the complex symbolic and economic exchanges taking place between the United States and
Senegal.[57] A discourse on slavery is always a commentary across regimes of value:
the interior emotional complexity of the human condition versus the economic
value that can potentially be extracted from laboring bodies.[58] To the extent that
many African nations are in the curious position of having to demonstrate a certain fitness for democracy as a prerequisite for foreign aid, these symbolic displays
have concrete economic consequences—and "legal implications." Countries like
Senegal cannot simply *promise* they are committed to democracy, their devotion to
this political ideal must be *demonstrated*, just as a legally binding promise cannot
simply be affirmed but requires the additional element of "material consideration"
(Drake 1905). William Pietz (2002, 38) offers an example that, in the interest of
brevity, makes this point forcefully. In referring to the cinematic reproduction of
John Grisham's novel *The Client*, Pietz describes a scene where a young boy who
will stand trial for his knowledge of a mob murder "hires a sympathetic lawyer
by handling her a crumpled dollar bill": "The real transfer of even this nominal
sum can cause a contractual relationship, in this case the lawyer-client relation,
to come into existence. . . . Although she accepts the dollar, the lawyer does not
regard it as a partial payment for her services. It is just [a retainer,] the technical
requirement for establishing the contractual relationship."

   On the face of it, the scenario Pietz discusses here has little to do with Senegal
or the events that took place on July 8, 2003. No formal legal document was
prepared to confirm Senegal's commitment to democracy or even to assure that
the United States would serve as the country's representative in any legal arena.[59]
Still, in being packaged and presented using the same techniques deployed by
those who enslaved Africans centuries before, the population of Gorée appears
as a bundle of value that, while not exchangeable in any concrete form, provides
Senegalese consent to the terms of this informal contractual arrangement. Which
is why, although, "The requirement to give up control over some material object
that in value might be a mere trifle might sound like a pointless formality or an
empty ritual left over from a more primitive age consideration . . . as a positive
doctrine and concrete legal reality . . . has [in fact] survived to the present day
both as a legal object in judicial decisions and," even more relevant to this case,
"as a practical reality in social transactions" (Pietz 2002: 39). In this "transaction,"
Bush traded on the spectral capital afforded by the virtual enslavement of Senegalese citizens for his own purposes: to cleanse the United States of any culpability

concerning its historic involvement in the Transatlantic slave trade, ironically, by re-staging the exchange of bonded Africans.[60]

Consideration might seem an inappropriate concept to invoke, since there was no formal exchange at Gorée like there is in the case of a legal retainer, as previously mentioned. Yet, at least in Anglo-American law, where the notion of "material consideration" operates, it is not crucial that the second party formally take possession of the object tendered. "Consideration is a social fact brought into being by the voluntary alienation of a valuable material object" (Pietz 2002, 38): "What is necessary for an enforceable contract is [only] that the person receiving the promise . . . effectively separates himself from something valuable under his control. The person making the promise need not actually take possession of it. She need only agree that this alienation of an object of material value [in this instance, a horde of 'enslaved' Africans] . . . is an acceptable consideration." Apparently Wade, who aspired to secure U.S. geopolitical might and financial resources found this informal social compact "acceptable," as did Bush, who stood to benefit from Senegalese cultural capital, through a spectacle staged in the crucible of national aspiration: a soccer field, some miles off the country's Atlantic coast.[61] This tension between slavery and neoliberal freedom, play, and imprisonment structures the contradictory experience of economic transformation in Senegal. "Sport," says Gerald Early, "is how human beings perform the art and craft of competition" (1996, 5). Neoliberalism, by contrast, is a game with more losers than winners. And, it's difficult to trust the referees. So even when people genuinely believe they are playing fair, they sometimes overlook the fact that it's hard to compete with shackles around your feet.

NOTES

1. "The reenactment of the event of captivity," as Saidiya Hartman (2002, 760) insists, "contrives an enduring, visceral, and personal memory of the unimaginable" and not simply for people from the African Diaspora returning to Senegal as part of a heritage crusade but also for this Senegalese population, many of whom acknowledged no previous link to the history of the transatlantic slave trade, despite inhabiting a region steeped in the history of its commercial transactions. Saidiya Hartman, *Lose Your Mother: A Journey along the Atlantic Slave Route* (New York: Farrar, Straus and Giroux, 2007).
2. BBC News, "Country profile: Senegal," *BBC News*, http://news.bbc.co.uk/1/hi/world/ africa/country_profiles/1064496.stm (accessed May 15, 2005).
3. Tidiane Sy, "Senegal opposition to amnesty law," *BBC News*, January 11, 2005.
4. Postcolonial "lawlessness," thus, often "turns out to be a complex north-south collaboration" (Comaroff and Comaroff 2006a, 8).
5. Seck was once so close to Wade that when the president appointed him prime minister of Senegal "he told the media that he knew the president so well that he did not even need directions to 'transform Wade's vision' into concrete acts." Seck, in fact, claimed he was "an embodiment of the President's vision." And yet, in the early months of 2005, rumors began circulating that Seck hoped first to have Wade displaced from the presidency through means as unspecified as they were clandestine, then to gain control over parliament in the elections of 2006, and ultimately to run for president in 2007. What

was once a political alliance disintegrated into a criminal conspiracy—a conspiracy theory, at least.

6. A short time later, opposition leader Abdourahim Agne was charged with "threatening state security" after making a speech that encouraged Senegalese people to "go into the streets by the millions and demonstrate peacefully for change," so they might be able to improve the country's "dire state." The director of a film "examining Abdoulaye Wade's election promises" Agne—his lawyer told reporters—"had been charged under laws banning attempts to overthrow the state." The state's charge against Agne seemed to render explicit what remained hidden behind the "more serious charges" added to the case against Seck after the fact of his arrest, that is, Wade's fear that he might be displaced either through a rival the people preferred or through a radical critique. For more on Agne's case, see "Senegal opposition leader charged," *BBC News*, May 31, 2005.

7. DJ Awadi's song "Sunuugal," whose name is inspired by a word that means "canoe" in Wolof, but which is also the etymological root of the word Sénégal, features this lyrical critique of Wade: You promised me I would have a job/ You promised me I would have food, You promised me I would have real work and hope/ But so far—nothing, that's why I am leaving, that's why I am taking off in this canoe . . . I would prefer to die than to live in this hell (Winter 2006).

8. "Alterity" is used here because forensic inquiries entail assumptions about human difference. They rely sometimes on historical patterns unique to specific demographics and nearly in every instance on bodily signatures like fingerprints (Beavan 2002), footprints (Abbott 1964), and even voices (Hollien 2002).

9. This extraordinary claim appears on the back cover of Colin Beavan's (2002) *Fingerprints: The Origins of Forensic Detection and the Murder Case that Launched Forensic Science.*

10. "In 1846," as Williams Pietz has shown, "the British Parliament abolished England's ancient quasi-religious law that compensated wrongful death accidents according to the money value of the lethal object. It was replaced with . . . [t]he Fatal Accidents Act of 1846 [which] established the modern method for determining the money value of human life that has been used ever since in Anglophone capitalist societies (Pietz 2002, 36)." The "quasi-religious" law of the deodand—which had been in operation for centuries—was finally eradicated, because it was believed, as noted legal reformer Lord Campbell argued, this doctrine was "not applicable to the present state of society." As far as he was concerned, in this period—which also witnessed "the legal revolution that created modern corporations" (Pietz 1997b, 105)—it should no longer be upheld "that the life of a man was so valuable that they could not put an estimate upon it" (*Hansard's Parliamentary Debates* 1846; also cited in Pietz 1997b). This was the same period in which clinical pathologists began routinely using invasive surgical procedures on a human body that was suddenly, neither in that context, too sacred for more rational and scientific treatment (Foucault 1975).

11. Methodologically and conceptually, forensics offers ethnographers a powerful set of tools. It is, of course, the science of human traces, of that which is "written in blood" (Wilson 2003; cf. Chaplan 1984; Walls 1974). Ideally, it connects us with the "'action' to be recounted" (Rhine 1998, xix). Thus while it is usually concerned with identifying and treating human remains (Nafte 2000; El-Najjar and McWilliams 1978; Weston and Wells 1974) or with assisting in legal investigations (Ferllini 2002; Rhine 1998; Ubelaker and Scammell 1992), it need not be restricted to these contexts (van Duyne et al 2003). In its inception, after all, the word has a rather broad field of signification. That which is "forensic" in Latin derives from the word "forum"—a meeting ground or market. *Forensis*, consequently, "means *belonging to a public space*, since in Roman times, legal trials, sentencing, and executions were . . . literally for the public to view" (Nafte 2000, 5). I ask

the reader to retain this sense of forensics as particular kind of perspective, as I consider the relationship between sports, spectatorship, and the fetishization of subject positions in athletic contests and in the political projects with which they share discursive thematics.

12. Wade's determined position on the matter did not reflect the nuanced exchanges taking place in Senegalese popular media. Critical of the way Bush's coalition was driven by what seemed to many to be a divine right to fight, Cheikh Bamba Dioum published an article entitled, "God bless the USA . . . *and* Afghanistan." (*Le Soleil*, Friday, 12 October 2001, emphasis added.) Malick Ndiaye, Leader of the Collective Social Forces for Change, declared that he supported "neither Bush nor Ben Laden" in his *L'info 7* piece ("Les partisans du 'Ni Bush ni Ben Laden' remittent ça aujourd'hui," *L'info 7*, Wednesday, 7 November 2001). Religious theorist Ebrahim Moosa (2006) has noted parallels between the two adversaries: "Both Bush and Bin Laden claim to have divine mandates, to have access to secret spiritual knowledge that obliges them to do certain things, even if those things run counter to their religions' most basic ethical teachings. Both men claim they're going to save people through their actions" ("In God's Name," *The Sun*, April 2006, Issue 364, 12). Moosa contends, furthermore, that both men "believe they have messianic missions to fulfill" (ibid.).

13. See "Pour la création d'un Pacte africaine contre le terrorisme," *L'info 7*, Jeudi September 20, 2001; and "Le président Wade propose un 'Pacte africaine contre le terrorisme,'" *Soleil*, Vendredi September 21, 2001.

14. No doubt this course of action was easier for Wade to pursue, because his Islamic nation is ruled by a secular state. "Islamic law [differs dramatically from] common law or civil law systems" (like the French codes Senegal uses); instead, it is structured largely by the "opinions of religious scholars, who argue, on the basis of the text of the Koran, the Prophetic *hadith* and the consensus of the first generations of Muslim scholars, what the law should be" (Peters 2005, 1); emphasis original.

15. "Les chefs d'État réaffirment l'engagement sans faille de l'Afrique," *Le Soleil* October 18, 2001, 3.

16. Ibid.

17. Ibid. "M. Bush également annoncé la création d'un fonds de soutien des investissements privés dans la région . . . des garantie et une couverture du risque politique pour leurs projets en Afrique sub-saharienne."

18. Greg Mills (2004) cites the preponderance of paramilitary groups operating within the African continent and the fact that NEPAD was ratified in October of 2001 as evidence that African leaders share with their Western counterparts an interest in fighting terrorism. It is doubtful that African heads-of-state would envision their struggle against militia groups and rogue soldiers in the same uncompromising way that Bush envisioned his crusade against terrorism. Mills' claim that, "had there never been a September 11, 2001, President George W. Bush arguably never would have made a visit to Senegal, South Africa" and other African countries (Mills 2004, 158), on the other hand, it is compelling to think about how U.S. Foreign Policy in Africa is, at times, structured by historical contingencies.

19. For more on this burgeoning alliance, see Lara Pawson's (2004) article, "France tackles U.S. trend," in *BBC News*. French President Jacques Chirac holds talks with Senegal's President Abdoulaye Wade in Paris on Thursday. The meeting comes as French-Senegalese ties appear to be under pressure, largely due to the West African state's more recent friend—the United States. France, the former colonial power, remains Senegal's biggest donor and trading partner. Dakar has played down the talks, insisting that relations are very good . . . But skeptics say the meeting is a chance for Paris to remind Senegal who pays out millions of dollars in donor assistance each year.

20. It should be noted that Senegal also started to gain favor with the United States during Reagan's presidency for its willingness to serve as a contingency landing site for space shuttle missions. During the presidency of George H. W. Bush, Senegal supported the United States by condemning the bombing of Pan Am flight 103, for which two Libyan men were convicted (although suspicions have been raised, subsequently, about the extent to which due process was upheld during the trial).

21. "Bush et Wade rêvent d'une finale Sénégal-USA," *Le Soleil* Wednesday June 19, 2002. The front-page headline translates as "Bush and Wade dream of a Senegal-USA final match." The article heading itself, on the newspaper's interior, carries a similar sentiment, "Bush et Wade souhaitent USA-Sénégal" ("Bush and Wade wish for a USA-Senegal final").

22. Admittedly, it is not altogether clear how this demographic label is being deployed. To the extent that "youth"—as a sociological moniker—often refers not simply to people of a certain age but also to those who have not yet accessed well-recognized institutions of social reproduction (marriage, parenthood, and stable employment in Senegal), it encompasses people from ages associated with adolescence up through age thirty—and beyond. What's more, the term is typically, as the Senegalese case emphasizes, gendered as male.

23. See statistics provided by the United States Central Intelligence Agency. See www .ciaworldfactbook.com/senegal.

24. See "Senegal celebrates Cup heroics," *British Broadcast Corporation Sport*, July 21, 2001, emphasis added.

25. "Senegal back to heroes welcome," *BBC Sport*, July 22, 2001.

26. Ibid.

27. "Lions players rule in France," *BBC Sport*, August 13, 2001.

28. "Africa's obsession with soccer," *BBC Sport*, August 18, 2001.

29. "Senegal in fever over World Cup debut," *BBC Sport*, May 14, 2002.

30. "Soccer fever," June 12, 2002.

31. See "Divided loyalties for Fadiga," *BBC Sport*, April 8, 2002; emphasis added.

32. "The whole of the Cameroonian squad plays abroad, along with 22 Senegalese, 21 Nigerians, 16 South Africans, and nine Tunisians." Isabelle Saussez, "Africa on the sidelines," *The Courier ACP-EU* (July–August 2002).

33. "Senegal's success story," *BBC Sport*, June 16, 2002.

34. This includes granting amnesty to Murids who, besides accounting for one of Senegal's largest incomes via remittance packages sent home from overseas, make valuable donations to the government in return for tax leniencies. In this way, their holy city of Touba can be considered, from a certain perspective, privatized.

35. Birdsall and Nellis (2003), for instance, assess the impact of privatization in Senegal by distinguishing between "winners and losers."

36. For elites, "the watchwords of the sporting canon—competition . . . 'fair play' . . . transpose quite naturally from his [or her] individual conduct to that of a company or indeed a nation. Here they may take on fancier, more imposing titles: Free Enterprise, Competitive Trading Position . . . the National Interest, Equality before the Law, etc., but the inherent ideas are still the same" (Brohm 1978, ix).

37. International lending agencies, North Atlantic economists, and donor nations seem agreed that good governance correlates with economic prosperity despite the fact that Angola, the Democratic Republic of the Congo, Equatorial Guinea, Nigeria, and, increasingly, the Sudan, have topped the list of African countries receiving foreign investment capital during the past decade (Ferguson 2006, 196, Reno 2001, 187) despite, for different reasons, evidencing some of the continent's most glaring forms of political and civil unrest. Development discourses still insist, however, that Cote d'Ivoire's recent political turmoil and

Sierra Leone's protracted social crisis surrounding fierce competition for diamonds are to blame for lackluster economic performance (African Development Report 2004).

38. In this regard, we might consider one of the World Bank's most recent poverty reduction efforts in Africa, the Comprehensive Development Framework (CDF). These measures were taken with the understanding that structural adjustment programs had been ill-conceived and seldom successful. CDFs promote autonomy for countries receiving assistance, but they contain a selectivity predicated on the presence of a "good policy environment" that is so ambiguously defined it tends to undermine the program's expressed guarantee of country ownership, restricting national autonomy (Pender 2001; Taylor 2005).

39. See "Pèlerinage à la Maison des Esclaves," *Le Soleil* Samedi le 3 Vendredi 1998.

40. See "Le Méridien Président sous haute surveillance," *L'info* 7 le 2 Juillet 2003. The writer who chronicled the modes of surveillance Bush's security forces used found it important to emphasize that not even Clinton's 1998 visit seemed to necessitate such precautions: "Il faut souligner que meme la visite que le prédéceseur de Bush, Bill Clinton, avait effectuée au Sénégal avant l'alternance, n'avait nécessité autant de precautions."

41. Bayo Holsey's 2003 work on slave tourism to Ghana reveals the cleavage between the absence of historical work on the slave trade in Ghanaian educational curricula and the diasporic discourses prevalent among African Americans that encourage them to visit Cape Coast and Elmina to confront a key point of departure for their ancestors who, it is believed, embarked on a transatlantic journey to the Americas from these sites (cf. Richards 2005).

42. Djibril Samb, ed. 1997. *Gorée et l'esclavage: Actes du Séminaire sur "Gorée dans la Traite atlantique: mythes et réalités."* Dakar: IFAN.

43. Scholars have not, as yet, considered how or why the experience of being corralled on a soccer field during the Bush visit provoked comparison with the Middle Passage. Despite inhabiting a different historical epoch, a good deal remains to be learned, I think, from studying the infrastructure(s) through which capital—symbolic and economic—continues to circulate in transatlantic circuits and the political networks that sustain these forms of exchange.

44. Bush here suggests the United States government is using its "power and resources" to introduce "peace" and "end conflict" in ways that are especially beneficial to African people. The millennial captiveries at Gorée, it hardly needs to be emphasized, tell a very different story (cf. Hesse 2002).

45. For the full transcript of his speech, see "President Bush Speaks at Gorée Island," Remarks by the President on Gorée Island, July 8, 2003, as posted on the official Web site of the White House, http://www.whitehouse.gov/news/releases/2003/07/20030708-1.html (accessed April 19, 2007).

46. And "criminality" has, for young actors in postcolonial and postindustrial contexts, provided if not exactly "a means of production" a repertoire of techniques for achieving the "productive redistribution" of resources among those people "alienated by new forms of exclusion" (Comaroff and Comaroff 2006b, 278; cf. Venkatesh 2006). Ironically, this was less true for Akon than it most likely was for his peers in the underworld. The top-selling recording artist was, by his own, admission raised "middle-class" though he "gravitated toward the 'hood, running with tough crews in all the cities where his parents moved" (Checkoway 2007, 94).

47. Laura Checkoway's (2007, 92) *Vibe* magazine article Akon quotes as saying, "Bend . . . over, look back, and watch me," to former "it" girl Tara Reid whose legs were then wrapped around his waist as the hip-hop balladeer performed, "Smack that," before an audience at the 2006 Sundance Film Festival. Crucial to the way Akon understands himself to be

racialized in ways consonant with African Americans is his concern with celebrating what I have elsewhere termed "hip-hop fantasy," an ideational frame that hinges on performances of material excess and the insistence—at least in the public persona of many hip-hop (and, in this case, hip-hop R & B) artists—of depicting the female body as a personal adornment and vehicle for sexual escapades (Ralph 2006a).

48. I use the term "African American" with a sense of its complicatedness given that I was born in Canada and raised in the United States in a family that is originally from Guyana, a Caribbean country which, technically, forms part of the South American mainland.

49. Hartman mentions encountering in Ghana "visiting scholars, artists, and journalists" who paid "*obruni* prices for rent and everything else" without "receiving the quality of goods and services that the powerful commanded and that Ghanaians exacted" (2007, 25). Yet her critique of the way Ghanaians skillfully exploit African Americans says nothing about opportunistic enterprises rendered in reverse. Looking in the opposite direction, what are we to make of Andrew Young, whose persistent interest in maintaining extensive diplomatic ties to Africa (especially Nigeria) throughout his tenure as a United Nations Ambassador subsequently translated into commercial interest in Nigerian oil reserves (Dart 2002, Ashley 1995) through his GoodWorks International consulting firm amidst local protest among students that oil exploitation had upset the local ecosystem. In the late 1990s GoodWorks was, in Nigeria, rumored to be behind the murder of local youth who protested oil deals that Young's company had helped to negotiate. It was never confirmed that Young nor any of his associates were guilty of such allegations. Yet one cannot but help notice the irony that a former U.S. Ambassador and freedom movement leader now finds himself in the curious position of having to defend corporate policies and profit-seeking projects linked to death and environmental destruction in Nigeria. Young has been soundly criticized for implicitly endorsing—or at least refusing to condemn—injustices perpetuated by Nigerian political officials, like President Olegun Obasanjo, whose regime remains mired in "corruption, crime, poverty, and violence" according to Human Rights Watch (Gentry and Poole 2007), forcing some to now question the former civil rights leader and Mayor of Atlanta whose political credentials no longer seem sufficient to defend his newest ventures in Africa (Meier 2007). Young, it should be noted, first met the Nigerian leader in his capacity as ambassador though he kept in touch with Obasanjo even after he resigned from the Carter Administration, not simply making it possible for his sons to attend to Morehouse College and Georgia Tech in Atlanta (Suggs 2000) but also providing him with "books, tapes, and a Bible" when Obasanjo was imprisoned under a previous president (ibid.).

50. If "slavery" ultimately "stripped your history to bare facts and precious details," Hartman (2007, 11) seems unable to imagine a history apart from any "trace" of ancestral evidence (7). I am sensitive to the formidable challenge of reckoning with one's past in the absence of documentation, of trying to form a narrative despite the innumerable "blank spaces" (12) that occasion the "bare bones" (11) recollections of familiar lore. Still it seems, paradoxically, evidence of too strong a romance with the violent dehumanization of enslavement to suggest "towns vanished from sight and banished from memory" are all any African American "can ever hope to claim" (9). What of the historical understandings African people developed beyond the rigid criteria of verifiable proof (cf. Brown 2003), the meaningful ties they manage to forge despite "the slipperiness and elusiveness of slavery's archive" (Hartman 2007, 17)? Whose ancestral connection, after all, is indubitably real?

51. The overlapping articulation of actors and constituencies that participated in the transatlantic slave trade (and domestic circuits connected to it) are difficult to discern given the

way Hartman (2007, 208) theorizes this historical matrix, "Simply put, slaves were stolen from one group, exchanged by a second group, and then shipped across the Atlantic and exploited in the Americas by a third group."

52. I find Hartman's quest to locate an appreciation for Africa steeped "in the efforts thwarted and realized, of revolutionaries intent upon stopping the clock and instituting a new order" promising, and think it not simply commendable but also appropriate that, instead of longing for the myth of a glorious past, she claims a position "articulated in the ongoing struggle to escape, stand down, and defeat slavery in all its myriad forms." Still, rendering this the preferred posture of a "lost tribe" (Hartman 2007, 234–35)—instead of theorizing the diasporic connection as an admittedly prosthetic but nevertheless consequential political alliance forged by people occupying similar structural positions linked through certain critical sensibilities who also share something of a historical trajectory (Edwards 2003,13–15, 143, 282, 303)—seems remarkably complicit with the peculiar institution's inevitable project to sever the histories of its captives. From whence this extraordinary efficacy?

53. Beyond the historical particulars of Senegalese social life, Jean Loup-Amselle (1990; repr. 1998) has convincingly argued that, instead of fixed ethnic distinctions, many African societies were marked by an originary syncretism that appears as a fixed tribal identity more readily in ethnographic accounts and, apparently, personal meditations than it does in reality.

54. "Jackson," said one senior official, "has given us a lot of help on this, and we'll all remember him for it," see "Clinton Opposes Slavery Apology," in Brooks, ed. (1999, 352); emphasis added.

55. In November 2006, British Prime Minister Tony Blair called the transatlantic slave trade a "crime against humanity" but, like Bush, "stopped short of issuing a full apology." See "Blair 'sorrow' over slave trade," *BBC News*, November 27, 2006.

56. As Malcolm Gladwell (2006: 138) is careful to remind us, "Crime . . . isn't a single discrete thing, but a word used to describe an almost impossibly varied and complicated set of behaviors." To the extent that Bush's speech reframed the legacy of U.S. activity in the Afro-Atlantic world by "silencing" so much of "the past," (Trouillot 1997) he produced a version of history at least as criminal as the virtual enslavement of the Senegalese people who, on July 8, 2003, understood their predicament through the lens of bondage.

57. Recall the scene in Voltaire's *Candide* when a mutilated slave who has escaped from a plantation in Surinam confronts the novel's young protagonist to display fleshy scars that remind him, "It is at this price that you eat sugar in Europe." It is at a similar price that Senegal might enjoy their slice of the neoliberal pie.

58. If "slavery" successfully "established a measure of man and a ranking of life and worth that has yet to be undone" (Hartman 2007, 6), forensic methods of calculation during the nineteenth century provided what was then considered to be objective criteria for determining "worth" in emergent forms of monetized value by drawing on a shift in Anglo-American legal perspective, which suddenly made human beings, previously considered more sacred than scientific, available for quantification (Pietz 1997a, 98) based on the physiological and demographic variables planters considered in the "slave pens" where they examined prospective chattel (Johnson 1999, 118).

59. That such a legal concept could nevertheless fit this scenario so appropriately makes one question whether this militaristic spectacle is evidence of the way "force 'trumps' law" or if "the very concept of law" if, indeed, "juridical reason itself, includes a priori a possible recourse to constraint or coercion and, thus, to a certain violence" (Derrida 2005, xi; cf. Bourdieu 1997, 95). Though, as the foregoing discussion suggests, I believe legal concepts

and categories are to be engaged and not dismissed as mere "ideologized obfuscations of how social life really comes about and operates . . . studied only in order to debunk them" (Pietz 2002, 35).

60. These Senegalese "slaves," then, might be considered "virtual commodit[ies]" (Trouillot 1994) in both senses of the term.

61. Apropos here, articles 1109 and 1117 of the *Code civil des français* (French civil code) indicate that a contract is not valid if consent derives from misapprehension (*erreur*), is achieved under duress (*violence*), or results from misrepresentation (*dol*). For the articles in question, see the *Code civil*: "Il n'y a point de consentement valable, si le consentement n'a été donné que par erreur, ou s'il a été extorqué par violence ou surpris par dol" (Article 1109). "La convention contractée par erreur, violence ou dol, n'est point nulle de plein droit; elle donne seulement lieu à une action en nullité ou en rescision" (Article 1117).

CHAPTER 22

NUCLEAR IMPERIALISM AND THE PAN-AFRICAN STRUGGLE FOR PEACE AND FREEDOM

GHANA, 1959–1962*

JEAN ALLMAN

> We face neither East nor West: we face forward.
> —Kwame Nkrumah, Positive Action Conference for
> Peace and Security in Africa, Accra, 2 April 1960

IN 1962 PROFESSOR ST. CLAIR DRAKE PREPARED A PAPER FOR THE ACCRA ASSEMBLY on the World without the Bomb—a high-profile international gathering in Ghana's capital of nearly a hundred activists, statesmen, scientists, teachers, and clergy opposed to nuclear armament.[1] Drake, a renowned Pan-Africanist and then professor of anthropology at Roosevelt University, served as the head of the Department of Sociology at the University College of Ghana from 1958–61. In his paper, "The African Revolution and the Accra Assembly" Drake predicted: "History will record a significant fact about the African Revolution, that it was led by men who always exhibited an unusual concern for minimizing the violence of the revolutionary struggle, for seeking solutions through the United Nations wherever possible; and who were always concerned to insulate the revolution from Cold War politics so that Africa would not run the danger of becoming the spot from which World

* Gwendolyn Mikell and Elliott P. Skinner, *Women and the Early State in West Africa*, Working Paper #190 (East Lansing, MI: Women in International Development, 1988).

War III—the nuclear war began."[2] Regrettably, Drake—an extraordinarily incisive and prescient social critic—was not on target with this particular historiographical prediction. By and large these are not the "facts" of the "African Revolution" that history has chosen to remember. While we have heretofore managed to avoid World War III (an accomplishment increasingly imperiled with each passing day), the achievements of the "African Revolution" have, for the most part, been buried beneath the detritus of coups and counter-coups, debt, civil war, and structural adjustment. What many have termed an "Afro-pessimism"—"nothing good ever comes out of Africa"—has erased the vision, the new world order that so many sought to build.[3]

What I would like to accomplish with this chapter is rather simple. I want to recount a story—for some it may be a familiar story—of one small episode in that African Revolution. I wish to focus on the movement against what was called by activists at the time "nuclear imperialism," as it emerged out of the Pan-African struggle for freedom from colonial rule in the late 1950s and early 1960s. It is a story I reconstruct out of the private papers and recollections of participants, government documentation, and newspapers, both in Ghana and the United States. I want to revisit this moment in Africa's past and in the history of Pan-African revolt for three basic reasons. First of all, and quite simply, it is important to remember in these days of war, torture, and U.S. imperialism run amok that there was a time when Africa was at the very center of the global peace movement and when radical visions of a new world order were being generated from the streets of Accra to the mountains of Kenya, from the townships of apartheid South Africa to the Qasbah in Algiers. Secondly, I offer these stories as a reminder, if not a corrective, to the new discourse on globalization that has washed across so many campuses in North America. New transnational and global studies programs and institutes are popping up everywhere, offering "new" ways to understand the world that transcend political and national borders. They are either oblivious to or strategically dismissive of the fact that interdisciplinary programs like African American studies and Peace studies have been "global"—in subject and method—from the outset.[4] The movement against nuclear imperialism that took root in the Pan-African freedom struggle showcases the "global" and the "transnational" in ways that need to be recovered and remembered. We must not forget that Pan-Africanists like W. E. B. Du Bois, Paul Robeson, George Padmore, Kwame Nkrumah, Walter Sisulu, and Patrice Lumumba were, in many ways, on the frontline in confronting the harsh realities of our current world order—the postwar imperial world that the United States has sought to make in its own racist image.

Thirdly and finally, this story stands as a counter-narrative, a corrective, I would argue, to the so-called afro-pessimism that has dominated scholarship on Africa since the 1980s. The focus of much of that pessimism has been on the failure of the African nation. Ever since the promise of the newly independent African nation states of the 1950s and 1960s and the celebratory historiography that heralded their rise crashed against the hard rocks of neocolonialism and neoliberalism, many scholars of African nations and nationalism have been immobilized by what has been widely deemed the failure of the nationalist and Pan-Africanist project in Africa. As a result, few, if any, have transcended the modernization-bound question, What went wrong?[5] Even when scholars have deployed counter-modernization theories of dependency and underdevelopment, the question has, in many ways, remained the same. This profound, almost discipline-defining pessimism—a pessimism that seems consistently fueled by horrors transpiring in places like Rwanda, Liberia, and more recently the Sudan—has prevented us from comprehensively recognizing, problematizing, and historicizing the legacies of nation and Pan-Africanism in late colonial and neocolonial Africa. Yet these are stories we need to remember, perhaps more so now than ever—"nation time," liberation times, times when Pan-Africanism recognized no boundaries and a United States of Africa was considered not a pipe dream but a plan just shy of a blueprint.[6]

BLACK INTERNATIONALISM AND EMERGENT GHANA

These kinds of stories always have multiple beginnings and manifold genealogies of origin.[7] The story I wish to tell here, for example—though it unfolds in West Africa during the first years of Ghana's independence—is inextricably connected to stories of radical black internationalism, especially in the United States. Thanks to the works of Horne, Plummer, Richards, Meriweather, Kelley, Von Eschen, Gaines, and many others we know that in the United States there was an efflorescence in black internationalism in the wake of Italy's invasion of Ethiopia in 1935—an internationalism that continued to insist on the inextricable connections between struggles for equality and racial justice in the United States and anticolonial resistance in Africa and Asia.[8] Through biographical accounts, through histories of organizations like the Council on African Affairs, and through careful readings of African American press, we can trace the paths and reconstruct the meetings and rallies that brought radical black internationalists into alliance, debate, and common struggle across the globe. We can follow Du Bois, Robeson, Max Yergan, and the Council on African Affairs from 1937 until the height of the cold war;[9] or we can look at the NAACP's colonial conference in April 1945, which brought together participants from throughout the colonial world, including Kwame Nkrumah.[10] In Manchester, England at the Fifth Pan-African Congress in October of that very same year, Padmore, Nkrumah, Jomo Kenyatta, Du Bois, Amy Ashwood Garvey, and many others made sure that Africa and an African agenda for liberation was front and center.[11] A year later, the Big Three Unity Rally in New York in 1946 emphatically placed an antinuclear agenda in the foreground of struggles against colonialism and racial oppression, as explicit connections were made between new U.S. investments in Africa (for example in the mining of uranium in the Belgian Congo for the construction of atomic bombs) and issues of social justice in the United States.[12] Those connections—between Pan-Africanism, anticolonialism, and global peace—continued to resonate throughout the Cultural and Scientific Conference for World Peace in New York (March 1949) and at the World Peace Conference in Paris (April 1949).[13]

But by the early 1950s, as many historians have argued, the Red Scare, McCarthyite repression, and the systematic persecution of black internationalists in the United States, including the confiscation of both Robeson's and Du Bois's passports, "limited the scope and capped the resources of many mass organizations that had been militant during the war years. Cold war rhetoric questioned the legitimacy of anticolonialism and pacifism in a world dominated by armed superpowers."[14] As a result, the cold war "severed the black American struggle for civil rights from the issues of anticolonialism and racism abroad." The politics of black internationalism in the United States, Von Eschen concludes, "did not survive the beginnings of the Cold War."[15]

Without minimizing the devastating impact of the cold war on black radical internationalism in the United States, it remains important to understand the ways in which that internationalist vision so deftly reconstructed by Horne, Plummer, Von Eschen, and others continued throughout the 1950s—albeit in very different forms and often in very different places.[16] For example, the so-called Bandung Conference in Indonesia in 1955, was one of the great watersheds in the history of struggles for peace, freedom, and nonalignment—and it provided an important context for radical black internationalists to continue to engage, on a global stage, with the newly emergent nonaligned world, despite the ravages of the cold war.[17] The famed African American writer, Richard Wright, was in attendance and, though Robeson was denied a passport for travel, he sent a long message of support that powerfully foregrounded the inextricable connections between colonialism, nuclear proliferation, and racial injustice.[18] After Bandung, the center of gravity for black internationalism, shifted decisively from what was fast becoming the center of a new global empire, that is, the United States, toward its margins—a small country in West Africa, under the leadership of a staunch Pan-Africanist, Kwame Nkrumah. Nkrumah spent a decade in the United States as a student from 1935–45. These were the very years that black internationalism was coming into its own and Nkrumah contributed directly to the forging of that radical internationalist agenda in the United States. In many ways, then, the shift from Harlem in 1945 to Bandung in 1955 to independent Ghana after 1957 was predicted by and directly predicated upon the Pan Africanist Congress in Manchester, which witnessed African leaders taking center stage for the first time.[19] Indeed, as Du Bois wrote to Immanuel Wallerstein in 1961, "Pan-Africanism was not dormant between the Manchester meeting in 1945 and the Accra meeting in 1958. It was alive in the plans of Nkrumah, Padmore and many others . . . but the question was where it could meet and how far its program could go."[20]

GHANA, THE SAHARA TEAM, AND THE PAN-AFRICAN STRUGGLE AGAINST NUCLEAR IMPERIALISM

Du Bois's second question—just how far did or might the radical Pan-Africanist agenda go—remains open. But his first question—the where—was addressed, at least until 1966, by Ghana's independence. The powerful initiatives of black internationalism, particularly with regard to anticolonialism, nonalignment, and peace resonated throughout the First Conference of Independent African States (Accra, April 1958), which demanded an end to the "production of nuclear and thermo-nuclear weapons" and the suspension of "all atomic tests in any part of the world and in particular the intention to carry out such tests in the Sahara."[21] Developing a position that he termed "positive neutrality," Nkrumah explained to Ghana's National Assembly in February of that same year that "Ghana has a vested interest in peace; our constant concern is national security, in order that we may get on with the job of economic and social reconstruction in an atmosphere of peace and tranquility."[22]

That African liberation was inextricably bound to struggles for peace, security, and nonalignment echoed throughout the continent in the months following Ghana's independence. At the All-African People's Conference (AAPC) in December 1958—a conference hailed by the *Chicago Defender* as proof that "Pan-Africanism is more than a vague dream of expatriates in London or Negroes in Harlem"[23]—the East-West conflict and the prospects for nonalignment were central to discussions.[24] Many black radicals joined Du Bois in considering the AAPC the direct successor of the 1945 Pan-African Congress—a sixth congress, as it were. Robeson was able to attend the historic gathering in Accra, though Du Bois was not. His message to participants was delivered by his wife Shirley Graham Du Bois.[25] Together conference members decried the ways in which imperialists "are now coordinating their activities by forming military and economic pacts such as NATO, European Common Market, Free Trade Area . . . for the purpose of strengthening their imperialist activities in Africa." Participants—Patrice Lumumba from the Congo and Tom Mboya of Kenya among them—pledged their "full support to all fighters for freedom in Africa . . . as well as to all those who are compelled to retaliate against violence to attain national independence and freedom for the people."[26]

The following year, when France made clear its intent to begin testing a new atomic bomb in the Sahara desert in the very near future, the connections between colonialism and nuclear proliferation took on an even greater urgency for African nationalists and Pan-Africanists. In August 1959 the Monrovia Conference on Algeria, which included foreign ministers from nine independent African states who gathered to discuss support for the Algerian nationalist struggle, passed a resolution denouncing the decision to conduct nuclear tests in Africa.[27] Two months later, when the AAPC steering committee met in Accra, Nkrumah demanded that the major world powers "stop all nuclear tests, stop research on, and manufacturing of nuclear weapons, destroy all existing stocks of atomic and hydrogen bombs and dismantle all rocket bases. . . . [N]uclear weapons," he declared, "constituted the 'sword of Damocles hanging over the head of mankind . . . which we must remove by positive action.'"[28]

But by the time Nkrumah addressed the AAPC steering committee, concrete plans were already well underway—for one of the first times anywhere in the world—to make manifest the combined struggle against imperialism and campaigns for nuclear disarmament. Ghana's central role in articulating a Pan-African vision, in reanimating black internationalism in the wake of systematic cold war repression, and in aiding anticolonial movements throughout the continent made it a logical place for mobilizing against France's nuclear imperial threat. And the Ghanaian government, as we will see, ended up providing significant logistical and practical support for the specific plans that began to unfold in April 1959. Yet the monumental work of engineering this crucial alliance between anti-imperial, Pan-African struggles for freedom and the antinuclear movement fell not to Nkrumah or his government but to two African American radical peace activists—Bill Sutherland and Bayard Rustin, who, like Nkrumah, were very much participants in and products of the efflorescence of black internationalism in the previous decade and were profoundly influenced by the Pan-African vision of Trinidadian revolutionary George Padmore.[29] Both Rustin and Sutherland resisted the draft during World War II and as a result of their actions were sentenced to federal prison—Rustin for three years, Sutherland for four.[30] Both were active in the Fellowship of Reconciliation[31] and in founding the Congress of Racial Equality in 1942 and had participated in the Committee for Non-Violent Action and a range of international campaigns for nuclear disarmament, including Britain's Direct Action Committee against Nuclear War. Indeed, Rustin was one of the main speakers at the Trafalgar Square launch of the famous Aldermaston March to protest nuclear armament on April 9, 1958.

Bill Sutherland settled in Ghana at the beginning of 1954.[32] On Bayard Rustin's recommendation, he had originally planned to travel to Nigeria to work with Nnamdi Azikiwe on the *West Africa Pilot*, but his visa was endlessly delayed by the British Colonial Office. Then one afternoon, as Sutherland recounts it, "Padmore had me meet him at a sidewalk café in Paris, where he was discussing politics with African American author Richard Wright. Wright had just begun working on the book that was to become *Black Power*, and was filled with exciting stories about his recent trip to the Gold Coast." Based on Padmore's recommendation, Sutherland headed for Ghana. After several years working on an educational project, he became the personal secretary of Ghana's first post-independence Finance Minister K. A. Gbedemah, who shared a background in international peace activism. According to Sutherland's recollections, he first learned of the French plans from April Carter and Michael Randle, who were centrally involved in the Direct Action Committee of the British antinuclear movement. Fortunately, because of his travels back and forth to Britain as Gbedemah's secretary, Sutherland was able to liaise closely with his British counterparts, with Africans residing in London, and with the Convention People's Party (CPP) in Ghana in order to work out plans for how peace activists should respond to the impending French threat to explode an atomic bomb in the Sahara.[33] In September, Sutherland also made his way to New York in an effort to build further support for direct action against the French threat. There he met with Bayard Rustin, who had played an absolutely critical role in organizing nonviolent protest during the Montgomery Bus Boycott and whom Sutherland considered crucial to any action against the French.[34] "Not only," as John D'Emilio writes, "did [Rustin] have more experience organizing complex direct action projects than did any of the British pacifists, but the fact that he was black and so deeply involved in the Southern freedom struggle would bring credibility among Africans." Ever the political tactician, Rustin was quickly persuaded by Sutherland and by the strategic possibilities of linking the struggles of African peoples to the campaign for nuclear disarmament. [35] By October, the War Resistors League agreed to fund Rustin's journey to Britain and Ghana to help develop a plan for direct action.

Kevin Gaines argues in his pathbreaking *American Africans in Ghana* that Ghana "provided an independent forum for black American radicals . . . offering them the opportunity to participate in a transnational culture of opposition to a Western culture seeking the preservation of colonial and neo-colonial dominance over the majority of the world's peoples."[36] Certainly Ghana provided an important forum for African American activists, like Rustin and Sutherland, but we must also appreciate the reciprocal processes at work: what black radicals from the United States and the broader Diaspora brought to Ghana. Ghana became the new focal point of a "transnational culture of opposition," in no small part because of the ongoing work of black internationlists like Du Bois, Robeson, Padmore, Rustin, and Sutherland. These activists were not simply accessing an existing (and heretofore separate) struggle for peace and freedom by locating their work in Ghana. They were generating and constituting that struggle—literally mapping it with their movements across borders and boundaries, as they forged links between pacifism, nuclear disarmament, and civil rights and reinvigorated the Pan-African struggle against colonial domination. In Ghana in 1959, it was Bayard Rustin and Bill Sutherland who sought to bring the lessons of Montgomery, Aldermaston, and the All African People's Conference on to the same page.

Yet the work of forging and maintaining transnational links—especially given the primacy of "nation" and "national interests" in a cold war world—was not always easy, even when operating from black internationalism's new center of gravity in an independent African state. Indeed, as D'Emilio points out, many in the civil rights movement, including Dr. Martin Luther King, Jr., and A. Philip Randolph, were not pleased with Bayard Rustin's decision to travel to Ghana during what was understood as a critical moment in the Black Freedom Movement. They urged Rustin not to participate in any action that might result in a prison term and thereby prevent him from returning to the United States. Meanwhile, Rustin's pacifist allies urged him on.[37] Unsure of how to proceed, Rustin—already in Ghana—requested that his allies from the civil rights and pacifist movements meet. The results of that meeting, held in Randolph's office in Harlem, were reported to Rustin by telegram: "'Randolph expressed firm view civil rights struggle paramount and decisively important to African colonial struggle as well as peace fight. Your indispensable role in domestic actions requires return. . . . Muste holds that Africa project potentially more important, capable of major contribution to civil rights struggle here as well as struggle against new nuclear colonialism.'"[38] In the end, Rustin found his own way, moving forward with his participation in the protest, while making arrangements, in consultation with Nkrumah and other protesters, for pacifist A. J. Muste (Fellowship of Reconciliation activist and then chair of the Committee for Non-Violent Action) to replace him in due course. [39]

Just what that protest would entail was "the culmination," according to April Carter, "of . . . negotiation and planning between Accra, New York, and London; journeys to France; deputations to African Embassies in London; and debate about the politics, route, personnel and financing of the team."[40] Rustin, Sutherland, Scott, Randle, Carter, and a handful of others ultimately decided that the protest team should assemble in Accra and then travel north through Upper Volta and the French Soudan, continuing directly to the test site in Algeria.[41] The plan had the full support of Ghana's ruling party—the Convention People's Party. In fact, the Ghana Council for Nuclear Disarmament (GCND), headed up by E. C. Quaye, chair of the Accra City Council, approached Ghana's Cabinet in October of 1959 with a request for assistance to send the protest team from Ghana "across West Africa into the atomic testing ground at El Hammoudia near Reggan"—approximately two thousand miles to the north of Accra some time in November.[42] After much discussion, the cabinet decided that it would provide a special grant of G£2,757 to the African Affairs Committee, "which in turn could make the sum available" to the GCND. Concerned about maintaining an image of government neutrality, the Cabinet directed the GCND "not to publicise the source of their income in order to avoid possible embarrassment to the Government."[43] Such budgetary laundering through African Affairs, however, could not have disguised government support for the protest. The African Affairs Committee met regularly at Flagstaff House, Nkrumah's residence in Accra, and included all of the top government ministers and CPP activists.[44]

By November 1959, international activists began to converge on Accra. Reverend Michael Scott, the renowned anticolonial, antiapartheid peace activist, who had been arrested for civil disobedience in South Africa in 1946 and would become a major figure in the movement for colonial freedom in southern Africa, arrived in mid-November.[45] Joining him was Muste (who founded the American Committee on Africa) and Randle (chair of the Direct Action Committee against Nuclear War from 1958–61). In addition to Bayard Rustin, Francis Hoyland (an art teacher and painter from Britain),and Esther Peter (a peace activist who worked for the Council of Europe in Strasbourg) made their way to Accra, although H. O. Hakansson, a professional dancer and lifelong pacifist from India who was supposed to join the team was not able to come.[46] As the team members assembled in Accra, they began the process of applying for travel documentation. The French government's immediate response to the activists' requests was to deny visas and entry permits to all team members, including Esther Peter, a French citizen. In a statement published in the *Ghanaian Times*, Rustin, serving as the team's secretary, responded by reaffirming the group's determination to proceed:

The French authorities, by imprisoning us, will hardly silence our voices. We cannot believe that the French people want to defy world opinion as expressed at the United Nations by exploding a bomb while the present nuclear powers are seeking an agreement to end all testing. We cannot believe that the French people want to perpetrate the infamy of violating and desecrating the soil of Africa in the interests of a new nuclear imperialism. . . . There is one thing that will cause us to abandon our mission—the abandonment of the Sahara Test.[47]

While diplomatic responses unfolded outside of Ghana, public fundraising for the campaign within Ghana reached a fever pitch with a mass rally at West End Arena on November 19 when Finance Minister K. A. Gbedemah appealed to the nation on behalf of the team and the GCND. "'We the people of Africa,'" Gbedemah declared, "can under no circumstances permit our God-given land to be used for the destruction of humanity.'" Over G£4,000 were collected at that rally.[48] Meanwhile, the Ghanaian government continued to battle at the United Nations. In New York, Ghana's Permanent Representative to the United Nations sharply questioned French sovereignty over the Sahara: "the whole question of French sovereignty in the Sahara is today being debated on the field of battle between the armies of France and the forces of the Provisional Government of the Algerian Republic. . . . If France must explode its bomb, they are quite welcome to do so somewhere in metropolitan France. . . . Those days when the destiny of Africa was decided at the conference tables outside Africa . . . are over."[49]

On December 5, 1959, the night before the team's departure, Rev. Scott told Radio Ghana listeners that the desert "'was being prepared as a base for nuclear war in North Africa and the Middle East.' Their journey would be 'a holy war, a non-violent war, against the inhumanity of nuclear war.'"[50] The following morning the Sahara Protest Team, which now included eighteen members, began its trek after a dawn farewell rally at the Arena. The members included eleven Ghanaians (C. Ablorh, B. M. Akita, K. M. Arkhurst, George Asante, K. A. Dornu, K. Frimpong-Manso, F. A. Koteye, P. G. Marshall, George Odoe and R. Orleans-Lindsay); a Nigerian student, H. Arinze; Ntsu Mokhekle, president of the Basutoland National Congress; Hoyland, Randle and Scott from Britain; Peter from France; and Rustin and Sutherland from the United States. Muste remained in Accra in charge of communications, though he planned to meet up with the team briefly when it reached the border. The team's departure captured world media attention the moment it left Accra.[51] By the afternoon of day one the team had reached Kumasi—the capital of the historical Asante empire—where it was met by an enthusiastic crowd. As Muste later wrote, "I wish that all activists in the United States and Britain could witness these scenes—the big rallies, the people lined along the streets and roads shouting, 'Freedom!' and 'Sahara Team!' as the huge truck and the Land Rovers rolled by."[52]

After leaving Kumasi, the team continued its journey north, reaching the town of Bolgatanga on December 7, where members then made the decision to try to cross the border at the frontier town of Bawku.[53] From Bawku, the team passed into French territory but had to travel sixteen miles before reaching the first

French government post at Bittou. There team members met three French officers and Michael Scott immediately set about explaining their mission. According to Muste's account, "one of the officers interrupted him and said: 'you do not need to speak at length. We know all about your group. . . . But we are here under instructions from Paris to forbid you to proceed.'"[54] And it was there, at Bittou, that the team sat for several days in a strange kind of limbo: they were not allowed to pass nor were they arrested. Sutherland recalls that they "maintained their presence in Bittou, handing out leaflets. . . . Once it was clear that the group was well received, evident by local offers of food and shelter, the French police surrounded them."[55] Esther Peter, as a French citizen, was the only one allowed to move about, but she could only do so under guard.[56] At times, the situation grew quite tense. As Sutherland recalls, Bayard Rustin at one point said, "'Let's start up our motors and see what happens.'" Apparently, seconds later, "the French paramilitary came and took their positions with their weapons at the ready and the team members stopped the motors." The team withdrew to Bolgatanga to reconsider their strategy.

Several days later, the *Ghanaian Times* announced the protesters' new strategy with a bold headline: "Protest Team Makes Second Dash." A smaller group of seven, led by Michael Scott, left Bolgatanga on December 17, this time heading toward Po in the Upper Volta. The team included Randle, Scott, Sutherland, and four others. While this group made their move, other members agreed to focus on propaganda in Ghana, the surrounding region, as well as Europe and the United States.[57] But the smaller version of the team was again halted at the end of December at the frontier post of Po. There they camped for some time, unable to move. Finally, as Sutherland recounts:

> After getting to know the African guards, the second team moved early one morning across the border. "The guards did not try to stop us physically," Bill remembered, "but they did alert their French superiors, who arrested us after we advanced about one mile into Upper Volta. We were put into jail. . . . The next morning, we were all put into a large van and the amiable military officer in charge led the way in a small Land Rover. We did not know where they were taking us, but we drove for several hours. Unfortunately, the officer's land rover skidded and turned over on the very sandy road. The team of seven of us was allowed out of our van to help put the overturned vehicle right and we continued on our way! Eventually . . . we . . . were dumped unceremoniously into Ghana."[58]

As all of the personal accounts of the protesters attest that the standoff between the team and the French officers was incredibly tense, though there were, at times, surprising moments of easy interaction with the soldiers. Michael Scott recounts several of these, including one on Christmas day when soldiers approached the camp and the protesters were sure they were about to be arrested. Instead, "the soldiers had come to share food and drink with us," Scott remembers, "and sing us French songs."[59] On another occasion, when the protesters were moving through French territory under French escort, one of their truck's tires was punctured and they were forced to stop. The soldiers who had been following them pulled up along side and set their rifles down, knowing that no one could flee. Then, according to Scott, "Michael Randle, who had a guitar, started singing and playing. The soldiers became interested and the whole thing ended up with them doing their dances and our people doing ours. The French officers, three of them, eventually arrived and found their own troops dancing with us, their rifles parked by the side of the road."[60]

Such light-hearted moments notwithstanding, by January team members were utterly exhausted, though they made several more attempts to cross the border, urged on, according to Sutherland, by the enthusiasm of Finance Minister Gbedemah.[61] Finally, in mid-January, the team again crossed the border—not at an official post—with the help of a "local guide along a path usually used by smugglers."[62] They hid in the bush and slept under blankets draped over tree branches for shade, and moved only at night, on foot, toward Ougadougou. Exhausted and without water, they eventually hitched a ride with a truck. But their relief gave way to dismay when, as Scott recalls, "Instead of being taken to the next town, we found ourselves being driven right into the compound of the French authorities. One of the French officers who had had the Christmas dinner with us lifted up the flap of the lorry and said, 'Bon jour. Good to see you again.'"[63] The team members were again arrested and again they refused to cooperate with the police. They were then carried, one by one, into a van, driven south, and dumped, for the last time, on the Ghana side of the border.[64]

After this failed attempt, the Sahara Protest Team, as it came to be known, returned to Accra where it now had a permanent office and was officially co-chaired by E. C. Quaye and Michael Scott, with Randle as secretary. Its working committee included Abdoulaye Diallo, E. J. Duplain, K. A. Gbedemah, R. T. Makonnen, Bayard Rustin, Bill Sutherland, and N. A. Welbeck.[65] For several months team members continued their struggle on a range of fronts. Scott left Accra for Tunis on January 25 for the AAPC, where he announced that he was "prepared to fly into the Sahara test area, if he could find a plane and a take-off strip."[66] Rustin and Muste returned to the United States where demonstrators, organized by the Committee to Support the Sahara Protest Team, marched in protest against the testing outside the French Tourist office in the heart of Manhattan.[67]

But as peace activists—African, American, and European—strategized around new transnational peace tactics, France's plans continued apace for the first atomic bomb testing. The only factor hindering French efforts in late January and early February was the weather and when those conditions improved, they exploded their first nuclear device at 6:00 a.m. on February 13, 1960. At an emergency cabinet meeting, Nkrumah proposed and cabinet members agreed that all of the assets of French firms operating in Ghana should be frozen and no permits for the transfer of assets issued.[68] Several weeks later, on April 1, the French exploded a second bomb—reported by the Ghana *Evening News* as "representing thrice the strength of the bomb which devastated Hiroshima."[69] The cabinet immediately met to discuss "stronger measures." The government then suspended diplomatic relations with France, recalled J. E. Jantuah, the Ambassador to France, refused visas to French citizens, and froze all assets and properties of French citizens residing within Ghana's borders.[70]

Outrage at France's unilateral disregard for African soil and African lives was expressed throughout the continent and beyond, although as David Birmingham rightly points out, there was "no outcry from the neo-colonial puppets"—most notably from Côte d'Ivoire's Felix Houphouet-Boigny.[71] The only act of colonial aggression to elicit a similar level of outrage in the Pan-African press was Italy's invasion of Ethiopia in 1935, which had galvanized black internationalism throughout the world. More than a few speakers and editorialists pressed the connection. "There is only one incident in living memory," an editorial in the *Ghana Evening News* exclaimed, "that compares in magnitude with the ephemeral, almost agonizing, triumph of Might over Right at Reggan last week—Mussolini's rape of Abbysinia with the gas bombs and mass slaughter that eventually sounded the death-kneel of the League of Nations. . . . Then as now, certain nations, loud in their profession of humanity, fraternity, justice and democracy sat complaisantly as Ethiopia, nailed to the wall, stretched her bleeding hands for succor."[72]

The most dramatic Pan-African-led response to the French tests occurred shortly after the second bomb exploded at the beginning of April. Building directly on the twinned legacies of Pan-Africanism and peace that were manifest at the 1945 Pan Africanist Congress, the 1949 Peace Conferences in New York and Paris, the Asian-African Conference in Bandung, as well as at the AAPC in 1958, Ghana hosted the Positive Action Conference for Peace and Security in Africa. From April 7 to 10 delegates gathered from throughout the continent (officials from independent states and representatives of liberation movements), as well as nonvoting observers from Japan, India, Britain, Sweden, the United States, Yugoslavia, and France.[73] Nonvoting delegates also represented the Committee for Non-Violent Action, the Montgomery Improvement Association (Ralph Abernathy), and the American Friends Service Committee. The conference was initially planned as a specific and direct response to the French testing but in 1960 history was moving faster than conference planners possibly could, and dramatic events of the moment demanded a rapid expansion in the conference agenda. On March 21, 1960, South African police fired on a group of unarmed demonstrators who were peacefully protesting the government's increasingly harsh and racist pass laws. In a matter of seconds, 63 lay dead and over 180 were wounded. The Sharpeville Massacre and the rapidly escalating violence and repression in Algeria that same month forced organizers to expand the conference mandate—a shift that generated some concern, especially from officials of invited governments. Nigerian Prime Minister Abubakar Balewa wrote to Nkrumah on March 25 that he was concerned not only by the expanded mandate but also by the expanded participant list: "I must also tell you that I find the proposed Conference very confusing in that delegations of Governments, of political parties, and of trade unions are all to join in together. I consider that this unorthodox procedure will make it very difficult for the delegations representing the elected Governments of their countries."[74] In the end, Nkrumah was able to assure concerned leaders that the threat of the atomic tests was paramount, but that

> on an occasion when so many of our brothers are able to meet together for an exchange of ideas in respect of the welfare of our motherland, advantage ought to be taken of any little opportunity that presents itself for the consideration of some of the most vital problems facing us as a people . . . it is precisely in the peculiar nature of our assemblies that we have an opportunity to provide new lessons to the older nations. The mobilization of the total African body politic in Conference is an achievement transcending the bounds of procedures, when the urgency of the matter and the colossal importance of protecting our interests are taken into consideration.[75]

In the end, the delegations that expressed concern about the conference's expanded mandate and the mixing of state and non-governmental agents agreed to attend.

Nkrumah carefully set out the conference's agenda in his opening address, as he explained to delegates that they had come to Accra "first to discuss and plan future action to prevent further use of African soil as a testing ground for nuclear weapons; secondly to consider effective means to prevent further brutalities against our defenceless brothers and sisters in South Africa, brutalities which are the result of the South African Government's racial policy of apartheid. Thirdly, this conference must consider the ways and means whereby Algeria can be helped to bring an end to this dismal flow of human blood consequent upon this lingering physical conflict."[76] At the start of the conference, the Sahara Protest Team, echoing the connections between the antinuclear struggle and African liberation, presented its manifesto to the delegates, calling not only for an end to nuclear tests and nuclear arms but also for a thousand volunteers for a renewed Sahara protest movement: "By joining with Africans from other parts of our Continent in positive non-violent action against nuclear imperialism they can make a decisive contribution to the liberation of all Africa."[77]

As the conference's aim was to mobilize the "total African body politic" in assemblies that did not privilege ministers and government representatives, delegates convened in separate committees to take on the specific questions of the conference's expanded mandate. Defying cold war rhetoric, they worked to bring substance to Nkrumah's famous dictum, "we face neither East nor West, we face forward," as they strategized about how to address the French atomic tests, the war in Algeria, apartheid South Africa, and the full liberation of the continent. In the end, resolutions were passed by all of the committees. The first committee called for total disarmament; the second (on Algeria) recommended support for the Algerian fight for independence and encouraged the independent African states to "consider formation of African volunteer corps to fight side by side with their Algeria brothers." The third committee called on African states to support victims of apartheid, to consider imposing sanctions on South Africa, and to demand that the South African mandate over South West Africa be revoked.

Finally the fourth committee on the liberation of Africa requested the United Nations to call a conference to consider a time table for the total liberation of all African countries.[78]

But perhaps what was more important than the resolutions passed at the Positive Action Conference was the process that had unfolded—that is, "the peculiar nature" of the assembly, as Nkrumah termed it. It was a process that was transnational, nonaligned, progressive, multi-pronged and not solely reliant on the power and expertise of the state. As Sutherland recalls:

> When Nkrumah asked the international grouping of mainly pacifist advisors to see what they could "come up with," even the most experienced amongst them found it to be a formidable task. Veteran US. Tacticians A.J. Muste and Ralph Abernathy worked long into the night with such figures as Madame Tomi Kora of Japan, Madame Asha Devi of India, Esther Peter and Pierre Martin of France and Britian's Michael Randle.[79]

Like its many precursors—from Manchester to Bandung—the Positive Action Conference sought to take on imperialism as a many-headed hydra—the forces of power that linked the atomic bomb with apartheid in South Africa and white rule in Algeria.

But if the conference demonstrated the common terrain upon which both Pan-Africanists and peace activists could work, even in the debilitating climate of the cold war, it also exposed the fissures that would eventually pull the movements apart. In this important way, then, it constituted a momentous turning point in the Pan-African struggle against nuclear imperialism. In the wake of Sharpeville and in the face of French intransigence in Algeria, the issue of nonviolence suddenly became a question. Among those attending the conference was Frantz Fanon. "He spoke in a quiet and sober voice," recalls Sutherland, "explaining his view of the regrettable necessity for armed struggle . . . 'we tried [nonviolent means, Fanon explained] . . . but the French came into the Casbah, broke down door after door and slaughtered the head of each household in the center of the street. When they did that about thirty-five consecutive times, the people gave up on non-cooperation.'"[80] And in many ways, the massacre at Sharpeville brought home a similar lesson: nonviolent direct action was only met with violence, murder, and massacre.

In the end, therefore, the Positive Action Conference brought the challenges and contradictions of the struggle against a nuclear imperialist world into sharp relief. Nkrumah, echoing the manifesto of the Sahara Protest Team, opened the conference with a call for continuing nonviolent positive action as the primary tactic against nuclear proliferation, apartheid, and entrenched settler rule on the continent. By the end of the conference, as Sutherland notes, there was "only a passing reference to the original proposals." While the conference ended with concrete plans for the setting up of a training center on nonviolent protest tactics, as a wing of the proposed Ideological Institute at Winneba, even those plans were off the table within a year.[81] It is hard to imagine how it could have been otherwise, given events unfolding in Algeria, South Africa, Angola, the Congo, Mozambique, and Zimbabwe. As Drake would later write, "Nkrumah, as a proponent of 'non-violent positive action' was committed to setting up a training center at Winnebah for activists from southern Africa until 1961 when the beginning of armed struggle made this an untenable position for a leading Pan-Africanist. Sharpville [*sic*] and Lumumba's murder also pushed Nkrumah away from a Gandhist position."[82]

Patrice Lumumba, the dynamic, progressive, and democratically elected leader of the Congo, was murdered by Congo secessionists with the complicity of Belgium and the U.S. Central Intelligence Agency on January 17, 1961. With Lumumba's assassination, the escalating violence in Algeria, and in the wake of Sharpeville, the transnational radical peace movement was severed from its Pan-African moorings. Ghana kept an official foot in the door of the international peace movement when it hosted the Accra Assembly in 1962—a gathering, which, as I noted at the outset, drew a cast of experts and activists from throughout the world.[83] But the parameters of that assembly were much narrower than those of 1960. The conference focused almost exclusively on the nuclear arms race—any radical transformative social agenda had disappeared.[84] The Assembly was official and in many ways predictable. As Sutherland recalled:

> Nkrumah was undoubtedly the leading voice of Pan-Africanism on the continent, and had become the leading African voice on the world scene. The nonviolence advocacy so prominent in the 1958 All African People's Conference, in the 1959 Sahara Protest Teams, and in the planning for the 1960 Positive Action Conference were all but eliminated from the mainstream political discourse. The Ghanaian government continued to provide support for the more conventional peace politicians . . . hosting the World Without the Bomb Accra Assembly . . . but there was little space or time for the radical experimentation of the previous years.[85]

In other words, as cold war lines were drawn ever more sharply, as the space to speak from a position of nonalignment grew narrower, and as evidence mounted—from Algeria, South Africa, Angola, and the Congo—of the lengths the neocolonial powers would go to preserve control and entrench their profits, time, experimentation, and the space to imagine new worlds must have appeared as luxuries no one could now afford.

LESSONS AND LEGACIES?

So, why remember the story of the Sahara Protest Team or recount the "peculiar assembly" of the Positive Action Conference that promised so much in 1960? The Sahara Team never made it more than a few kilometers north of the Ghana border and the "radical experimentation" of the Positive Action Conference dissolved in a matter of months. Certainly contemporary observers and participants considered the dramatic efforts of the Sahara Protest Team of immense historical significance. "It would not matter if not a single person ever reached the site," Nkrumah wrote, "for the effect of hundreds of people from every corner of Africa and from outside it crossing the artificial barriers that divide Africa to risk imprisonment and arrest, would be a protest that the people of France . . . could not ignore. Let us remember that the poisonous fall-out did not, and never will, respect the arbitrary and artificial divisions forged by colonialism across our beloved continent."[86] Sutherland's more recent recollections were just as positive: "this joining up of the European anti-nuclear forces, the African liberation forces, and the U.S. civil rights movements could help each group feed and reinforce the other. Both the civil rights struggle and the CND were on a high at that time; they were really strong, people's movements. Then, to be sponsored by a majority political party in government clearly marked a unique moment in progressive history."[87] And radical pacifist Muste echoing Sutherland's enthusiasm, characterized the Sahara protest as "an immense propaganda job for the idea of nonviolence . . . among the masses."[88] The reflections of Nkrumah, Sutherland, and Muste remind us, I would argue, of the ways in which national liberation and nation-building, pan-Africanism and the radical, transnational peace movement were constitutive political struggles. Building upon two decades of sustained black internationalist struggle, Ghana's leadership role in the Pan-African movement of the postwar era was forged in the context of the antinuclear movement and, just as importantly, that radical, direct action struggle for peace gained some of its most dramatic impetus from the leadership and participation of Pan-Africanists from throughout Africa and the Diaspora.

Their "global emancipatory vision," as Gaines has termed it, and their "radical experimentation" toward realizing that vision in the first years of Ghanaian independence is not without its legacies—legacies that we would do well to recall, especially when we hear that "nothing good ever comes out of Africa." [89] Indeed, in many ways one of the most significant legacies of those years is the very nation of Ghana—a nation that has managed to endure despite the fact that its disintegration, like that of so many African countries, appeared overdetermined by the antidemocratic colonial legacies of dependence, uneven development, kleptocracy, and outsized militaries. "What is astonishing," as C. L. R. James reminded us in 1972, "is not the failures but the successes. When did so many millions move so far and so fast?"[90] The stories of the Sahara Protest Team and the accounts of the radical experimentation that defined the Pan-African struggle in those critical years force us to remember not only the persistence and resilience of black internationalism during the worst ravages of the cold war but also the possibilities and promise of the African Revolution in ways that fulfill Drake's prediction of nearly a half century ago. They expose the inextricable connections between war, racism, and empire (whether it be the imperialism of DeGaulle's France or of Bush's U.S. hegemony) and remind us not only of the necessity of combining struggles for peace with movements against racism, economic injustice, and imperialism but also of the absolute centrality of Africa to each and every one of those struggles. As Nkrumah wrote nearly a half century ago, "the future of the world will be decided in Africa."[91]

NOTES

1. In the end, Drake was unable to attend, but his paper is included in the volume produced out of the assembly. A list of those who accepted invitations, along with their affiliation, can be found in the *Ghanaian Times*, June 14, 1962. See also, Accra Assembly, *Conclusions of the Accra Assembly* (Secretariat of the Accra Assembly, 1962); and Julian Mayfield, ed. *The World Without the Bomb: Selections from the Accra Assembly Papers* (Accra: Government Printer, 1962).

2. St. Clair Drake, "The African Revolution and the Accra Assembly," St. Clair Drake Papers, SC MG 309, Box 67, Schomburg Library. Also reprinted in, Mayfield, ed., *World Without the Bomb*, 33–37.

3. For an excellent overview of afro-pessimism, see Waful Okumo, "Afro-Pessimism and African Leadership," *The Perspective* (April 5, 2001), http://www.theperspective.org/afro_pessimism.html (accessed March 2005), which makes special mention of the works of Paul Kennedy, David Lamb, Peter Marnham, and Robert D. Kaplan. See also Manthia Diawara, *In Search of Africa* (Cambridge, MA: Harvard University Press, 1998); and Goren Hyden, "African Studies in the Mid-1990s: Between Afro-Pessimism and Amero-Skepticism," *African Studies Review* 39, no. 2 (September, 1996): 1–17. For a defense of afro-pessimism, see David Rieff, "In Defense of Afro-Pessimism," *World Policy Journal* 15, no. 4 (1998–99): http://www.worldpolicy.org/journal/rieff.html (accessed March 2005). Disillusionment set in early with scholars of Ghana, see especially, Dennis Austin, *Ghana Observed: Essays on the Politics of a West African Republic* (Manchester: Manchester University Press, 1976), 2–5.

4. See the special issue of *Radical History Review* 87 (2003), "Transnational Black Studies," ed. by Lisa Brock, Robin D.G. Kelley, and Karen Sotiropoulos.

5. Among the many scholars who have grappled with this question for Ghana since 1966 are: Bob Fitch and Mary Oppenheimer, *Ghana: End of an Illusion* (New York: Monthly Review, 1967); Samir Amin, *Neo-Colonialism in West Africa* (New York: Monthly Review, 1973); Dennis Austin and Robert Luckham, *Politicians and Soldiers in Ghana, 1966-1972* (London: Frank Cass, 1975); Dennis Austin, *Ghana Observed*; C. L. R. James, *Nkrumah and the Ghana Revolution* (London: Allison and Busby, 1977); Manning Marable, *African and Caribbean Politics: From Kwame Nkrumah to Maurice Bishop* (London: Verso Books, 1987); Kofi Buenor Hadjor, *Nkrumah and Ghana: The Dilemma of Postcolonial Power* (London: Kegal Paul International, 1988); David Rooney, *Kwame Nkrumah: The Political Kingdom in the Third World* (New York: St. Martin's, 1988); and most recently, Kevin Gaines, *American Africans in Ghana* (Chapel Hill: University of North Carolina, 2006).

6. The promise of those years is eloquently captured in Gaines, *American Africans*, esp. 19–26.

7. The difficulty of finding beginnings and examining "intricate connections" is described in the preface to Marable's *African and Caribbean Politics*, vii.

8. See firstly, Paul Robeson, *Here I Stand* (1958; repr. Boston: Beacon, 1988); and W. E. B. Du Bois, *The World and Africa* (New York: International Publishers, 1979). The literature produced by U.S.-based scholars, particularly over the last decade, on the development of black internationalism during World War II and the fate of that radical agenda during the cold war is quite substantial. See, for example (by date of publication): Gerald Horne, *Black & Red: W.E.B. Du Bois and the Afro-American Response to the Cold War, 1944–1963* (Albany, NY: State University of New York Press, 1986); Brenda Gayle Plummer, *Rising Wind: Black Americans and U.S. Foreign Affairs, 1935–1960* (Chapel Hill: University of North Carolina Press, 1996); Robin D.G. Kelley, *Race Rebels: Culture Politics and the Black Working Class* (New York: Basic Books, 1996); Penny Von Eschen, *Race against Empire:*

Black Americans and Anticolonialism, 1937–1957 (Ithaca, NY: Cornell University Press, 1997); Yevette Richards, "African and African-American Labor Leaders in the Struggle over International Affiliation," *International Journal of African Historical Studies* 31, no. 2 (1998): 301–34; Gerald Horne, *Race Woman: The Lives of Shirley Graham Du Bois* (New York: New York University Press, 2000); Yevette Richards, *Maida Springer: Pan-Africanist and International Labor Leader* (Pittsburgh, PA: University of Pittsburgh Press, 2000); Robin D. G. Kelley, "Stormy Weather: Reconstructing Black (Inter) Nationalism in the Cold War Era," in *Is It Nation Time? Contemporary Essays on Black Power and Black Nationalism*, ed. Eddie S. Glaude, Jr. (Chicago: University of Chicago Press, 2002), 67–90; James H. Meriwether, *Proudly We Can Be Africans: Black Americans and Africa, 1935–1961* (Chapel Hill: University of North Carolina Press, 2002); Penny Von Eschen, *Satchmo Blows Up the World: Jazz Ambassadors Plays the Cold War* (Cambridge, MA: Harvard University Press, 2004); Brenda Gayle Plummer, ed., *Window on Freedom: Race, Civil Rights, and Foreign Affairs, 1945–1988* (Chapel Hill: University of North Carolina Press, 2003); Nikhil Pal Singh, *Black is a Country: Race and the Unfinished Struggle for Democracy* (Cambridge, MA: Harvard University Press, 2004); and Gaines, *American Africans*. In addition to these, see the following, which look specifically at U.S. foreign policy and race within the United States: Mary L. Dudziak, *Cold War Civil Rights: Race and the Image of American Democracy* (Princeton, NJ: Princeton University Press, 2000); Azza Salama Layton, *International Politics and Civil Rights Policies in the United States, 1941–1960* (Cambridge, MA: Cambridge University Press, 2000); Thomas Borstelmann, *The Cold War and the Color Line: American Race Relations in the Global Arena* (Cambridge, MA: Harvard University Press, 2001); and George White, Jr., *Holding the Line: Race, Racism, and American Foreign Policy toward Africa, 1953–1961* (Lanham, MD: Rowman and Littlefield, 2005).

9. On the Council on African Affairs, see Robeson, *Here I Stand*, esp. Appendix D, "A Note on the Council on African Affairs," by Alphaeus Hunton, 117–21; Horne, *Black and Red*, esp. Chapter 11, *passim*; Plummer, *Rising Wind*, esp. 116–18; Von Eschen, *Race Against Empire*; Meriwether, *Proudly We Can Be Africans*, esp. 59–68.

10. Horne, *Red and Black*, 28–30; Meriwether, *Proudly We Can Be Africans*, 63–64.

11. On the Pan African Congress and its connections to U.S. black internationalism, see Horne, *Black and Red*, 29–47; Plummer, *Rising Wind*, 154–61; Von Eschen, *Race Against Empire*, 45–53; Meriwether, *Proudly We Can Be Africans*, 182–85. See also Kwame Nkrumah, *Ghana: The Autobiography of Kwame Nkrumah* (London: Thomas Nelson and Sons, 1957), 52–55.

12. Von Eschen, *Race Against Empire*, 103–7

13. See Horne, *Black and Red*, 119–26; and Robeson, *Here I Stand*, 41–47.

14. Plummer, *Rising Wind*, 213. The argument was first set out in Horne, *Red and Black*.

15. Von Eschen, *Race Against Empire*, 2–33.

16. For a similar argument, but utilizing the life and work of Lorraine Hansberry, see Fanon Che Wilkins, "Beyond Bandung: The Critical Nationalism of Lorraine Hansberry, 1950–1965," *Radical History Review* 95 (Spring 2006): 191–210.

17. The Asian-African Conference received wide coverage in the African American press. See, for example, Horne, *Black and Red*, 190–91.

18. Robeson, *Here I Stand*, 45–47. See also Richard Wright, *The Color Curtain* (New York: World, 1956).

19. Von Eschen points out that in Drake's subsequent overviews of Pan-Africanism, "The work of Paul Robeson, Alphaeus Hunton, and the CAA was invisible." Even in discussions of Padmore and Nkrumah, there is "scant mention of their roots in the left and without reference to their early alliances with African Americans or the repercussions of the

Cold War for that generation of anticolonial activists." Von Eschen, *Race Against Empire*, 176. See St. Clair Drake, "Diaspora Studies and Pan-Africanism," in *Global Dimensions of the African Diaspora*, ed. Joseph Harris (Washington, DC: Howard University, 1982), 451–514.

20. Du Bois, letter to Wallerstein, May 3, 1961. Cited in Horne, *Black and Red*, 336.

21. Text reproduced in Colin Legum, *Pan-Africanism: A Short Political Guide* (New York: Praeger, 1963), 147. See also Kennett Loves, "African Nations Ask Nuclear Ban," *New York Times*, April 23, 1958; and Kwame Nkrumah, *I Speak of Freedom: A Statement of African Ideology* (New York: Praeger, 1961), 130. As Kwesi Armah points out, Ghana's commitment to a nuclear-free world actually pre-dated independence: "Ghana was consistent in this policy as expressed in the position taken on the eve of independence with regard to the Bandung Conference of Afro-Asian Countries in 1955." Kwesi Armah, *Peace without Power: Ghana's Foreign Policy, 1957-1966* (Accra: Ghana Universities Press, 2004), 137. See also Michael Dei-Anang, *The Administration of Ghana's Foreign Relations, 1957-1965* (London: Athlone Press, 1975), 19–20.

22. Nkrumah, *I Speak of Freedom*, 151.

23. Homer A. Jack, "African Confab Results," *Chicago Defender*, January 17, 1959.

24. See, for example, Bill Sutherland and Matt Meyer, *Guns and Gandhi in Africa: Pan African Insights on Nonviolence, Armed Struggle and Liberation in Africa* (Trenton, NJ: Africa World, 2000), 35.

25. Horne, *Black and Red*, 342; and Horne, *Race Woman*, 156.

26. *All-African People's Conference News Bulletin* 1, no. 4 (Accra: AAPC, 1959): 2.

27. "Special Conferences," *International Organization* 16, no. 2 (Spring 1962): 445.

28. Reported in the *Ghanaian Times*, October 3, 1959.

29. See Nkrumah, *Ghana*, 49–63; Gaines, *American Africans*, 34–39. As Gaines writes, "Padmore was the leading theoretician, strategist, and publicist of anticolonialism and African liberation, linking metropolitan agitation to the nationalist movements on the African continent." (34). He first met Nkrumah in London in 1947. After Ghanaian independence, he relocated to Ghana and worked as Nkrumah's key advisor on African Affairs. Padmore died suddenly in London on September 23, 1959. See also Marable, *African and Caribbean Politics*, 120–23; Von Eschen, *Race against Empire*, 13–15, 45, as well as Padmore's *Africa and World Peace* (London: M. Secker and Warburg, 1937), and *Pan-Africanism or Communism* (London: Dobson, 1956).

30. For Rustin's life story, see John D'Emilio's authoritative, *Lost Prophet: The Life and Times of Bayard Rustin* (Chicago: University of Chicago Press, 2003). For Sutherland, see Sutherland and Meyer, *Guns and Gandhi*. See also Gaines, *American Africans*, 103–6. To situate Rustin and Sutherland in the broader context of African American international activism, see Brenda Gayle Plummer's introduction to her edited, *Window on Freedom: Race, Civil Rights, and Foreign Affairs, 1945–1988* (Chapel Hill: University of North Carolina, 2003), 1–16.

31. Nkrumah joined the Fellowship while he was teaching philosophy at Lincoln University in Pennsylvania. See Gaines, *American Africans*, 43.

32. Sutherland and Meyer, *Guns and Gandhi*, 5–7, 21–22. See also, Ernest Dunbar, *The Black Expatriates: A Study of American Negroes in Exile* (New York: E.P. Dutton, 1968), 88–109.

33. Sutherland and Meyer, *Guns and Gandhi*, 36. See also D'Emilio, *Lost Prophet*, 280.

34. For an account of Rustin's work in Montgomery, see D'Emilio, *Lost Prophet*, 223–48.

35. D'Emilio, *Lost Prophet*, 280.

36. Gaines, *American Africans*, 13.

37. D'Emilio, *Lost Prophet*, 283–86.

38. D'Emilio, *Lost Prophet*, 285, citing Muste and Levison to Bayard Rustin, 11/14/59, Bayard Rustin Papers, University Publications of America.

39. D'Emilio, *Lost Prophet*, 286.

40. April Carter, "The Sahara Protest Team," in *Liberation without Violence*, ed. A. Paul Hare and Herbert H. Blubert (London: Rex Collings, 1980), 130.

41. When Rustin arrived in London he began to meet with officials and various embassies, including the Moroccan Embassy, which offered to secretly provide financing for the team. According to D'Emilio, Rustin was concerned about the strings that might be attached to such an offer. See D'Emilio, *Lost Prophet*, 281.

42. National Archives of Ghana, Accra [NAG], ADM 13/2/65, Cabinet Memorandum, 23 October 1959.

43. NAG, ADM 13/1/28, Cabinet Minutes, October 23, 1959

44. NAG, Special Collections, Bureau of African Affairs [SC/BAA] 251, African Affairs Committee Minutes, 19 November 1959. [File is in the process of being renumbered as RG 17/1/465.] Committee members included: Nkrumah himself, Ako Adjei, N. A. Welbeck, Kofi Baako, Abdullai Diallo, Tawiah Adamafio, T. R. Makonnen, E. J. Du Plain, John Tettegah, Kwaku Boateng, Yankeh, Joe Fio Meyer, Eric Heymann, A. K. Barden, Kojo Botsio, Amoah Awuah, A. Y. K Djin, and Mbiyu Boinange.

45. For Scott's life story, see Michael Scott, *A Search for Peace and Justice: Reflections of Michael Scott*, ed. Paul Hare and Herbert H. Blumbert (London: Rex Collings, 1980).

46. *Ghanaian Times*, October 1, 1959 and October 22, 1959. For reconstructions of the protest based on participants' accounts, see Bayard Rustin, *The Reminiscences of Bayard Rustin*, no. 8: Interview of Bayard Rustin by Ed Edwin, November 6, 1985 (Alexandria, VA: Alexander St., 2004), 341–44; D'Emilio, *Lost Prophet*, 279–88; Scott, *Search for Peace*, 131–37. For brief secondary accounts of the Sahara protest, see also, Armah, *Peace without Power*, 138. Gaines, *American Africans*, 103–6.

47. *Ghanaian Times*, November 30, 1959.

48. *Ghanaian Times*, November 21, 1959.

49. Full text reprinted in *Ghanaian Times*, November 23, 1959.

50. *Daily Graphic*, December 5, 1959.

51. Many scholars have noted the decline in coverage of African affairs in African American newspapers in the wake of the cold war. See esp. Meriwether, *Proudly We Can Be Africans*, 153–57; and Von Eschen, *Race Against Empire*, 107, 118, 152. Ghana's leadership role in the antinuclear movement received some coverage in the African American press before Rustin, Sutherland, and others assembled in Accra. See, for example, *Chicago Defender*, September 13, 1958, January 17, 1959, August 8, 1959; and September 16, 1959. The Sahara protest itself was covered in the mainstream U.S. press: *New York Times*, October 15, 1959, November 5, 1959; November 20, 1959; January 5, 1960; January 31, 1960; and February 10, 1960; *Washington Post*, October 15, 1959; and November 20, 1959. However, the Sahara protest did not receive coverage in the *Chicago Defender* or the *Pittsburgh Courier*—two of the leading African American papers.

52. Muste, *Essays*, 398.

53. For a full account see Muste, *Essays*, 398–400.

54. Ibid., 400.

55. Sutherland and Meyer, *Guns and Gandhi*, 38.

56. *Daily Graphic*, December 14, 1959.

57. From late December until early January, Pierre Martin, a peace activist and teacher in a mass education unit in North Africa fasted outside the French Embassy in Accra to protest the proposed French tests. See *Daily Graphic*, January 5, 1950; and *Ghana Evening News*, January 1, 1950. In an interview with Mabel Dove, Martin explained, "There are two

Embassies in Ghana today. I, Pierre Martin, I am the Ambassador presenting the people of France and the Embassy on the fifth floor of Ghana House represents French Officialdom" *Ghana Evening News*, January 1, 1950.

58. Sutherland and Meyer, *Guns and Gandhi*, 38–39. For Scott's account see his *Search for Peace*, 132.

59. Scott, *Search for Peace*, 131.

60. Ibid., 132.

61. In early January, Sutherland and Randle traveled to Accra to confer with government officials and activists there. See *Daily Graphic*, January 7, 1960.

62. Sutherland and Meyer, 39.

63. Scott, *Search for Peace*, 131.

64. See Sutherland and Meyer, *Guns and Gandhi*, 39–40; and *Daily Graphic*, January 19, 1960.

65. See NAG, SC/BAA 514, Sutherland to Nkrumah, dd. Accra, August 26, 1960.

66. *Ghana Evening News*, 1, 29–60.

67. Ibid.

68. NAG, ADM 13/1/29, Cabinet Minutes, February 13, 1960. See also, "Ghana to Freeze Assets of French," *New York Times*, February 14, 1960. Over the next ten days, Professors R. W. H Wright, A. H. Ward, and John Marr of the University College of Ghana conducted extensive testing to measure fallout. They noted that there was a dramatic increase in levels of radiation throughout Ghana, all the way to Accra, but nowhere did measures exceed "what is considered to be the maximum safe dose." See ADM 13/2/69, "Confidential: Radioactive Fallout Effect in Ghana Following the French Atom Bomb Test," February 23, 1960.

69. *Ghana Evening News*, April 4, 1960

70. NAG, ADM 13/2/71, Cabinet Memorandum, "Second French Nuclear Test in the Sahara," April 1, 1960. See also, Armah, *Peace without Power*, 138–39.

71. David Birmingham, *Kwame Nkrumah: The Father of African Nationalism* (Athens: Ohio University Press, 1998), 104.

72. *Ghana Evening News*, February 17, 1960.

73. A pamphlet, "Positive Action Conference for Peace and Security, April, 1960," which includes Nkrumah's opening and closing speeches, resolutions of the conference, and the manifesto submitted by the Sahara Protest Team can be found in NAG, ADM 16/24.

74. NAG, SC/BAA 68, Abubakar Balewa to Nkrumah, dd. Lagos, March 25, 1960.

75. NAG, SC/BAA 68, Nkrumah to Balewa, dd. Accra, April 2, 1960.

76. For Nkrumah's opening address, see ADM 16/24, "Positive Action Conference." A slightly different version of the speech is reprinted in Nkrumah, *I Speak of Freedom*, 211–22.

77. ADM 16/24.

78. Ibid.

79. Sutherland and Meyer, *Guns and Gandhi*, 40.

80. Ibid., 41.

81. Ibid. In August 1960, after conversations with Scott, Muste, and Rustin, Sutherland wrote to Nkrumah to follow up on plans for the "Non-Violent Positive Action Training Centre" but realized that "the Congo Crisis naturally took precedence over everything else." See NAG, SC/BAA 514, Sutherland to Nkrumah, dd. Accra, 26 August 1960. [File to be renumbered as RG 17/1/411]

82. St. Clair Drake, "Nkrumah's 'Ban the Bomb' Conference,' (handwritten note, no date), St. Clair Drake Papers, SC MG 309, Box 67, Schomberg Library. Sutherland also believed the murder of Lumumba was crucial: "After the 1961 murder of Lumumba—a murder that was originally planned by the U.S. Central Intelligence Agency (CIA)—Nkrumah's

disillusionment with nonviolent strategies became solidified." Sutherland and Meyer, *Guns and Gandhi*, 47.

83. The papers and proceedings of the Assembly were edited and introduced by playwright Julian Mayfield. See Mayfield, ed., *World Without the Bomb*. For a brief account, see Gaines, *American Africans*, 164.

84. See NAG, SC/BAA 137, "The Accra Assembly, 10–17 May 1962," (pamphlet). Invitees to the Assembly were to come as individuals, not as delegates or representatives. As the preliminary program explained, "The Assembly will be composed of about one hundred individuals invited in their personal capacity who have been active in putting forward proposals independent of the policies of any Power Bloc."

85. Sutherland and Meyer, *Guns and Gandhi*, 42.

86. Nkrumah, *I Speak of Freedom*, 215.

87. Sutherland and Meyer, *Guns and Gandhi*, 59, citing a 1993 conversation in Brooklyn, New York.

88. Muste, *Essays*, 397. Many years later, as Scott reflected on both the significance of the Sahara protest and its limitations, he considered the government's support of the protest or rather the reliance on the government a problem: "any time we wanted to go further, we had to send people back to Accra to find out whether the government would be willing to put up the necessary funds to allow the action to continue. Bayard Rustin and Michael Randle had to go back to Accra for several days in order to keep us 'on location', so to speak. In the end the decision was really made by the Ghanaian government, that they were willing to risk their trucks." Scott, *Search for Peace*, 134.

89. Gaines, *American Africans*, 285. On "radical experimentation," see Sutherland and Meyer, *Guns and Gandhi*, 42.

90. C. L. R. James, *At the Rendezvous of Victory: Selected Writings* (London: Allison and Busby, 1984), 184.

91. Nkrumah, *I Speak of Freedom*, 220.

About the Editors

Dr. Manning Marable is one of the United States' most influential and widely read scholars. Since 1993 he has been professor of public affairs, political science, history, and African-American studies at Columbia University in New York City. From 1993 to 2003 Dr. Marable served as the founding director of the university's Institute for Research in African-American Studies, one of the nation's most prestigious academic centers of scholarship on the black American experience. Since earning his PhD three decades ago, Dr. Marable has written over two hundred articles in academic journals and edited volumes, and has written or edited twenty-five books and scholarly anthologies. In 2002 Dr. Marable established the Center for Contemporary Black History (CCBH) at Columbia University, an advanced research and publications center that examines black leadership and politics, culture, and society. CCBH produces *Souls: A Critical Journal of Black Politics, Culture, and Society*, a quarterly academic journal of black studies, which is published and distributed internationally by Taylor and Francis Publishers. CCBH also directs the *Amistad Digital Resource Project*, funded by the Ford Foundation, which is developing multimedia resources for teaching African American history in secondary schools. He is currently completing a biography of Malcolm X, to be published by Viking Press.

Vanessa Agard-Jones is the managing editor of *Souls: A Critical Journal of Black Politics, Culture, and Society*. She is a PhD candidate in anthropology and French studies at New York University, where she works on the politics of sexuality, nationalism, and *la mémoire de l'esclavage* (the memory of slavery) in France and its Départements d'Outre Mer (Martinique, Guadeloupe, Guyane, and Réunion). She holds a Master of Arts degree from Columbia's Institute for Research in African-American Studies and a Bachelor of Arts degree from Yale University. She has worked as both the director of a prison abolitionist nonprofit in California (the Prison Activist Resource Center) and as a teacher in elementary and middle schools in Georgia and is the former Chair of the Board of Directors of New York's Audre Lorde Project: A Community Organizing Center for Lesbian, Gay, Bisexual, Transgender, Two Spirit, and Gender Non-Conforming People of Color.

About the Contributors

Jean Allman teaches African history at Washington University and has authored and edited numerous works on West African history, including *The Quills of the Porcupine: Asante Nationalism in an Emergent Ghana*, *"I Will Not Eat Stone": A Women's History of Colonial Asante* (with V. Tashjian), and *Fashioning Africa: Power and the Politics of Dress*. Allman co-edits the *New African Histories* book series at Ohio University Press and is the co-editor of the *Journal of Women's History*.

Asale Angel-Ajani is a professor in the Gallatin School at New York University. She is the author of numerous articles on the transnational traffic in African women and women's roles in drug trafficking. Angel-Ajani has also published widely on human rights, migration, displacement, conflict, and postconflict. Most recently, she is co-editor of the book *Engaged Observer: Anthropology, Advocacy, and Activism* (Rutgers University Press, 2006) with Victoria Sanford. Angel-Ajani earned her MA and PhD in Anthropology from Stanford University.

Kevin Gaines is the director for the Center for Afroamerican and African Studies and professor in the Department of History at the University of Michigan. He is author of *Uplifting the Race: Black Leadership, Politics, and Culture in the Twentieth Century* and *American Africans in Ghana: Black Expatriates and the Civil Rights Era* (University of North Carolina Press, 2006). Professor Gaines is a past winner of the John Hope Franklin Book Prize of the American Studies Association and has received fellowships from the National Humanities Center and the Center for Advanced Study in the Behavioral Sciences.

Faye V. Harrison is director of African American studies and professor of anthropology at the University of Florida. Her publications include *Resisting Racism and Xenophobia: Global Perspectives on Race, Gender, and Human Rights* (ed.) and *Outsider Within: Reworking Anthropology in the Global Age* (University of Illinois Press, 2008).

Scholar and filmmaker **Robin J. Hayes** is an assistant professor of ethnic studies at Santa Clara University. While completing a combined doctorate at Yale University in African American studies and political science she served as co-founder and facilitator of the Black Resistance Reading Group and program coordinator for the Center for the Study of Race, Inequality and Politics. Robin has published and presented several articles about black politics and culture and served on the faculty of Williams College and Northwestern University. In addition, she produced, directed, and co-wrote *Beautiful Me(s): Finding our Revolutionary Selves in Black Cuba*, a documentary film about the historical relationship between African Americans and Afro-Cubans. Her current book project, *African Liberation, Black Power and the Diasporic Underground*, generates a theory about transnational exchanges between social movements in the African Diaspora by exploring the relationship between African anti-colonialism and the Black Panther Party, Organization of Afro-American Unity, and Student Nonviolent Coordinating Committee. Robin thanks the Ford Foundation for their support of the research that appears in her contribution.

Gerald Horne has published twenty books including, most recently, *The White Pacific: U.S. Imperialism and Black Slavery in the South Seas after the Civil War* (University of Hawaii Press, 2007), and *The Deepest South: The U.S., Brazil and the African Slave Trade* (New York University Press, 2007).

Joseph Jordan is associate professor of African/Afro-American studies and director of the Sonja Haynes Stone Center for Black Culture and History at the University of North Carolina at Chapel Hill. He previously served as director of the Auburn Avenue Research Library on African-American Culture and History, was founding chair of African/African-American Studies at Antioch College, and taught at Howard and Xavier Universities. He is editor of a forthcoming collection on African Americans and Native Americans and is a contributor to *No Easy Victories*, a collection of essays on the U.S. antiapartheid movement (Africa World, 2008). His curatorial work includes *Without Sanctuary: Lynching Photography in America*. He currently serves as Senior Policy Advisor at TransAfrica Forum.

Ricardo Rene Laremont is professor of political science and sociology at the State University of New York at Binghamton. He is the author of *Islam and the Politics of Resistance in Algeria, 1783-1992* (Africa World, 1999). He is editor of *The Causes of War and the Consequences of Peacekeeping in Africa* (Greenwood, 2001) and *Borders, Nationalism, and the African State* (Lynne Rienner, 2005).

Clarence Lusane is an author, activist, scholar, and well-respected expert in the areas of human rights, global racism, and international relations. He is the author of seven books on human rights, political rights, and global black politics. He has lectured on these topics in numerous foreign nations, including Colombia, Cuba, England, France, Germany, Guadeloupe, Guyana, Haiti, Japan, the Netherlands, Panama, Switzerland, Turkey, and Zimbabwe, among others. His latest book is *Colin Powell and Condoleezza Rice: Foreign Policy, Race, and the New American Century* (Praeger, 2006). His current research is focused on the intersection of jazz and international relations, the impact of the Hurricane Katrina crisis on black politics, and new grassroots human rights movements in the global South.

Anthony W. Marx is the President of Amherst College. Before coming to Amherst he was on the faculty at Columbia University where he was professor and director of undergraduate studies in political science. Marx is a respected teacher and an internationally recognized scholar who has written three books on nation-building, particularly in South Africa but also in the United States, Brazil, and Europe. He also has established and managed programs designed to strengthen secondary school education in the United States and abroad. Marx is the author of a dozen substantive articles and three books: *Lessons of Struggle: South African Internal Opposition, 1960-1990* (Oxford University Press, 1992), *Making Race and Nation: A Comparison of the United States, South Africa and Brazil* (Cambridge University Press, 1998), and *Faith in Nation: Exclusionary Origins of Nationalism* (Oxford University Press, 2003). *Making Race and Nation* received the American Political Science Association's 1999 Ralph J. Bunche Award (co-winner for the best book on ethnic and cultural pluralism) and the American Sociological Association's 2000 Barrington Moore Prize (for the best book of the preceding three years in comparative-historical sociology).

Maureen Mahon is an assistant professor of anthropology at the University of California, Los Angeles. Her research interests include African American expressive culture and the intersection of music, race, and identity. *Right to Rock: The Cultural Politics of Black Music and Black Authenticity*, her book about the Black Rock Coalition, is published by Duke University Press (2004). She earned a PhD in anthropology from New York University.

Elizabeth Mazucci has been affiliated with the Malcolm X Project at Columbia University since 2003. As researcher and former project manager, she has focused on the discovery and analysis of primary source material, particularly government records. She holds a BA in religion and archaeology from the University of Rochester and an MA in anthropology (archaeology) from Columbia. Her master's thesis examines what Malcolm X carried in his pockets at the moment he was assassinated; it is currently being adapted and expanded for publication. Mazucci has worked in the archives at the American Red Cross September 11 Recovery Program, Bausch & Lomb, the New York City Department of Records, and the New York City Department of Parks and Recreation.

Irma McClaurin is an activist anthropologist, writer, teacher, and administrator, who has had a commitment to eradicating social inequality for over thirty years. She was associate professor of anthropology at the University of Florida (1995–2006), the Mott distinguished professor of women's studies at Bennett College for Women (2004), deputy provost at Fisk University (2002–04), and an American Association for the Advancement of Science Diplomacy Fellow at USAID (2000–2001). She is currently a program officer at the Ford Foundation. McClaurin has a BA in American studies, an MFA in English, and holds both the MA and PhD in anthropology. She is the author of *Women of Belize: Gender and*

Change in Central America (Rutgers 2001 [1996]) and the editor of *Black Feminist Anthropology: Theory, Politics, Praxis and Poetics* (Rutgers 2001), a 2003 Choice Outstanding Academic Title. She is also a poet and has published *Black Chicago* (1971), *Song in the Night* (1974), and *Pearl's Song* (1988).

Brian Meeks is professor of social and political change in the Department of Government at the University of the West Indies, Mona. He is the director of the Centre for Caribbean Thought and a former chairman of the Michael Manley Foundation. He has published five books and many articles on Caribbean politics and political theory. Among them are *Envisioning Caribbean Futures: Jamaican Perspectives*; *Caribbean Revolutions and Revolutionary Theory: An Assessment of Cuba, Nicaragua and Grenada* (University of the West Indes Press, 2001) and *Radical Caribbean: from Black Power to Abu Bakr, Narratives of Resistance: Jamaica, Trinidad, the Caribbean*, and he is editor of *Culture, Politics, Race and Diaspora: The Thought of Stuart Hall* (Ian Randle, 2007). His first novel, *Paint the Town Red*, was published in 2003 by Peepal Tree.

Leith Mullings is a distinguished professor of anthropology at the Graduate Center of the City University of New York. She received her PhD in anthropology from the University of Chicago. Her books include: *Therapy, Ideology and Social Change: Mental Healing in Urban Ghana* (University of California Press, 1984), *Cities of the United States* (Columbia University Press, 1987), *On Our Own Terms: Race, Class and Gender in the Lives of African American Women* (Routledge, 1996), *Let Nobody Turn Us Around: Voices of Resistance, Reform and Renewal, An African America Anthology* (Rowman and Littlefield 2000, co-edited with Manning Marable), *Stress and Resilience: The Social Context of Reproduction in Central Harlem* (Springer Press 2001, with Alaka Wali), *Freedom: A Photohistory of the African American Struggle* (Phaidon 2002, with Manning Marable), and *Gender, Race, Class and Health: An Intersectional Approach* (Jossey Bass 2005, co-edited with Amy Schulz). She has written articles on such subjects as stratification, ethnicity, race, gender, health, globalization, participatory research, and public policy. In 1993 Professor Mullings was awarded the chair in American Civilization at the Ecole des Hautes Etudes en Sciences Sociales in Paris, France, and in 1997 she received the Prize for Distinguished Achievement in the Critical Study of North America from the Society for the Anthropology of North America. She is currently serving as a member of the Executive Board of the American Anthropological Association.

Gina Perez is assistant professor of comparative American studies at Oberlin College. Her research interests include U.S. Latinos, gender, political economy, migration, transnationalism, urban anthropology, poverty, militarization and Latino youth, Latina feminisms. She earned a PhD in anthropology in 2000 from Northwestern University.

Michael Ralph completed his PhD in anthropology at the University of Chicago. He is now an assistant professor in the Department of Social and Cultural Analysis at New York University.

Besenia Rodriguez earned a PhD in African American and American studies from Yale University. She lives in Los Angeles, California.

Mark Q. Sawyer currently holds appointments as an associate professor with the Department of Political Science and with the Bunche Center for African American Studies at UCLA. He is also the director of the Center for the Study of Race, Ethnicity & Politics at UCLA. His book *Racial Politics in Post Revolutionary Cuba* (Cambridge University Press, 2005) received the Du Bois Award for the best book by the National Conference of Black Political Scientists 2006. He earned a PhD from the University of Chicago.

T. Denean Sharpley-Whiting is professor of African American and Diaspora studies, French director of African American and Diaspora studies, and director of the W. T. Bandy Center for Baudelaire and Modern French Studies at Vanderbilt University. She has recently published a book on young black women, feminism, and hip-hop culture, *Pimps Up, Ho's Down: Hip Hop's Hold on Young Black Women* (NYU, 2007). Her other works include *Négritude Women* (University of Minnesota Press, 2002), *Black Venus: Sexualized Savages, Primal Fears, and Primitive Narratives in French* (Duke University Press, 1999), and *Frantz Fanon: Conflicts and Feminisms* (Rowman and Littlefield, 1998). She has co-edited three volumes, the latest of which includes *The Black Feminist Reader* (Blackwell, 2000).

Howard Winant is professor of sociology at the University of California, Santa Barbara, where he is also affiliated with the Black Studies and Chicano Studies departments. He founded and directs the UCSB Center for New Racial Studies. Winant's work focuses on the historical and contemporary importance of race in shaping economic, political, and cultural life, both in the United States and globally. He is the author of *The New Politics of Race: Globalism, Difference, Justice* (University of Minnesota Press, 2004), *The World is a Ghetto: Race and Democracry Since World War II* (Basic Books, 2001), *Racial Conditions: Politics, Theory, Comparisons* (University of Minnesota Press, 1994), *Racial Formation in the United States: From the 1960s to the 1990s* (co-authored with Michael Omi, Routledge 1994 [1986]), and *Stalemate: Political Economic Origins of Supply-Side Policy* (Praeger, 1988).

Lisa Yun is associate professor of English and Asian and Asian American studies at the State University of New York at Binghamton. She is the author of *The Coolie Speaks: Chinese Indentured Laborers and African Slaves in Cuba* (Temple University Press, 2007).

INDEX

CPSIA information can be obtained
at www.ICGtesting.com
Printed in the USA
LVOW13s1624260217
525473LV00010B/77/P